# The Student Laboratory and the Science Curriculum

**CURRICULUM POLICY AND RESEARCH SERIES**
Edited by William Reid and Ian Westbury

EVALUATING CURRICULUM PROPOSALS
*A Critical Guide*
Digby C. Anderson

THE CLASSROOM SOCIETY
*The Construction of Educational Experience*
Herbert A. Thelen

INNOVATION IN THE SCIENCE CURRICULUM
*Classroom Knowledge and Curriculum Change*
Edited by John Olson

SCHOOL SUBJECTS AND CURRICULUM CHANGE
*Case Studies in the Social History of Curriculum*
Ivor Goodson

TEACHER THINKING
*A Study of Practical Knowledge*
Freema Elbaz

TIME AND SCHOOL LEARNING
Lorin W. Anderson

LEARNING MATHEMATICS
*The Cognitive Science Approach to Mathematics Education*
Robert B. Davis

CLASSROOM ENVIRONMENT
Barry J. Fraser

LEARNING ABOUT TEACHING THROUGH CLINICAL SUPERVISION
Edited by W. John Smyth

QUESTIONING AND TEACHING
*A Manual of Practice*
J. T. Dillon

# The Student Laboratory and the Science Curriculum

Edited by
**ELIZABETH HEGARTY-HAZEL**

ROUTLEDGE
London and New York

First published 1990
by Routledge
11 New Fetter Lane, London EC4P 4EE

Simultaneously published in the USA and Canada
by Routledge
a division of Routledge, Chapman and Hall, Inc.
29 West 35th Street, New York, NY 10001


Printed and bound in Great Britain by Mackays of Chatham



*British Library Cataloguing in Publication Data*

The Student laboratory and the science curriculum.
1. Education. Curriculum subjects. Science.
Laboratory techniques
I. Hegarty-Hazel, Elizabeth
507'.8

ISBN 0-415-00118-8

*Library of Congress Cataloguing in Publication Data*

The Student laboratory and the science curriculum.
(Curriculum policy and research series)
1. Science–Experiments–Study and teaching.
2. Physical laboratories. 3. Curriculum planning.
4. Science–Study and teaching. I. Hegarty-Hazel, Elizabeth. II. Series.
Q183.AlS78 1990 507.8 89-10462

ISBN 0-415-00118-8

CONTENTS

# FIGURES

# TABLES

# THE CONTRIBUTORS

This is a group of authors with long experience as well as specialized qualifications to bring to the task. Almost all hold higher degrees or doctorates in both science and science education and hold academic positions in universities. (Others hold such positions as senior research fellows or are senior administrators.) As a group, they are active in research on the role of the student laboratory in science at secondary school or university level and on various areas which significantly impinge: the ways in which science students develop concepts, technical skills, enquiry skills and attitudes, as well as evaluation, policy-oriented research, curriculum research, and the history and philosophy of science.

With surnames in alphabetical order, the authors are shown below together with their major institutional affiliations at the time of going to press.

## AUTHOR AND INSTITUTIONAL AFFILIATION


John Ainley
Australian Council
for Educational Research
9 Frederick Street
Melbourne 3122, AUSTRALIA

Elaine Atkinson
Education Department of Victoria
234 Queensberry Street
Melbourne 3053, AUSTRALIA

John Baird
Faculty of Education
Monash University
Melbourne 3168, AUSTRALIA

Joan Bliss
Centre for Educational Studies
Kings College, University of London
London SW10 0UA,
ENGLAND

Audrey Champagne
American Association for the
Advancement of Science
1776 Massachusetts Avenue, NW
Washington, DC 20036
UNITED STATES OF AMERICA

John Edwards
Department of Education
James Cook University
Townsville, AUSTRALIA

Contributors

Peter Fensham
Faculty of Education
Monash University
Melbourne 3168, AUSTRALIA

Yael Friedler
Amos de-Shalit Science Teaching Centre
Hebrew University
Jerusalem, ISRAEL

Paul Gardner
Faculty of Education
Monash University
Melbourne 3168, AUSTRALIA

Colin Gauld
School of Education
University of New South Wales
Sydney 2033, AUSTRALIA

Richard Gunstone
Faculty of Education
Monash University
Melbourne 3168, AUSTRALIA

Elizabeth Hegarty-Hazel
School of Microbiology
University of New South Wales
Sydney 2033, AUSTRALIA

Leopold Klopfer
Learning Research & Development Center
University of Pittsburgh
Pittsburgh, PA 15260
UNITED STATES OF AMERICA

David Layton
School of Education
University of Leeds
Leeds LS2 9JT, ENGLAND

Joseph Novak
Department of Science Education
Cornell University
Ithaca, NY 14850
UNITED STATES OF AMERICA

John Olson
Faculty of Education
Queens University
Kingston, Ontario, CANADA

Colin Power
Faculty of Education
Flinders University
Adelaide 5042, AUSTRALIA

Michael Prosser
Centre for Teaching & Learning
University of Sydney
Sydney 2006, AUSTRALIA

Pinchas Tamir
Amos de-Shalit Science Teaching Centre
Hebrew University
Jerusalem, ISRAEL

# FOREWORD

For a hundred or so years, the science laboratory has been seen as perhaps the key place for the teaching of 'real' science both in schools and universities. These labs have not always been well-equipped or well-used and, when the improvement of science teaching is considered, the laboratory is usually assumed to have a special significance. When, for example the curriculum reform projects of the 1960s sought to symbolize what was new and different about what they were doing, it was their lab experiences which seemed most central to their intentions. This, in its turn, led to a larger focus on the improvement of labs as places in which the most effective and important 'new' science teaching would take place. In the early 1970s, for example, when the Australian federal government first considered whether it had a role to play in what had been to that time state-administered elementary and secondary education, the improvement of science education by way of the improvement of science laboratories was the first task that the government undertook. The decision did not seem to be one that could or should be questioned and, as a result, the many sparsely equipped and often spartan schools that had been built in the 1950s found themselves equipped with elaborate labs. The curriculum projects that the Australian federal government suppported at that time, in common with other similar projects elsewhere, had the idea of 'inquiry' at their core and this ideology and rationale, in its turn, provided a further, more concrete and immediate rationale for the importance of the laboratory.

This book had its beginning in this Australian experience. As part of its programme of support for science education, the federal government supported evaluation of its programmes; Australia's science educators were prepared to ask some basic questions about what was happening in the labs that were so intimately associated with the new courses that were being developed and initiated a set of research programmes that have continued to the present. This book with its core of chapters reporting this work shows how significant this research and evaluation has been. Building from around the group that undertook this work, Elizabeth Hegarty-Hazel, a microbiologist working in a university milieu, has gathered a distinguished group of educational researchers from Great Britian, the United States, Canada, and Israel and asked them to join their Australian colleagues and evaluate the curriculum and teaching practices of the school and university teaching laboratory. The result is a wide-ranging, comprehensive review of research and thinking on all aspects of teaching, and the curriculum, of the science laboratory in both the secondary school and also the university.

*The Student Laboratory and the Science Curriculum* is the only book of its kind and as such will be the starting point for all thinking and research on labs for many years. It is a book which is both exciting and deeply disturbing at the same time: it shows quite conculsively that the unexamined belief in the significance of the lab as a place for science teaching must be evaluated in the light of the realities of practice found in many schools and settings but at the same time it shows what the laboratory can and might achieve when curricula and teaching practices are thought through and appropriately supported by school systems.

But in addition to the immediate relevance of this book to those with professional interests in the teaching lab, *The Student Laboratory and the Science Curriculum* has a deeper significance to all who are interested in the curriculum. The 'teaching' that is discussed in most educational research is usually *classroom* teaching. But there are other kinds of settings in which teaching takes place and for which curricula are developed. The laboratory is perhaps the most immediately visible of these different settings from the viewpoint of the secondary schools, but the library has a similar place in the school and carries its own freight of symbolic meaning both for teachers and for publics. In universities and professional schools, in addition to their labs and libraries, there are also studios and practicum sites. All of these are places in which goals and curricula, and the forms of teaching and assessment, are quite different from those found in classrooms or lecture halls.

What does this variety of sites with their varying traditions in which teaching takes place mean for curriculum theory and research? In his *The Design Studio: An Exploration of its*

*Traditions and Potential*, Donald Schön suggested that the focus of educational thinking on the classroom has limited conceptions of curricula, teaching and assessment. He goes on to argue strenuously for analyses of the design studio as a unique site in which important, but from the point of view of the classroom, unconventional kinds of teaching and learning take place. The science laboratory is not, of couse as invisible to conventional educational theory and research as the studios of artists and architects, but it too is a place that contains many lessons for those interested in alternative forms of teaching and their possible implications for curriculum, teaching and schools. And because of its scope, and the unity of the structure with its constant focus on basic curriculum elements and processes, *The Student Laboratory and the Science Curriculum* offers an excellent starting point for those who want to pursue the line of thinking sketched by Schön - as well as the related problem of evaluating teaching settings other than the classroom. We wondered, for example, what would emerge if the secondary school library, or the university practicum site, were to be subjected to as rigorous, self-critical, and searching analysis as this book gives to the teaching laboratory. There are several chapters which provide explicit and provocative exemplars for such an anlysis, but all of the chapters provide grist for the mill, and models, for such thinking. For this reason we believe that this book will and should have a wider readership than its authors may have suspected as they prepared their chapters.

**W.A. Reid**
University of London Institute of Education

**Ian Westbury**
University of Illinois at Urbana-Champaign

# Part 1

# THE MODEL

# 1.1

# The Student Laboratory and the Science Curriculum: An Overview

**Elizabeth Hegarty-Hazel**

## INTRODUCTION

Since professional scientists work in laboratories at some or all stages of their careers, then student scientists must also work in laboratories. So runs the rationale for laboratory work as a unique feature of science education at secondary and tertiary levels. Student laboratory work has been regarded as an essential step in the socialization of students into the state of scientific literacy, or even into professional science. It has been likened to the role of sacraments in religious observances, i.e. an outward and visible sign that students are acquiring inward (scientific) grace. Like the sacraments it has also had its vehement critics who have claimed student laboratory work is simply an empty ritual, time wasting and expensive. Certainly the laboratory is not only unique in kind but uniquely expensive of resources including time, space, equipment, personnel, expertise and money.

The combination of expense with the suggestion of uniqueness has prompted decades of questioning as to what is achieved or what can best be achieved in the laboratory. The cumulative index to *Science Education* (Champagne and Klopfer, 1978) covered the sixty years to 1976 and showed scores of research studies directed towards answering these questions. Now, with common sense, the clear vision of hindsight, and ample reviews (Cunningham, 1946; Bates, 1978; Bradley, 1968; Black and Ogborn, 1979; Blosser, 1980;

Garrett and Roberts, 1982; Hegarty, 1982; Hofstein and Lunetta, 1982), we can see that there are no adequate answers. Part of the problem is that scientific literacy and induction into professional science do not always go together. In this book, those writing about higher education tend to stress induction into professional science whilst those writing about secondary education may espouse both aims or either one (for a variety of reasons including conflict between them for the 'average' student).

Another part of the problem is that the term 'science laboratory work' is too loosely defined and thus can seem both too broad or too specific. For the purpose of this book we will take student laboratory work to be a form of practical work taking place in a purposely assigned environment where students engage in planned learning experiences, and interact with the materials to observe and understand phenomena. (Some other forms of practical work such as field trips are thus excluded.)

The problem of interpreting the research literature is not simply one of definition but is also conceptual. Garrett and Roberts (1982) introduced some useful terminology which helps explicate the difficulty. According to these authors *strategies* are the highly complex instructional procedures which reflect the overall approach employed by a teacher or course and *tactics* are less complex procedures used within the primary strategy. Different teaching methods such as laboratory classes or lectures are, by these definitions, tactics. At a lower level of detail within a tactic is a set of *techniques*, the activities employed by the teacher, the fundamental units of any teaching situation. Examples would be the types of laboratory equipment handled, the procedures used, the experiments undertaken and the questions posed by the teacher. To the practising teacher, these techniques (like the grand strategies) are also likely to be more important and interesting than the tactic of laboratory work itself.

Using this terminology, it emerges that research studies, and especially comparative learning studies, have often mismatched instructional procedures which are at different levels. Thus a Method A vs Method B study may be comparing two strategies, or one strategy vs one tactic, or two tactics, and so on. It is not surprising the results have been ambiguous.

Thus laboratory work not only provides the setting for students to acquire technical expertise but can also be used to achieve a myriad of other educational objectives. The quotation and comments below give an idea of this complexity.

(Student, tertiary level microbiology, referring to laboratory exercises requiring experimental design intended to provide experience of scientific enquiry)

> You make a plan and as far as you know it's logical and consistent before you get into the lab. but when I get into the lab. and really see things and that they aren't as I expected, even little things, they kind of wreck your plans - or they can do ... and that's half the trouble when you're not familiar with the gear - that's why it does take you four times as long as it should ... you know it would be a lot easier next time ... well, a training in logistics as much as anything else. (Hegarty, 1979, p. 284)

It is a most obvious observation to make that in an adequately equipped laboratory, students can learn manipulative skills and can acquire technical expertise in the use of instruments and apparatus. In the quotation above, we see that this can be successfully accomplished or can result in more of a comedy of errors. As we shall see later, the aim of acquiring technical skills is customarily more valued at tertiary level than secondary and even at tertiary level is valued by students but is downplayed by staff.

(Referring to graduate students, chemistry majors)

> 'There was even more evidence of lack of understanding of electric current: 'transformers convert AC to DC', 'an ionic solid is a good conductor', 'resistors slow down electric current' (a common view), 'insulators concentrate electricity so that more energy is obtained from it', 'the flow of electrons is by convention from positive to negative but is really the other way around'. Several students said that either some or all domestic current was DC. Two students argued that the insulation around a conducting wire serves the function of holding the current in and directing it to where it is needed 'rather than let it come out along the way'. (Gunstone and White, 1981, p. 6)

In this quotation we are reminded that curriculum planners and teachers have often regarded the laboratory as a place where important science concepts will be learned or reinforced by processes of illustration, concrete exemplification of abstract principles, induction, deduction or discovery. Sometimes students can have the 'Aha!'/'it all clicks'/'penny drops' reaction where lab work makes everything come clear. At other times we hear the question 'How could these good university students develop such poor and mixed-up ideas about areas of content that they have studied extensively?' (West, 1984).

(Student, tertiary level, microbiology)

> I thought it was an exercise on designing and running the prac ourselves as contrasted to the kind of prac ... where everything was there and we just had to follow the recipe more or less ... I thought it was not so much the content which was important but the exercise of carrying it out. (Hegarty, 1979, p. 285)

(Researcher, secondary level, chemistry)

> To take what might be thought to be a simple problem ... A student is to learn that the higher the temperature of water, ... the more sodium chloride will dissolve. Almost any number of things can go wrong. The student might not understand that the solution has to be saturated. He might not be interested enough to take care that the solution is at the same temperature throughout ... He/she might not realise that the temperature thermometer takes some time to register the temperature or that indeed it is broken ... (Newman, 1981, p. 216)

In the quotations above we can see that students can certainly be aware of the difference between cookbook and enquiry exercises and that the success of the latter depends on many factors, including the technical expertise of the students. In some of the 'new' secondary school science curricula (e.g. BSCS, PSSC), understanding of scientific enquiry is highly regarded. The nature of such a school version of scientific enquiry is debated (Herron, 1971) and the extent to which any such aims are desirable for all students is also debated (Welch *et al.* 1981). And while participation in some student version of the scientific endeavour is deemed feasible in school laboratories (Tamir, 1983) it has been found often to fail to materialize (Stake and Easley, 1978). As we shall see later, at tertiary level, there is little uniformity of opinion on the importance, desirability or feasibility.

## THE FOCUS OF THIS BOOK

The examples above have illustrated that there will be a few simple questions worth asking or answering about the science curriculum and the student laboratory. It will be the burden of this book to examine this web of people, places, events, equipment and time which make up the laboratory experience.

We are aware that there are innumerable questions to be asked, both general and very specific. We are aware that the questions are different for science teachers and lecturers, for researchers, curriculum specialists and theorists, and for administrators. We shall try to ask and answer the various ranges of questions and address the different audiences at appropriate times.

A model was used to provide the overarching structure for this book and it is presented briefly in Chapter 1.2. The grandest claim which will be made of the model is that it serves as a form of middle range theory, that it stimulates the generation of good questions and that it has useful descriptive value. The remainder of this chapter will be devoted to introducing the key areas which will be incorporated into the model and which have served as a basis for later chapters.

## ELEMENTS OF A SCIENCE DISCIPLINE

Although in the wider sense, cultural content is seen as a source of curriculum, we shall be concerned only with the elements of a science discipline as curricular resources (Schwab, 1974). We suggest that choices from the elements of a science discipline are potential resources which will have a cascading effect on all other stages of the educational process to the extent that the potential is realized. It may be observed most directly in statements of intended learning outcomes. For questions on the role of laboratory work, one needs to examine the results of instructional planning decisions about which goals are to be met in laboratory classwork vs goals met in other classwork.

The elements of a science discipline are many, but Schwab identified the structure of science knowledge as the dominant influence on science education practices to date. This knowledge is of two kinds - 'knowledge that' and 'knowledge how' (labelled substantive knowledge and syntactic knowledge by Schwab). In a very important sense the laboratory in professional science is where much of the structure of scientific knowledge is made, modified or radically changed. Questions follow about the extent to which student science can or should reflect professional science (Chapters 3.2 and 5.4 show contrasting arguments).

To some extent, chapters in the book reflect this dominant emphasis on the structure of scientific knowledge. Thus we have chapters reflecting on learning the substantive structure of a science in the laboratory (see Chapters 3.1 and 3.3 on technical skills and scientific knowledge). We also have a chapter on syntactic structure (see Chapter 3.2 on scientific enquiry - the modes of questioning and interpreting in the sciences). The relationship between theory and methodology is attacked in both theoretical and practical ways in Chapters 2.1 and 2.2.

However, we have understood the need to move beyond the traditional elements of a scientific discipline into one of the areas described by Schwab as significantly neglected: the community of scientists/enquirers. As members of this community, scientists must possess personal and professional commitments for employing both types of knowledge ('that' and 'how') in a scientifically appropriate manner. Part of the scientist's drive may come from a deep love of the subject area being investigated, a desire to find out what makes things work, or a belief that scientific procedures provide at least one avenue by which dependable knowledge can be generated. Within the cognitive apparatus of a scientist these dispositions or commitments provide the motivation to use successfully what skills and knowledge are possessed. Such dispositions are often called attitudes, and in Chapter 3.4 some aspects of the development of two of these attitudes, namely 'attitude to science' and the group constituting 'scientific attitudes', are discussed. One consequence of the lack of attention in the curriculum to the community of scientists/enquirers may be the dominance in the curriculum of the empirical view of public science, the tensions described as public science vs private science and the implications of these on students' scientific attitudes (Gauld, 1982).

With mention of scientific knowledge and the nature of scientific activity, there arises also the question of the extent to which philosophy of science constitutes one of the elements of a discipline and hence a potential resource for curriculum planning. In the most obvious sense, philosophy of science is or could be in this position. However, Chapter 2.1 reminds us of the competing interpretations of the nature of science which philosophers have provided as they engage in their second-order, meta-enquiry. From that perspective, philosophy of science is more about science rather than being of science. Layton claims that its use in prescriptions for the science curriculum has been selective in accordance with the diverse purposes of those attempting to influence the conduct of science education. Thus, it is appropriate that several chapters (5.1 to 5.4) deal with the policy context of the curriculum and its influence on curriculum development, instructional planning and instruction itself. The politics of curriculum making and specifically the exercise of power and control in relation to the philosophy of science are further pursued in Chapter 2.1.

## LEARNERS, TEACHERS, SUBJECT MATTER, CONTEXTS (FOUR EDUCATIONAL COMMONPLACES)

Despite the realization that the term 'commonplaces' grates with some readers, it has been retained for historical reasons (notably its consistent use by Joseph Schwab) and because it serves as a convenient shorthand to cover learners, teachers, subject matter and contexts. Each of these commonplaces will be discussed separately in the following pages.

### Learners

Here we examine the effects on the curriculum development and instructional planning systems of knowledge about student learners: not only learning theory and findings from educational psychology but knowledge about students' readiness to undertake laboratory work of certain kinds, their interests, motivations and career aspirations. What do we know about the ways students might learn science concepts, scientific enquiry, technical skills and attitudes in the science laboratory? To read many student laboratory manuals one would imagine teachers assume all students to be identical. Yet nothing could be further from the truth. Students have had a variety of laboratory experiences before, some successful, some disastrous, some enjoyable, some dull. Students therefore hold many different views about science and scientists and different images of labs and labwork. Those who have only experienced illustrative/recipe style exercises may find discovery learning or participation in scientific enquiry unusual and perhaps exciting, threatening or repellent. There are curriculum implications in each alternative. Some of the 'new' school science curricula have explicitly taken account of learners' needs (for example, 'readiness' in Piagetian terms, in ASEP, the Australian Science Education Project). Others have not. A reappraisal of the 'new' curricula and their influence on students' understanding of scientific enquiry suggested that needs of learners had received far less attention than demands of subject matter (Welch *et al*. 1981). In turn, this must be one contributing factor to the reported widespread failure to materialize of enquiry-oriented laboratory work in USA schools and indeed, in some cases, the absence of any laboratory work whatsoever (Stake and Easley, 1978). This issue is addressed in the chapter on scientific enquiry (3.2) and, in Chapter 5.4, the conflict between the aims of scientific enquiry and science for all is raised as a policy issue. There is, however, no suggestion that laboratory work be allowed to disappear with science for all. Indeed, in 5.4 an increased role is advocated.

It is being recognized increasingly that the process of learning with understanding places complex and stringent demands on both teachers and learners. Particularly, recognition of the fundamental importance of the constructivist nature of learning, whereby the learner actively generates and manipulates conceptual frameworks in memory in order to make sense of formal and informal learning experiences, requires reappraisal of laboratory purposes and curricula. In Chapters 4.1 and 4.2, some findings regarding students' conceptual development and standard of metacognition (personal knowledge, awareness and control of learning) in science laboratories are presented. These findings have implications for attempts to make the laboratory a context for purposeful and productive enquiry. In Chapter 4.2, it is asserted that such enquiry requires that students master the requisite cognitive skills and that they practise strategies for fostering metacognitive awareness and control over their learning.

## Teachers

In a review of the theory and practice of practical work,

Newman (1983, p.10) wrote, 'Teachers are often the forgotten element in research; it would appear that only student differences and needs are important. Perhaps the time has come for increased attention in this area'. Whilst not entirely lacking, research on the teacher in the student science laboratory is sparse and focuses almost entirely on teachers' aims. We can use the literature to obtain a sense of what is missing.

Welch *et al.* (1981, p. 37) stated, 'Many teachers are ill-prepared, in their own eyes and in the eyes of others, to guide students in inquiry learning and over one-third feel they receive inadequate support for such teaching'. Tamir (1983) outlined a number of possible strategies for improving teacher education for enquiry (including those of an historical-philosophical approach, audio-tutorial individualized courses, enquiry-oriented laboratory investigations, current research papers and field work). Many prospective science teachers have not had such experiences and need to rely on curriculum guides. When Ben-Peretz and Tamir (1981) investigated what teachers want to know about curriculum materials, items such as requirement for laboratory work and stress on enquiry skills were found to be important but were outweighed by teachers' concern about subject matter (emphasis on principles and concepts, up-to-date, new developments, broad knowledge). A different approach to the question of teachers' knowledge was taken by Posner *et al.* (1982), who differentiated between teacher tasks and student tasks when resources such as laboratory equipment are deployed. They called this a cognitive science conception of curriculum and instruction.

Reminding us emphatically of the importance of teachers' knowledge, Schwab wrote:

> ... teachers will not and cannot be merely told what to do. Subject specialists have tried it. Their attempts and failures I know at first hand. Administrators have tried it. Legislators have tried it. Teachers are not, however, assembly line operators, and will not so behave. Further, they have no need, except in rare instances, to fall back on defiance as a way of not heeding. There are a thousand ingenious ways in which commands on what and how to teach can, will, and must be modified or circumvented in the actual moments of teaching. Teachers practise an art. Moments of choice of what to do, how to do it, with whom and at what pace arise hundreds of times a school day, and arise differently every day and with every group

> of students. No command or instruction can be so formulated as to control that kind of artistic judgement and behaviour, with its demand for frequent, instant choices of ways to meet an ever varying situation.
>
> (Schwab, 1983, p. 245)

Olson (1982) described the notion of teachers' practical knowledge and in Chapter 4.3 he explores the demands of curriculum upon teachers and the mechanisms developed by teachers to cope with science laboratory classes. He continues the argument of Brown and McIntyre (1982a) that if curriculum designers fail to take account of teachers' views then there is likely to be a gap between official intentions and teachers' actions.

## Subject matter

One of the commonplaces is certainly the subject matter (whether it be microbiology, chemistry or astrophysics) and we must be concerned with established knowledge, its modes of enquiry, the state of stable and fluid enquiry in the field (Schwab, 1960, 1962, 1964a, b). However, on numerous occasions Schwab warned of the tendency of subject matter to overwhelm consideration of the other commonplaces during curriculum deliberations. Descriptive studies of actual deliberations have borne this out (some of these were summarized and compared with Schwab's prescriptions by Hegarty, 1978).

We have selected for inclusion some of the important research on students' cognitive structures because of the extent to which it has stimulated discussion of subject matter issues. It has done so in a framework which acknowledges students' understanding of subject matter as the foremost issue whilst constantly referring also to teachers' and scientists' understanding of subject matter. One branch of this research concerns students' beliefs about natural phenomena.

At the moment, the implications of these findings for the use of the laboratory are more conjectural than the product of research and development. Nonetheless it is argued in Chapter 4.1 that the work has such implications, and that these can be argued from two perspectives. Firstly, it has been frequently suggested that the naive systems which students have arise largely from the attempts by students to make sense of their observations of the world around them. Secondly, many of the probes which have been used to explore student theories about the world have used tasks and approaches which are consistent with concrete laboratory experiences.

Some of the suggestions which impinge on laboratory work include the importance of using everyday equipment and phenomena in science teaching; the role of qualitative thought in the development of concepts; the ways in which cognitive discussions can assist in understanding ways of interpreting observations; the ways students conceptualize what they do in the laboratory. In addition, since many of the focus-problems and tasks used in research involve manipulation of science materials and equipment, common sense suggests a connection with the role of laboratory teaching. Some, like the physics velocity tasks for university students (McDermott, 1982) seem developed to the point where they could be used as student exercises in the laboratory for large groups.

## Contexts

Writing of the educational commonplaces, Schwab (1973) referred to the contexts in which a student's learning will take place and in which the fruits of teaching are born. He stated that the relevant contexts are manifold and used the metaphor of nesting one within another like Chinese boxes. He exhorted curriculum workers to consider contexts ranging from classroom environment to level of federal government funding. Numerous questions are called to mind concerning the contexts of science laboratory work. For example, what linkages are there (or might there be) between staff development programmes and the introduction of curriculum changes? Can funding facilitate these linkages? What support systems are there (and how might choices be made amongst them) which might more effectively liberate the creativity of educators and help the institutionalization of their ideas? How best should funds be distributed between equipment, materials and salaries in order to promote beneficial long-term influence on a programme? Policy questions of this kind are not usually thought of as relevant by curriculum developers (McKinney and Westbury, 1975) and have, thus, been largely ignored by workers in the field of curriculum studies. Yet such factors play an important role in deliberations or negotiations in schools and universities when curricula are planned within institutions or when institutions implement the work of outside planning groups (Reid, 1975). They have been taken more seriously as 'frame factors' (Bernstein, 1971) by workers in the sociology of education (e.g. Young, 1971; Eggleston, 1974). One of the arguments of the sociologists is that curriculum processes cannot be considered independently of decisions made at higher levels ranging from institution to government. To give some examples from work in schools, such decisions may involve such apparently disparate constraints as funding (Dahllöf, 1971; Lundgren, 1972; Wiley and Harnischfeger, 1974; Erickson, 1976; Hallinan, 1976; Harnischfeger and Wiley, 1976), the level of classroom teaching (Arlin and Westbury, 1976) and time (Westbury, 1977).

Chapter 5.2 illustrates the importance of the context in a description of the effects of the practical laboratory matriculation examination in Israel on the place that laboratory has occupied in the high school biology curriculum in Israel. The impact of these enquiry-oriented problem-solving examinations has been twofold: it stimulated and forced schools to build and operate well equipped laboratories run by specially trained laboratory technicians, and, more importantly, it has resulted in students and teachers actually spending a substantial portion of their time working in the laboratory and performing genuine investigations. Products have been a novel form of practical test as well as a standardized scoring instrument which has already been adopted by researchers and practitioners in countries other than Israel.

In considering policy contexts, the level of impact constantly arises. The level may be national, as illustrated in Chapter 5.1 on provision of laboratory resources and Chapter 5.2 on institution of practical laboratory examinations. Though sometimes referring to the utilization of nationally available enquiry-oriented science curriculum materials, the case studies reported by Stake and Easley (1978) were more concerned with the level of school district or individual school. Walker (1982) discussed school community and outsider and Brown and McIntyre (1982b) differentiated between organizational innovations (on which there is more consensus and more success) and pedagogical innovations (rely on individual teacher classroom practice with less consensus and less success). It is helpful to view Chapter 5.3 in this light. Innovations in the use of computers have been at the level of an individual teacher or, more importantly, at the level of a science department (especially in higher education). Questions of implementation across organizations or at a national level have seldom arisen.

## INTENDED LEARNING OUTCOMES

Important determinants of implementing a science curriculum are the goals which are selected as major learning outcomes to be achieved via each type of instruction (such as laboratory work, lecture and assignment). For laboratory classwork, the goals selected will influence the nature of the laboratory manuals provided, the nature of laboratory teaching and the activities of students in laboratories and periods of allied study.

Reviews of comparative learning studies (Cunningham, 1946; Bradley, 1968; Bates, 1978; Blosser, 1980) suggest that choices of goals for laboratory classwork should be made from amongst the following:

1. Teaching manipulative skills and increasing understanding of the apparatus;
2. Fostering understanding and experience of scientific enquiry;
3. Practice in designing and executing experiments, generating data for analysis and interpretation;
4. Developing attitudes to science laboratory, resourcefulness, creativity;
5. Introducing a new discipline, providing for individual differences, providing concrete learning experiences;
6. Fostering a sense of success, motivation and control in science.

In the research literature, the main approach has been to try to determine the relative importance of different goals for laboratory work by compiling a suitable list of goals and asking a group of expert judges (e.g. practising teachers or scientists) to rate each. This procedure has been followed at both secondary and tertiary levels for chemistry (Henry, 1975; Ben-Zvi, Hofstein and Samuel, 1977; Lynch and Gerrans, 1977), physics (Boud, 1973), microbiology (Hegarty, 1979). Repeatedly these lists showed rather routine goals such as those related to observation and measurement, accuracy and errors, analysis and interpretation as being the most important and those with most consensus. Goals related to manipulative skills were thought important by students but much less so by staff.

Some goals central to the notion of scientific enquiry were accorded lower importance and less consensus, e.g. recognition of problems, design of experiments.

Overall we now have some beginning studies on the nature of goals sought. The goal statements and methodology could be considerably improved. The studies could be extended to other disciplines and far more remains to be done at the tertiary level. However, more important than these criticisms is the lack of study of the constraints and mitigating factors which affect the selection and achievement of goals.

## INTENDING LEARNING OUTCOMES AND ACTIVITIES

### Technical skills

If one argues that the laboratory is the *sine qua non* of science, then technical skills are the *sine qua non* of the laboratory. If students' technical skills fail then it can be difficult to retrieve the experiment. Chapter 3.1 notes that technical skills may be broken down into categories such as methodical working, experimental technique, manual dexterity and orderliness. If this is done then it becomes easier to establish performance criteria, to plan teaching permitting practice and feedback and to plan evaluations.

Several research-based studies described in Chapter 3.1 show that improvement of technical skills may be obtained via manipulation of the conditions of learning (such as provision of criteria, practice or feedback). Schemes for content analysis showed a great variation in the extent to which different aspects of technical skills are emphasized in science curriculum materials. This can be linked to the earlier observation that secondary and, to a greater extent, tertiary teachers (and course designers?) regard technical skills as of relatively low priority and perhaps not fully worth systematic attention in curriculum materials. Students, on the other hand, seem to regard technical skills as scientifically and vocationally very important.

### Scientific enquiry

Here we focus first on student learning activities to be provided when intended learning outcomes stress scientific enquiry.

Lucas (1971) differentiated four different meanings for the term scientific enquiry: (1) scientists' techniques and procedures for enquiry, (2) scientific logic, (3) teaching-learning emphasizing probing, questioning, designing of techniques to gather information, (4) combinations, e.g. using technique (3) to teach about enquiry as in (1). Tamir (1983) reported considerable confusion by teachers amongst these different meanings. In laboratory work, what is needed is for course planners to choose or design exercises which require students to utilize enquiry skills such as formulating hypotheses and designing and executing experiments. As well as such implicit teaching, there should be explicit teaching about enquiry. This issue is fully explored in Chapter 3.2.

Content analysis techniques provide course planners with a tool for monitoring the level of scientific enquiry in intended learning activities (and thus the match between intended outcomes and activities). The scheme developed by Herron (1971) described laboratory exercises for their levels of openness for enquiry.

Reports of content analysis research based on Herron's technique or expanded versions of it (Herron, 1971; Hegarty, 1978; Tamir and Lunetta, 1978; Tamir, 1983) show that the majority of student exercises in published laboratory manuals are at low levels.

## Scientific knowledge

Science educators are increasingly seeking to understand how laboratory experiments actually influence the learning of scientific knowledge - particular symbols or facts, concepts, laws, principles, theories and generalizations. With a greater understanding of this, teachers will be able to develop more effective instructional environments that should result in intended student learning outcomes being more likely achieved.

It is now recognized that different instructional strategies are needed to promote the learning of different types of scientific knowledge. For example, learning a new concept requires the prior learning of subordinate concepts, whereas facts are more readily learned if associated within a meaningful context and need not be subordinately related. Simply including a laboratory experiment in the instructional process is insufficient. Of significant importance is the need to clearly identify the intended learning outcomes for students. This influences how the experiment is to be conducted, the nature of the questions to be asked and the kind of student involvement. Recent studies, as described in Chapter 3.3, have explored student perceptions of science laboratory experiments and have highlighted that often only able students appear to strongly link the appropriate scientific knowledge with the laboratory experiment. Other students generally recall the 'doing of the experiment' but not the associated scientific knowledge. Teachers need to emphasize more the links between a laboratory experiment and the scientific knowledge associated with this, rather than leaving students to make such links themselves. When students conduct an experiment in a small group, each student has a somewhat different experience. There is no one standard laboratory experiment. Exploring the impact of such differences on the learning of scientific knowledge is a significant challenge.

## Attitudes

In order to deal systematically with the affective outcomes of science education, Klopfer (1976) developed a taxonomy of educational objectives for this area. These objectives include outcomes related to a student's attitude to science and those related to the student's scientific attitude (or the tendency for the student to adopt scientific enquiry as a way of thought). Major reviews of research into attitudes to science (Gardner, 1975) and scientific attitudes (Gauld and Hukins, 1980) do not refer specifically to laboratory work and indicate the poor quality of much of the work in this area.

Attitudes are related to a willingness or a preference to behave in a particular way. For example, one might express a preference for (or spontaneously engage in) science-related activities such as reading science books or carrying out experiments; when carrying out experiments one might habitually check observations carefully before proceeding. These activities are consistent with the possession of attitudes which are thought to be an important aspect of scientific enquiry.

The types of activities most suitable for developing student attitudes are those which encourage students to be satisfied with behaving in a manner appropriate to doing good science. Thus one might argue that successfully solving real-life scientific problems is likely to promote satisfaction and that this satisfaction thus becomes associated with behaving in a scientific manner and enjoying doing science. The evidence for such a supposition and for other means by which such attitudes may be developed is discussed further in Chapter 3.4. But in spite of the emphasis placed on science-related attitudes in the science education literature (Fraser, 1978) there is some evidence that this emphasis is not reflected in classroom practice (Welch *et al*. 1981). One likely cause of this discrepancy is that classroom attitudes are difficult to assess in a valid manner.

We end this section by stating that where we suggest there should be congruence between intended learning outcomes concerning scientific enquiry and intended activities, evidence suggests that congruence can be obtained but not without special effort.

## ACTUAL LEARNING ACTIVITIES

### Life in classrooms

Here we are concerned with what actually happens in the classroom as distinct from goals, instructional plans as embodied in curriculum materials, or teachers' hopes. There is a growing literature on life in laboratory classrooms in different disciplines and institutions and at different levels, with reviews by Power (1977), Hegarty (1982), Dunkin and Barnes (1985) and White and Tisher (1985).

Chapters 6.1 and 6.2 deal with junior and senior levels of secondary science. They show that activities-based and enquiry-oriented laboratories are significantly different from conventional ones: teachers are less direct, more planning takes place, processes of science receive more emphasis, there is more post-lab discussion and teachers give fewer instructions in front of the whole class, but move around more, checking, probing and supporting. Students are usually more active and they initiate ideas more frequently. Unexpectedly, the verbal interaction among students in their working group is not related to the level of enquiry. Interactions are predominantly on a low cognitive level, featuring techniques and procedures. The enquiry-oriented curriculum appears to impose an overload on the average and slow learners at the secondary level, possibly as a result of the need to attend simultaneously to a wide range of objectives.

Recognizing this, Chapter 6.1 points out that teachers in

enquiry-oriented programmes adapt their teaching goals and style in accordance with the perceived ability level of the class. Indeed, Tisher and Power (1973, 1975) found that ASEP teachers spent much time in laboratory settings giving directions and information in low ability classes and much less in fostering enquiry than in high ability classes, while Edwards and Marland (1984) report a high incidence of interactions with neighbours in ASEP classrooms involving comparing one another's work, cooperating in assigned tasks, discussing results - a reasonable amount of non-enquiry-oriented thinking and discussion. While it is clear that enquiry-oriented curricula do call forth much more interaction with materials, experimentation, observation and recording of results than conventional curricula, both students and teachers develop strategies for coping with the multiple demands of such, which may lead to gaps between the intended and actual learning activities and outcomes.

At the tertiary level, Chapter 6.3 shows that teaching staff seem to spend most of their time concentrating on development of subject matter, content and laboratory procedures. Some staff talk nearly all the time, leaving little time for listening. The talk is generally at low cognitive levels; there is little emphasis on scientific enquiry or reflections of it in problem identification or design of experimental approaches. However, there is some evidence of emphasis on data interpretation and the formulation of conclusions. Supervision and management are time consuming. Talk and activities unrelated to class are noticeable. By comparison with staff, university students seem to spend far less time talking and then mostly on questions about techniques and procedures and responses characterized by facts and definitions. Most time is spent silently working on task related laboratory activities and organization. In Chapter 6.4, students' reactions are compared across the different years of study.

## Effects of enquiry-oriented curriculum materials

Given the emphasis so far on scientific enquiry, it is appropriate that the effects of enquiry-oriented curriculum materials on behaviour in laboratories be considered separately. We will be looking to see the extent to which the effects are in the desired direction, i.e. the extent to which intended learning outcomes and activities are reflected in actual learning activities.

Chapter 6.2 uses three main data sources about the

activities of students and teachers in secondary science laboratories: (1) students' reports, (2) teachers' reports, (3) direct observations (either structured or unstructured) by independent observer/s. Using the most common technique, direct observation, results show that the teacher is still the dominating figure who initiates and manages secondary laboratory classroom transactions. On the average, the students talk about 13 per cent of class time and only about 6 per cent of the time they are off task. About half the teachers' talk is devoted to content associated with the students' work. Relatively little time is devoted to the nature of scientific enquiry, to the process of science and to the assessment of limitations of investigations performed. Enquiry and the time allocated to performing experiments hardly exist at the ninth grade and become more noticeable in higher grades.

Chapters 6.3 and 6.4 show little evidence of enquiry-oriented science in higher education unless special exercises or projects are designed and are used with special teaching methods. Expected results of using enquiry-oriented curriculum materials may include increases in time spent by students talking on matters related to scientific enquiry. However, this may be with fellow students rather than with staff and in all there is a strong sense of staff acting as laboratory managers but not as questioners, challengers or promoters of scientific enquiry.

## CONCLUSION

This chapter conveys the structure of the book. The process of translating elements of a science discipline into curriculum and instruction is shown to be influenced by many factors - policies and constraints as well as considerations of learners, teachers, subject matter and context. These influences and their effects are taken up in later chapters where it is found helpful to monitor the process by examining contrasts between intended and actual learning activities. The structure of the book is also represented as a model (Chapter 1.2) which is used throughout.

## REFERENCES

Arlin, M. and Westbury, I. (1976) The leveling effect of teacher pacing on science content mastery, *Journal of Research in Science Teaching*,

13, 213-219.
Bates, G.C. (1978) The role of the laboratory in secondary school science programs, in *What Research says to the Science Teacher*, Volume 1 (Ed. Rowe, M.B.), Washington, D.C.: National Science Teachers' Association.
Ben-Peretz, M. and Tamir, P. (1981) What teachers want to know about curriculum materials, *Journal of Curriculum Studies*, 13, 45-54.
Ben-Zvi, R., Hofstein, A. and Samuel, D. (1977) Use of objectives questionnaire as a method for curriculum evaluation, *Science Education*, 61, 415-422.
Bernstein, B. (1971) On the classification and framing of educational knowledge, in *Knowledge and Control: New Directions for the Sociology of Education* (Ed. Young, M.F.D.), London: Collier-Macmillan.
Black, P.J. and Ogborn, J. (1979) Laboratory work in undergraduate teaching, in McNally, D. *Learning Strategies in University Science*, University College Cardiff Press, ICSU Committee on the Teaching of Science, pp. 161-201.
Blosser, P.E. (1980) A critical review of the role of the laboratory in science teaching. ERIC ED 206445.
Boud, D.J. (1973) The laboratory aims questionnaire - a new method for course improvement? *Higher Education*, 2, 81-94.
Bradley, P.L. (1968) Is the science laboratory necessary for general education science courses? *Science Education*, 52, 58-66.
Brown, S. and McIntyre, D. (1982a) Costs and rewards of innovation: taking account of the teachers' viewpoint, in Olson, J. (Ed.). *Innovation in the Science Curriculum*. Ch. 4, pp. 107-139. London: Croom Helm.
_____ and McIntyre, D. (1982b) Influences upon teachers' attitudes to different types of innovation: a study of Scottish Integrated Science, *Curriculum Inquiry*, 12, 35-51.
Champagne, A.B., Gunstone, R.F. and Klopfer, L.E. (1983) Naive knowledge and science learning, *Research in Science and Technological Education*, 1, 173-183.
_____ and Klopfer, L.E. (1978) *Cumulative Index to "Science Education": Volumes 1 through 60, 1916-1976*. New York: Wiley.
Cunningham, H.A. (1946) Lecture demonstration versus individual laboratory method in science teaching - a summary, *Science Education*, 30, 70-82.
Dahllöf, U.S. (1971) *Ability Grouping, Content Validity, and Curriculum Process Analysis*, New York: Teachers' College Press.
Devonport, J., Lazonby, J.N. and Waddington, D.J. (1979) Attitudes to practicals. *Education in Chemistry*, 16, 188-190.
Dunkin, N.J. and Barnes, J. (1985) Research on Teaching in Higher

Education, in *Handbook of Research on Teaching* (Ed. Wittrock, M.C.) Third edition, New York: Macmillan.

Edwards, J. and Marland, P. (1984) A comparison of students' thinking in a mathematics and a science classroom, *Research in Science Education*, 14, 29-38.

Eggleston, J. (Ed.) (1974) *Contemporary Research in the Sociology of Education*, London: Methuen.

Erickson, D.A. (1976) Implications for organizational research of the Harnischfeger-Wiley model, *Curriculum Inquiry*, 6, 61-68.

Fraser, B. (1978) Use of content analysis in examining changes in science education aims over time, *Science Education*, 62, 135-141.

Gardner, P.L. (1975) Attitudes to science - a review, *Studies in Science Education*, 2, 1-41.

Garrett, R.M. and Roberts, I.F. (1982) Demonstration versus small group practical work in science education. A critical review of studies since 1900, *Studies in Science Education*, 9, 109-146.

Gauld, C. (1982) The scientific attitude and science education: a critical reappraisal, *Science Education*, 66, 109-121.

_____ and Hukins, A.H. (1980) Scientific attitudes - a review, Studies in *Science Education*, 7, 129-161.

Gunstone, R.F. and White, R.T. (1981) Bringing students' conceptions into accord with scientists'. *The Australian Science Teachers' Journal*, 27(3), 5-7.

Hallinan, M.T. (1976) Salient features of the Harnischfeger-Wiley model, *Curriculum Inquiry*, 6, 45-59.

Harnischfeger, A. and Wiley, D.E. (1976) The teaching-learning process in elementary schools: a synoptic view, *Curriculum Inquiry*, 6, 5-43.

Hegarty, E.H. (1978) Levels of scientific enquiry in university science laboratory classes: Implications for curriculum deliberations, *Research in Science Education*, 8, 45-57.

_____ (1979) The Role of Laboratory Work in Teaching Microbiology at University Level. Unpublished Ph.D. thesis, University of New South Wales.

_____ (1982) The role of laboratory work in science courses: Implications for college and high school levels, Chapter 7 (pp. 82-105 and 162-167) in *Education in the 80s: Science* (Ed. Rowe, M.B.) Washington, D.C. National Education Association.

Henry, N.W. (1975) Objectives for laboratory work, in *The Structure of Science Education* (Ed. Gardner, P.L.), Hawthorn, Victoria: Longman.

Herron, M.D. (1971) The nature of scientific enquiry, *School Review*, 79, 171-212.

Hofstein, A. and Lunetta, V.N. (1982) The role of the laboratory in science teaching: neglected aspects of research. *Review of*

*Educational Research*, 52, 201-218.

*How Can Written Curriculum Guides Guide Teaching?* (1983) Papers by Westbury, I., Anderson, D.C., Olson, J.K., Harris, I.B. and Reid, W. are revised versions of papers presented at a symposium held at the AERA Annual Meeting 1981. Published in *Journal of Curriculum Studies*, 15.

Kerr, J.F. (1964) *Practical Work in School Science: an Account of an Inquiry sponsored by the Gulbenkian Foundation into the Nature and Purpose of Practical Work in School Science Teaching in England and Wales* (revised edition), Leicester: Leicester University Press.

Klopfer, L.E. (1976) A structure for the affective domain in relation to science education, *Science Education*, 60, 299-312.

Lucas, A.M. (1971) Creativity, discovery and inquiry in science education, *Australian Journal of Education*, 15, 184-196.

Lundgren, U.P. (1972) *Frame Factors and the Teaching Process: A Contribution to Curriculum Theory and Theory on Teaching*. Stockholm: Almqvist and Wiksell.

Lynch, P.P. and Gerrans, G.C. (1977) The aims of first-year chemistry courses, The expectations of new students and subsequent course influences, *Research in Science Education*, 7, 173-180.

McDermott, L.C. (1982) Problems in understanding physics (kinematics) among beginning college students - with implications for high school courses, Chapter 8 (pp. 106-128 and 167-168) in *Education in the 80s: Science* (Ed. Rowe, M.B.) Washington D.C.: National Education Association.

McKinney, W.L. and Westbury, I. (1975) Stability and change: the public schools of Gary, Indiana, 1940-1970, in *Case Studies in Curriculum Change: Great Britain and the United States* (Ed. Reid, W.A. and Walker, D.F.) London: Routledge and Kegan Paul.

Newman, B.C. (1981) Practical work in secondary school chemistry, in Fogliani, C.L. and McKellar, J.R. (Eds.) *The 1980s: a challenge and responsibility for chemical education*, pp. 209-224, Bathurst: RACI Chemical Education Division.

_____ (1983) The theory and practice of practical work, *South Australian Science Teachers' Association*, No. 833, 4-14.

Olson, J. (1982) Classroom knowledge and curriculum change: an introduction, in Olson, J. (Ed.) *Innovation in the Science Curriculum*, Ch. 1, pp. 3-33, London: Croom Helm.

Posner, G.J., Strike, K.A., Hewson, P.W. and Gertzog, W.A. (1982) Accommodation of a scientific conception: towards a theory of conceptual change. *Science Education*, 66, 211-227.

Power, C. (1977) A critical review of science classroom interaction studies, *Studies in Science Education*, 4, 1-30.

Reid, W.A. (1975) The changing curriculum: theory and practice, in *Case*

*Studies in Curriculum Change: Great Britian and The United States* (Ed. Reid, W.A. and Walker, D. F.) London: Routledge and Kegan Paul.

Schwab, J.J. (1960) What do scientists do? *Behavioral Science*, 5, 1-27.

_____ (1962) The concept of the structure of a discipline, *Educational Record*, 43, 197-205.

_____ (1964a) Structure of the disciplines: meanings and significances, in *The Structure of Knowledge and the Curriculum* (Eds. Ford, G.W. and Pugno, L.), Chicago: Rand McNally.

_____ (1964b) The structure of the natural sciences, in *Yhe Structure of Knowledge and the Curriculum* (Eds. Ford, G.W. and Pugno, L.), Chicago: Rand McNally.

_____ (1969) The practical: a language for curriculum, *School Review*, 78, 1-23.

_____ (1973) The practical 3: translation into curriculum, *School Review*, 81, 501-22.

_____ (1974) Decision and choice : the coming duty of science teaching, *Journal of Research in Science Teaching*, 11, 309-317.

_____ (1983) The Practical 4: something for curriculum professors to do, *Curriculum Inquiry*, 13, 239-265.

Stake, R.E. and Easley, J.A. (1978) *Case Studies in Science Education*, University of Illinois at Urbana-Champaign: Center for Instructional Research and Curriculum Evaluation and Committee on Culture and Cognition. Volumes I and II (prepared for National Science Foundation Directorate for Science Education; Office of Program Integration).

Tamir, P. (1983) Inquiry and the science teacher, *Science Education*, 67, 657-672.

_____ and Lunetta, V.N. (1978) An analysis of laboratory inquiries in the BSCS Yellow Version, *American Biology Teacher*, 40, 353-357.

Tisher, R. and Power, C. (1973) The effects of teaching strategies where ASEP materials are used. University of Queensland: Report to Australian Advisory Committee on Research and Development in Education.

_____ and Power, C. (1975) The effects of classroom activities, pupils' perceptions and educational values in lessons where self-paced curricula are used. Monash University: Report to Australian Advisory Committee on Research and Development in Education.

Walker, D.F. (1976) Toward comprehension of curricular realities, in *Review of Research in Education 4* (Ed. Shulman, L.S.) Itasca, Illinois: F.E. Peacock.

Walker, R. (1982) The school, the community and the outsider: case study of a case study, in Olson, J. (Ed.) *Innovation in the Science Curriculum*, Ch. 2, pp. 34-71. London: Croom Helm.

Welch, W.W., Klopfer, L.E., Aikenhead, G.S. and Robinson, J.T. (1981) The role of inquiry in science education: Analysis and recommendations, *Science Education*, 65, 33-50.

West, L. (1984) Learning large bodies of discipline knowledge. *HERDSA News*, 6 (3), 8-11.

Westbury, I. (1977) Education policy-making in new contexts: the contribution of curriculum studies, *Curriculum Inquiry*, 7, 3-18.

White, R.T. and Tisher, R.P. (1985) Natural sciences in *Handbook of Research on Teaching*, Third edition, (Ed. Wittrock, M.C.) New York: Macmillan.

Wiley, D.E. and Harnischfeger, A. (1974) Explosion of a myth: quantity of schooling and exposure to instruction, major educational vehicles, *Educational Researcher*, 3 (April), 7-12.

Young, M.F.D. (Ed.) (1971) *Knowledge and Control: New Directions for the Sociology of Education*, London: Collier-Macmillan.

# 1.2

# The Student Laboratory and the Science Curriculum: A Model

Elizabeth Hegarty-Hazel

## INTRODUCTION

In this book it is argued that the laboratory ought to be an integral part of the science curriculum and that it ought to raise key curriculum questions to anyone developing or teaching in a science programme. In many of the chapters it is shown that the student laboratory has not fulfilled its potential, is not well integrated, is often excluded from certifying examinations and so on. It is suggested that what is needed is not just more and better laboratories but rather a much better understanding of the interactions between learning in the laboratory and elsewhere.

The purpose of this brief chapter is to present a model of the student laboratory and the science curriculum and to redress the situation where few models have focused on the science curriculum and few or none on the laboratory. Models have often been the province of curriculum theorists and researchers, or of educational developers. However, this model is presented to the general reader as a summary of the underlying arguments of this book and of its structure. All chapters were invited on the basis of the model and authors used it as best served their purposes in their chapters.

Since models are always a contentious issue (some love them, some hate them), some defence is in order. The usual debates continue among protagonists about exactness of terms and definitions, and the positioning of components of models. Antagonists claim that models invite over-simplifications and over-generalizations and that models are seldom explicitly useful to curriculum developers in exactly their published form.

For the present model it is not relevant to state anything more ambitious than that it is a pictorial representation of an important educational issue and that it is a logically derived systematic method of interrelating ideas, propositions, theories and bodies of knowledge. Using Pratte's (1981) definitions, it would be described as a metaphorical model, constructed as a way of dismembering or picturing phenomena and representing them via a simplification. It meets his description (from Toulmin) of a valuable model as one which suggests further questions, taking its users beyond the phenomena from which they began and encouraging them to formulate hypotheses. Suggestiveness and deployability elevate such a model beyond the level of simple metaphor.

The model displayed in Figure 1.2.1 is logical not chronological and thereby makes no suggestion that curriculum development should occur in the order depicted. As common sense and experience have shown, the processes of curriculum reflection and deliberation take place, chronolo- gically, in a back-and-forth manner between means and ends (Schwab, 1983). This however does not reduce the value of a good model as a conceptual tool and a way of depicting interrelationships. The model is presented in a way which is suggestive of new questions, fresh interpretations, and different techniques. It should help illuminate complex systems and events and help workers in the field to find a focus amongst the seamless web of events. It is irrelevant if the model is not used in its entirety by either researchers or practitioners in curriculum. Ample descriptions exist for successful and productive use of parts of models (e.g. Posner and Rudnitsky, 1982). However, more often questions of usefulness arise because models are entirely technical in nature (Walker, 1976), ignoring the everyday constraints ofthe world of curriculum decision making. The present model is emphatically set in the domain of the practical (in the sense used by Schwab, 1969, 1971, 1973, 1974, 1983).

## EARLIER MODELS

Early curriculum models (Tyler, 1949; Taba, 1966; Wheeler, 1967) became increasingly descriptive but were criticized for starting with aims and thereby lacking the dimension of values

(Macdonald, 1975). Johnson's (1967; 1969) curriculum model made it clear that value choices would be involved in the development of a curriculum from 'available cultural content'. It also had the advantage of explicitly separating the processes of curriculum and instruction. Disadvantages were that the model relied on a rather theoretical and technical approach without reference to practical considerations, i.e. there was no representation of the realities of curriculum or instruction. The student was not represented in the model but learning outcomes were shown. The ways in which curriculum and instruction may result in learning outcomes were unclear.

Several other curriculum models were developed which did seek to represent students. For example Saylor and Alexander (1974) showed both 'student inputs' and 'student outputs'. However, the model of school effects developed by Harnischfeger and Wiley (1976) suggested that care is required in the positioning of these variables - in their model, the only variables which exerted *direct* influence on student achievement were those described as student pursuits, i.e. task related activities inside and outside the classroom.

## THE MODEL OF STUDENT LABORATORY AND SCIENCE CURRICULUM

The main areas of development in the model (Figure 1.2.1) may be summarized briefly:

1. Use of a *logical* (rather than chronological) sequence with means-ends reasoning.
2. *Arrows* or connecting points in the model which serve as the focus for questions of matching.
3. Inclusion of the notion of *cultural content* (Johnson, 1969) as the source of curriculum since this implies a curriculum process of significantly wide scope, e.g. encompassing an entire school or higher education system. When the scope is more limited (e.g. study of a single science discipline at school or higher level) a more circumscribed source of curriculum content is appropriate. In the present model, the notion of *elements of science* (Schwab, 1974) is also used.
4. With Johnson (1969) the separate representation of the *processes* of curriculum development, instructional planning and instruction. The same convention is

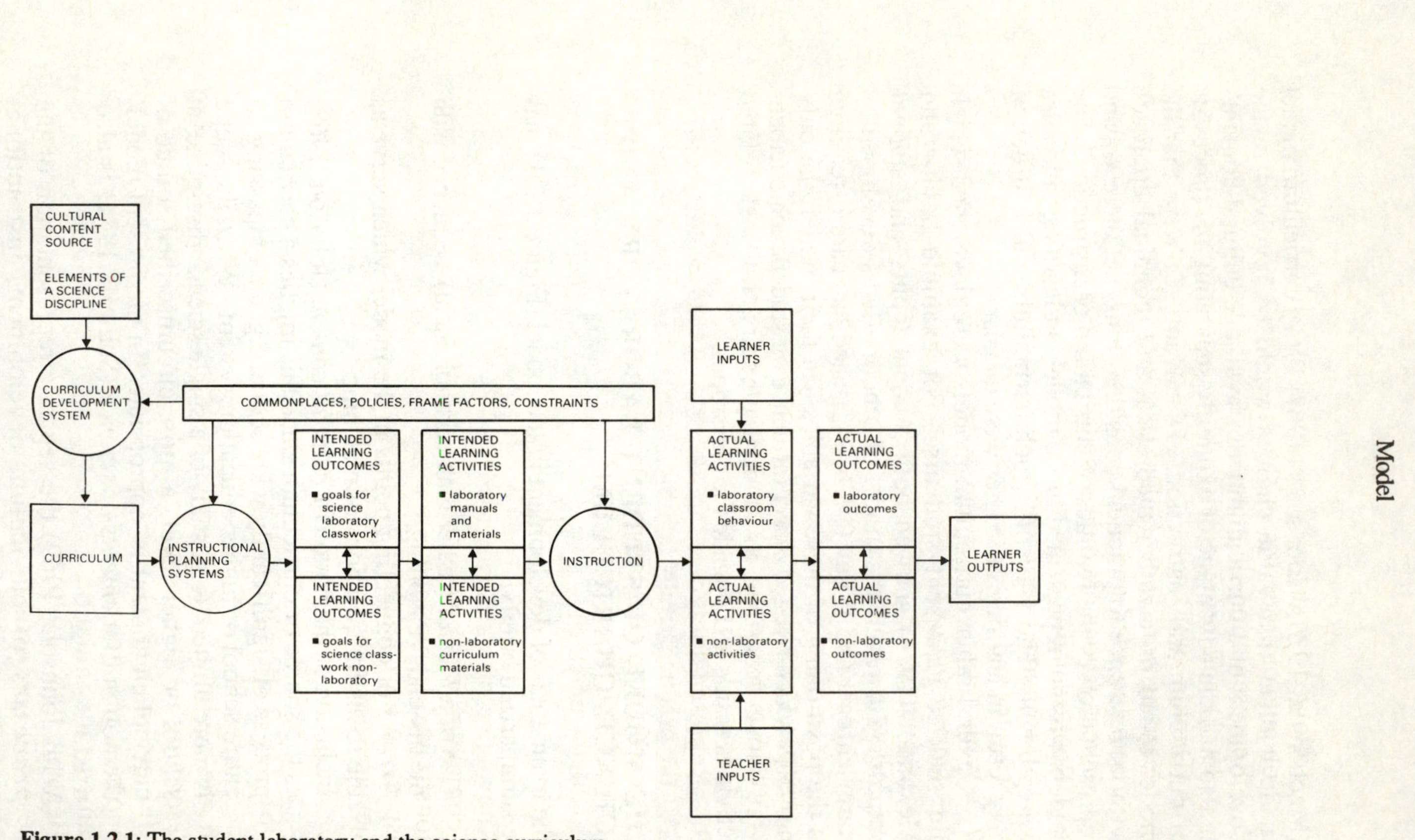


**Figure 1.2.1:** The student laboratory and the science curriculum

followed by showing processes as circles and inputs or outputs of processes as blocks.

5. Representation of intended learning *activities* (Herron, 1971) and representation of actual learning *activities* (Dunkin and Biddle, 1974) as well as intended and actual learning *outcomes* (Johnson, 1967; 1969). Inclusion of both *intended* and *actual* invites both comparisons and investigations of congruence (Stake, 1967).
6. Representation of *goals* of both the curriculum overall and of a planned series of learning activities and outcomes (Saylor and Alexander, 1974).
7. With Harnischfeger and Wiley (1976) representation of *learner inputs* as directly affected only by actual learning activities. Learner outputs follow logically. Knowledge *about* students exerts an indirect effect and is represented in the model as one of the commonplaces (see the section on 'learners' in Chapter 1.1).
8. In parallel with (7), *teachers* are shown as extending direct effects on actual learning activities. Again, knowledge *about* teachers is one of the commonplaces represented.
9. Representation of all those forces which, in the realm of the practical, must be brought to bear on the judgement of curriculum alternatives (Schwab, 1974). This is the point at which the emphasis is clearly changed from the technical to the practical.

Knowledge of learners, teachers, subject matter and contexts ('commonplaces') was introduced in Chapter 1.1. Although constraints, frame factors and policies may be considered as aspects of the context in which a science curriculum operates, they are shown separately in the model for emphasis.

## USE OF THE MODEL

In the realm of the theoretical, the model aspires to be *descriptive* of the educational processes in a science discipline. All elements and relationships in the model are amenable to empirical research and the model can be judged partly on its value to the initiation and execution of investigative studies (Hegarty, 1978, 1979). Where research findings are available, these will be discussed in later chapters. In the realm of the practical, the model is *prescriptive* and thus could be used by curriculum workers involved in developing new science curricula, planning methods and materials of instruction or in reviewing and re-planning established curricula. The model pays suitable tribute to the complexities of the practical domain. Its value as a conceptual scheme can be judged partly on the quality of the reviews and recommendations based on it and authors of later chapters have accepted a specific brief to tease out practical consequences of their investigations. Ultimately, however, work in the domain of the practical cannot be judged in the abstract - its value must be judged by its usefulness to curriculum deliberation groups in a specific situation (e.g. Hegarty, 1979, who reported deliberations on laboratory work in microbiology where the model was used to guide the structure of the deliberative group and the processes used).

## REFERENCES


Dunkin, M.J. and Biddle, B.J. (1974) *The Study of Teaching*, New York: Holt, Rinehart and Winston.

Harnischfeger, A. and Wiley, D.E. (1976) The teaching-learning process in elementary schools: a synoptic view, *Curriculum Inquiry*, 6, 5-43.

Hegarty, E.H. (1978) Levels of scientific enquiry in university science laboratory classes: implications for curriculum deliberations, *Research in Science Education*, 8, 45-57.

_____ (1979) The Role of Laboratory Work in Teaching Microbiology at University Level; unpublished Ph.D. thesis, University of New South Wales.

Herron, M.D. (1971) The nature of scientific enquiry, *School Review*, 79, 171-212.

Johnson, M. (1967) Definitions and models in curriculum theory, *Educational Theory*, 17, 127-140.

_____ (1969) The translation of curriculum into instruction, *Journal of Curriculum Studies*, 1, 115-131.

Macdonald, J.B. (1975) Curriculum theory, in *Curriculum Theorizing: the Reconceptualists* (Ed. Pinar, W.) Berkeley, California: McCutchan.

Posner, G.J. and Rudnitsky, A.N. (1982) *Course Design*, Second edition, New York: Longman.

Pratte, R. (1981) Metaphorical models and curriculum theory, *Curriculum Inquiry*, 11, 307-320.

Saylor, J.G. and Alexander, W.M. (1974) *Planning Curriculum for Schools*, New York: Holt, Rinehart and Winston.

Schwab, J.J. (1969) The Practical: a language for curriculum, *School Review*, 78, 1-23.

_____ (1971) The Practical 2: arts of eclectic, *School Review*, 79, 493-542.

_____ (1973) The Practical 3: translation into curriculum, *School Review*, 81, 501-522.

_____ (1974) Decision and choice: the coming duty of science teaching, *Journal of Research in Science Teaching*, 11, 309-317.

_____ (1983) The Practical 4: something for curriculum professors to do, *Curriculum Inquiry*, 13, 239-265.

Stake, R.E. (1967) The countenance of educational evaluation, *Teachers' College Record*, 68, 523-540.

Taba, H. (1966) *Learning Strategies and Cognitive Functioning in Elementary School Children*, U.S.O.E. Cooperative Research Project No. 2404, San Francisco State College.

Tyler, R.W. (1949) *Basic Principles of Curriculum and Instruction*, Chicago: University of Chicago Press.

Walker, D.F. (1976). Toward comprehension oxf curricular realities, in *Review of Research in Education 4* (Ed. Shulman, L.S.) Itasca, Illinois : Peacock

Wheeler, D.K. (1967) *The Curriculum Process*, London: Hodder and Stoughton.

# Part 2

# ELEMENTS OF A SCIENCE DISCIPLINE

# 2.1

# Student Laboratory Practice and the History and Philosophy of Science

**David Layton**

## INTRODUCTION

The aim of this chapter is to explore the relationship between the practices in which students engage whilst learning science in laboratories and our understandings of the nature of scientific activity. The general thesis is that the philosophy of science has rarely been used in a systematic and deliberate manner as a prime source of objectives for student laboratory work. Instead, it has been a resource, drawn upon selectively - raided, even - to underwrite purposes and practices which have their origins in considerations remote from philosophy. As a result, student laboratory practices have frequently been reflective of 'dubious or discarded philosophies of science' (Martin, 1979, p. 331); questions about the desirability and feasibility of students' laboratory experiences being congruent with the discipline, as practised by research scientists, have rarely been addressed, much less answered.

First, an account will be given of the origins and institutionalization of laboratories for the teaching of science. Second, some broad features of the philosophy of science as it was expounded in the period when student laboratories were first designed will be identified. Third, the educational objectives of laboratory work will be briefly reviewed in the light of evidence from the history of student laboratories and the philosophy of science. The purpose here will be to evaluate the extent to which views about the nature of scientific activity have, in the past, served as a major influence on the intended learning outcomes of students' laboratory work. Finally, consideration will be given to the implications of contemporary views about the nature of scientific activity for the laboratory practices in which students engage when they are learning science.

It will be seen that there is a political dimension to curriculum making and that the model (Figure 1.2.1) needs complementing along other dimensions in order to understand the shaping of student laboratory practice. The model includes reference to policy constraints and frame factors and to some extent these issues are taken up in Chapters 5.1 to 5.4. However, the excercise of power and control needs to be specifically addressed. There is here a suggestion of one limitation of the model: whilst identifying and articulating the pieces of the game, it is silent on the identity and powers of the players.

## THE ORIGINS OF LABORATORIES FOR THE TEACHING OF SCIENCE

The teaching, as opposed to the non-teaching, science laboratory was a product of the nineteenth century and a concomitant of the professionalization of scientific activity in that period. Laboratories for scientific enquiry have existed from the seventeenth century (if not before), especially analytical chemical laboratories in institutions such as universities and mining academies. But it was not until much later that, to borrow the words of one of the pioneers - Thomas Thomson of Glasgow University - an attempt was made to 'establish a real chemical school ... and to breed up a set of young practical chemists' (Morrell, 1969, p. 249). The pedagogy developed in such centres, and institutionalized there in laboratory designs, equipment and technique, became a powerful means of uniting the diverse chemical community of the mid-century and of controlling the definition of chemistry thereafter (Morrell, 1972; Bud and Roberts, 1984).

A major obstacle to the establishment of practical laboratory classes in universities at this time was the direct dependence of the salary of many professors upon fees paid by students. As a result, efforts were directed to recruiting large classes and the style of teaching adopted, frequently involving dramatic, eye-catching demonstration experiments in a lecture to some hundreds of students, was an outcome of the economic context in which the teacher had to work (Simpson, 1982, p. 49). Outside the universities, private, entrepreneurial teaching laboratories arose as a means of providing for medical students and others a training in those practical chemical skills necessary for their intended profession. In contrast, however, one great attraction of Liebig's laboratory at Giessen in the 1830s, perhaps the most celebrated and successful nineteenth-century institution for the breeding of chemists, was the low level of fees charged. With salaried status, and independence of student fees for his livelihood, Liebig could offer laboratory chemical education at a cost which no English institution could match. Another inducement was the mastery of new techniques of organic analysis which could be applied to the investigation of a very wide range of substances and practical problems in fields such as agriculture, manufacturing industry, medicine, pharmacy and public health.

It was one of Liebig's former students, A. W. Hofmann, who was recruited to become the first professor of the Royal College of Chemistry, founded in London in 1845. Perhaps sensing a threat in this development, both King's College and University College London decided in the same year to build chemical laboratories and institute analytical courses, in each case appointing a Giessen-trained man to teach the new practical schemes (Bud and Roberts, 1984, p. 63). Indeed, according to J. B. Morrell, 'by the 1850s the Liebig clan had assumed most of the "plums" available in British university chemistry', eleven chairs then being occupied by former students at Giessen (Morrell, 1972, p.19).

What accompanied this expansion of laboratory teaching in chemistry has sometimes been described in terms of the importation of the German ideal of *Lernfreiheit* (Cardwell, 1972, p. 87). Certainly there was a progressive curriculum shift in emphasis from science as useful knowledge to science as an activity which required no extrinsic justification. In both the UK and the USA the ideology of pure science was burnished strongly in the second half of the nineteenth century (Williamson, 1870; Gore, 1882; Daniels, 1967), so much so that by the opening of the twentieth century a President of the Royal Society could state that the 'supreme value of research in pure science ... could no longer be regarded as a question open to debate' (Huggins, 1906, p. 21).

The reasons for this change in objectives in the case of chemistry illustrate something of the complexity of the macro-level influences which can contribute to the social shaping of a curriculum. One strand of debate concerned the extent to which the laboratory course for students should embrace knowledge and techniques relevant to particular industries; specifically should it include trade skills and manufacturing 'know-how'? On this point industrial opposition was general, both manufacturers and employees being unwilling to participate in such an education lest trade secrets were revealed to competitors and jobs were lost (Science and Art Department, 1866).

There was also a point of principle about state-supported chemical education, the Royal College of Chemistry having been 'nationalized' in 1853. The idea of such an institution producing, or contributing to the production of, marketable products which might compete against the products of private enterprise was totally unacceptable. As one nineteenth- century civil servant expressed it, 'A state pottery school and Messrs Minton could not exist side by side'. Such a situation was incompatible with the prevailing economic doctrine of a 'fair field and no favours'.

Most crucially, however, the institutional contexts of the new chemical laboratories - University College London, King's College London, the Royal College of Chemistry and, from 1851, Owens College, Manchester (the future University of Manchester) - obliged chemists to construct justifications for the study of their subject in terms identical to those applicable to the established components of the curriculum, classics and mathematics. It was important to demonstrate the educational value, as well as the practical utility, of chemistry.

From an educational point of view, chemistry without laboratory work was seen as a body of factual information and general laws which conveyed nothing of lasting power to the mind. Its educational potential could only be realized in the laboratory. As John Gardner, another of Liebig's students and the first secretary of the Royal College of Chemistry expressed it, 'In a university education, the information obtained is generally held to be inferior in importance to the discipline of the mind, the attainment of habits of continued attention, and the exercise of reasoning power and memory. The deep study of classics and mathematics teaches men *how to learn*, as well as furnishes them with instruments available on all occasions and on all subjects ...'. Similarly, the study of chemistry, when pursued practically as in Liebig's laboratory, could be 'an instrumental means of attaining an intellectual habit and qualification for the observation and interpretation of Nature. He who has thus become an expert analyst, has obtained a power capable of application in any direction for advancing his own knowledge' (Gardner, 1846, p. 296). In other words, the twin aims of the advancement of chemical knowledge and the acquisition of a liberal education could be made compatible and simultaneously achieved if student chemical laboratories were conducted on Giessen lines.

Despite a lack of warranty for his claim about unlimited transfer of learning, Gardner had articulated the *leitmotiv* of justifications for the laboratory teaching of chemistry throughout the century. When, some forty years later, a committee of the British Association for the Advancement of Science reported on methods of teaching chemistry it was categorical in its assertion that 'science should be taught from the first as a branch of mental education, and not mainly as useful knowledge'. There was no more effective method of training 'the logical faculties' than a course which illustrated 'the scientific method' by means of 'observation, experiment, and reasoning with the aid of hypothesis' (British Association, 1889, pp. 228-9). Operationalized in terms of content, apparatus and technique, this led to the dominance of qualitative chemical analysis as the staple of the introductory student laboratory course in universities (Roscoe, 1906, p. 248).

When pressure grew in Britain in the late nineteenth century to introduce science into the secondary school curriculum and when the limitations of mere textbook knowledge of science had become appreciated, school laboratories were built at a rapidly increasing rate, those for chemistry far outnumbering those for physics and biology (Table 2.1.1).

**Table 2.1.1:** School laboratories recognized by the Science and Art Department (Abney, 1904, p. 875)

| Year | Chemistry | Physics | Biology |
|---|---|---|---|
| 1880 | 133 | - | - |
| 1900 | 669 | 219 | 17 |
| 1901 | 722 | 291 | 26 |
| 1902 | 758 | 320 | 34 |

Indeed, the comparative ease with which chemistry, as a science subject, could be shaped to satisfy the requirements of a liberal education, especially with regard to mental training, had led to a situation in which science was regarded by many classical headmasters at the end of the century 'as summed up in Chemistry' (Ashford, 1902, p. 2). Those wishing to argue a case for the teaching of physics in boys' schools found, amongst numerous other problems, the need to overcome a curriculum tradition that chemistry was the most educational of the sciences.

It is important to note here that in girls' schools the situation was very different, botany, especially in its systematic aspects, emerging as the staple of the curriculum. Again, one reason was that it could readily be presented as a study which provided a mental training. However, considerations of expense were also significant. Schools for girls throughout this period were much less well resourced than schools for boys and it was notably more costly to build, equip and maintain a laboratory for chemistry (or physics) than one for botany.

Three points are worth emphasizing about student chemical laboratories before turning to the case of physics. First, the influence of institutional context on the version of laboratory chemistry which became dominant is difficult to overemphasize. Views of what constituted a liberal education, although not entirely static over the period under consideration (McPherson, 1959), were a powerful determinant of the curriculum and obliged chemists to incorporate in schools and universities a version of their subject which was fitted for survival in those environments. Other versions of chemistry than that associated with mental training had been developed and were available. For example, an early catalogue of chemical apparatus offered for sale by Richard Griffin and Co. included details of a kit of apparatus for experiments 'to demonstrate the chemical facts contained in Professor Johnston's Catechism *Agricultural Chemistry and Geology, adapted for the use of schoolmasters*', purchase of the boxed collection being assisted by government grant (Griffin, 1844, p. 85; Layton, 1973, pp. 48-9, 98). By the closing decades of the century such items had vanished from the schools section of catalogues of apparatus manufacturers. Demonstration apparatus for the preparation and properties of chemicals remained plentiful, but a major provision at the level of the individual student was for analytical procedures. Once laboratories were built with accessible reagents for wet analysis and equipment for dry methods, class work in qualitative analysis was easy to organize and examine. Inorganic exercises, especially, could be set which matched the constraints of time and resources in schools and colleges.

The second point is that 'academic drift' towards a pure chemistry curriculum was a reflection of the hegemony which the relatively small band of research chemists had acquired in the chemical community by the mid-century. The reasons for this and the mechanisms by which it was achieved have been touched on above and are treated in detail by Bud and Roberts (1984). Once in power, what constituted an acceptable chemistry curriculum was controlled, and the primacy of pure science was sustained, by their influence over syllabuses, examinations and other factors such as the grant regulations of the Department of Science and Art. There existed, therefore, an institutionalized procedure for replicating the principles of laboratory work which had come to assume educational importance for academic chemists.

The third point relates to micro-level influences on the curriculum and particularly its determination in specific laboratory contexts. It is well known that the intended curriculum and its manifestation in practice - the curriculum as experienced by students - frequently differ. This was the case with student laboratory work in chemistry. Liebig's laboratory curriculum was profoundly different from the school laboratory exercise of 'taking the solution through the charts'. His pedagogy might be summarized as peripatetic tutoring; once the basic procedures of qualitative and quantitative analysis had been acquired, using known compounds, students were launched into original research. His autobiographical account records that:

> I gave the task and supervised its execution ... There was no actual instruction; I received from each individual every morning a report upon what he had done on the previous day, as well as his views on what he was engaged upon. I approved or made my criticisms. Everyone was obliged to follow his own course. In the association and constant intercourse with each other, and by participating in the work of all, everyone learned from the others. (Campbell Brown, 1914, p. 40)

For Liebig, successful experimentation was an art which could not be learnt other than by close apprenticeship. In the laboratory, the student acquired a knowledge of the vocabulary and grammar of chemistry and, hopefully, an ability to ask productive questions. All this was a far cry from the teaching of a formalized version of scientific method by qualitative analysis.

Turning now to the design and development of student physics laboratories, quite different problems were posed from those associated with chemical laboratories. Some were severely practical. Thus, much available physics apparatus was expensive, so that a substantial financial investment was needed to equip a teaching laboratory: there was also a danger of costly damage to delicate instruments in the hands of inept students. In universities, if not in schools, the concept of one general physics teaching laboratory was invalidated because many experiments could not be conducted 'in close contiguity owing to their mutual interference' (Adams, 1868). The sensitivity of certain instruments to mechanical vibrations also necessitated special rooms. Even within a single laboratory the provision of sufficient pieces of the same apparatus to enable each student in a class to perform the experiment at the same time was out of the question and a different style of pedagogy had to be evolved. In addition, many experiments required a longer time than was available in those schools whose timetable was constructed on the basis of a literary curriculum.

Other problems were of a different order. Physics experiments of a 'merely qualitative' kind were deemed insufficiently educational, 'only leading to play' (Worthington, 1881, p. 1). It was the precision of statements about physical phenomena and general laws that enabled the subject to make a distinctive contribution to education. Compared with other sciences, physics was a more severe discipline; conclusions were of 'a stricter character than ... in any other branch of natural science', their truth or falsehood being easier to test (Carey Foster, 1874, p. 507). What was needed, therefore, was a student laboratory course which provided a strict training in accuracy of observation, precision of measurement, and the 'combination of deductive and inductive reasoning'. The first successful attempts in England to provide such a course of carefully graded experiments, matched to the manipulative and intellectual abilities of students, occurred in the late 1860s and early 1870s in university contexts (Sviedrys, 1976; Shepherd, 1979). Comparable pioneering developments at the Massachusetts Institute of Technology and the Royal Technical School, Chemnitz were alike in stressing the importance of qualities such as perseverance and manipulative skill in the education of a physicist (Pickering, 1871; Weinhold, 1875). Again institutional context and resources were significant determinants of the curriculum experienced by students. With one 'demonstrator' for some twenty students, the 'circus' arrangement of experiments made it difficult, if not impossible, to keep theory and laboratory work in step. Despite exhortations to 'Read all about the experiment, so as to have a clear idea of what you are going to do before you begin' (Worthington, 1896, p. 11), much practical work was reduced to following instructions in a laboratory manual. Similarly, the common stress upon presentation of results in graphical form was not unconnected with the ease with which the instructor of a large class could see at a glance how carefully the work had been done by each student (Pickering, 1871, p. 241).

The setting up of such courses frequently required great ingenuity on the part of instructors and their technical assistants in constructing simplified and inexpensive apparatus (Glazebrook and Shaw, 1885, p. vii). In many laboratories the construction of equipment by the students themselves was later to assume importance (Guthrie, 1886, p. 661). Occasionally the ease with which a particularly sensitive instrument might be put out of adjustment became a significant determinant of course content and student curriculum experience. In the later years of his long tenure of the chair of natural philosophy at Oxford (1865-1921), R. B. Clifton is reputed to have demonstrated the solar spectrum with a diffraction grating spectrometer once every summer term. No one but he was allowed to touch the instrument, a student having previously left it out of adjustment, so obliging Clifton to spend many hours re-setting it (Brown, 1968).

The educational arguments for student physics laboratories in universities transferred readily to the context of secondary schools. As with chemistry, the stress was on mental discipline. Much pioneering work here was undertaken by a schoolteacher, A. M. Worthington (later professor at the Royal Naval Engineering College, Dartmouth). The need was perceived to be for experiments which involved inexpensive apparatus only, which could be performed and recorded within a one-hour lesson, and which yielded measurements sufficiently precise to illustrate physical principles convincingly. Worthington claimed for his course that it provided a good training in '(1) skilful manipulation; (2) exact observation; (3) intelligent and orderly recording of observations; (4) principles of indirect measurement; (5) the application and intelligent use of Arithmetic, Geometry, and easy Algebra; (6) the varying of experimental combination; (7) *Common Sense'* (his emphasis) (Worthington, 1881, p. 3). It was also a sound foundation for subsequent work with costlier physical instruments. The physics laboratory curriculum which Worthington developed was published in 1886 as *A First Course of Physical Laboratory Practice*. It did for physics what qualitative analysis schemes had previously done for chemistry and became the model for many subsequent courses, achieving a sixth edition by 1903. Its opening sentence, 'Physics is essentially the science of measurements', both captured the flavour of the book and provided an indication of the view about the nature of science which it transmitted.

In practice, and especially in the hands of teachers less imbued with the spirit of research than Worthington, his course degenerated into mechanical manipulation and observation without understanding. Even his successor, David Rintoul, a science teacher of outstanding ability, found it necessary to revise the course to avoid 'all incentive to independent thought' being lost. In Rintoul's view, a prime objective was the fostering of 'a spirit of scientific curiosity'; hence experiments were presented in the form of a problem which the student had to solve (Rintoul, 1898, p. vi).

Also of significance for the fate of courses such as Worthington's, the closing years of the century witnessed momentous developments in physics leading to quantum theory and relativistic ideas which made it impossible to retain a view of physics as compatible with 'common sense'. They also underscored the role of imagination in the conceptualizing of physical phenomena. It is appropriate at this point, then, to step back from an account of the origins and development of student science laboratories to examine the extent to which the philosophy of science exerted an influence.

## PHILOSOPHY OF SCIENCE AND STUDENT LABORATORY PRACTICE

According to Landau, in an extensive review of theories of scientific method from Plato to Mach, philosophy of science became a discipline in its own right in the early nineteenth century (Landau, 1968, p. 29). Significant products of this academic emancipation - works such as Herschel's *Preliminary*

*Discourse* (1829), Ampère's *Essai sur la philosophie des sciences* (1834), Whewell's *Philosophy of the Inductive Sciences* (1840) and J. S. Mill's *System of Logic* (1843) - were, therefore, available to scientists and science educators at the time when student laboratories were being established. To what extent did they influence the objectives, content and pedagogy of student laboratory practice?

In order to answer this question it is necessary to examine some issues which writers on the philosophy of science saw as important in this period. A major concern was the context of discovery and the process by which scientific knowledge was won. At one end of a spectrum of opinion were those who adopted a narrow inductivist position, their aim being to formulate logical procedures whereby observations could be made to generate laws and theories. Others, such as Whewell, were severely critical of the view that scientific enquiry entailed the passive observation of phenomena and the mid-century debate between Whewell and J. S. Mill on the nature of induction has been well analysed by Strong (1955). For Whewell, 'discoverers' induction', the process of scientific discovery, could not be captured in formalized, explicit canons of procedure. An inductive inference was always more than a logical colligation of facts; not only were the facts bound together, but they were 'seen in a new point of view. A new mental Element is superinduced' (Whewell, 1838, p. 71).

The distinction being drawn is well illustrated by passages from the writings of two of the foremost experimentalists of the period. Humphry Davy's *Elements of Chemical Philosophy* opens with the assertion that: 'By experiment, new facts are discovered; ... observation, guided by analogy, leads to experiment, and analogy, confirmed by experiment, becomes scientific truth' (1812, p. 2). In contrast Liebig held that: 'Experiment is only an aid to thought ... (which) ... must always and necessarily precede it if it is to have any meaning. An empirical mode of research, in the usual sense of the term, does not exist. An experiment not preceded by theory, i.e. by an idea, bears the same relation to scientific research as a child's rattle does to music' (Liebig, 1863, p. 49). Of course it is possible to argue about the validity of these two positions. As Ian Hacking has pointed out, in an original philosophical treatment of experiments in science, an experiment might still be significant even if it was not derived from a theory about the phenomenon under scrutiny (Hacking, 1983, p. 154). What concern us here, however, are the educational implications of the two positions, the classical inductivist approach to discovery, a developed version of which was expounded by J. S. Mill (1843), and the more Kantian approach of Whewell, according to which 'ideas are prescribed to, and are not derived from, sensations' (Losee, 1972, p. 121). In essence, if scientific discovery could be reduced to a set of explicit logical procedures it could be taught; if not, it couldn't.

Taking the negative proposition first, what seems to follow from the accounts of those like Whewell and Liebig is an association of scientific investigation with 'craftsman's work' (Beveridge, 1961; Ravetz, 1973, pp. 75-108). On this view, 'the informal and largely tacit precepts of method' are learnt mostly by imitation and experience and not by formal instruction (Ravetz, 1973, p. 103). In writing about university education, Whewell was explicit on this point. 'A practical instruction in Inductive Reasoning is not possible', he asserted. 'We may lead men to feel the force of demonstration, but we cannot teach them to discover new truths' (Whewell, 1838, p. 43). So far as formal education was concerned 'scientific method', in a strict sense, was unteachable. Todhunter, who edited Whewell's considerable correspondence, was clearly of the same mind on this point. 'Little of what is characteristically valuable in experimental philosophy is susceptible of transmission', he wrote (Todhunter, 1873, p. 19).

This was a most unpalatable conclusion for those trying to establish science in school and university curricula. The educational claims for science rested heavily upon its ability to provide a mental training. Subjects like mathematics and classics were as able as science to provide opportunities for the exercise of deductive thinking: the unique contribution of science stemmed from its purported inductive procedures. If these were unteachable, the case for its inclusion in the curriculum fell.

It is not surprising, therefore, that proponents of science in education, in so far as they appear to have been influenced by philosophy of science, very largely rejected the Whewellian version of scientific method in favour of the logicians' version, exemplified in the writings of J. S. Mill. An explicit acknowledgement of a debt to Mill occurs in T. H. Huxley's influential lecture on the educational value of the natural history sciences, his debut as a speaker on educational matters, in 1854. It was in this lecture that Huxley gave his celebrated definition of science as 'nothing but *trained and organised common sense*', outlining a general scientific method involving *'observation and experiment, comparison and classification, deduction and verification'* (Huxley, 1893, pp. 45, 52-3, 56).

To evaluate the significance of Huxley's reference to Mill it has to be realized that the main purpose of his lecture was to elevate the status of his subject, biology, and to emphasize its similarities to chemistry and physics by invoking a common 'scientific method' (Layton, 1975). To this end, Mill was a resource to be pressed into service, rather than a prime determinant of Huxley's thinking. In the 1854 lecture we can detect the outlines of that mythical view of science 'erected more or less deliberately in the late nineteenth century, which purports to tell how various scientific discoveries were made' (Price, 1966, p. 247). It implied that students could be educated in the processes of new knowledge production and it overloaded science with educational responsibilities for the development of problem-solving abilities and critical thinking. So far as the student laboratory was concerned, the logician's view of scientific method provided a warranty for practices which had been shaped less by philosophy of science than by institutional ideals about the purposes of education and the resource constraints under which these ideals were operationalized.

## PHILOSOPHY OF SCIENCE AND CURRICULUM CHANGE

The declared aims of science teaching since the late nineteenth century have not remained constant. Standing back from the fine detail of classroom and laboratory practices, a number of concerted attempts to bring about strategic shifts can be discerned. The advocacy of heurism in the early twentieth century, associated with the name of H. E. Armstrong, would be one such endeavour; the emphasis in the inter-war years on the cultural value of science, with a playing-down of formal training, represents another; whilst the attempts to place students in the position of 'a scientist for the day' in the reforms of the early 1960s was a third turning point. Each of these has had significant implications for the role of experimental work in student laboratories.

In the case of Armstrong, his own account of the source of his interest in 'the practice of scientific method' acknowledges 'a purely literary origin', R. C. Trench's *Study of Words,* which included a discussion of the discovery of relationships between words (Brock, 1973, p. 9). A later experience of the workings of a court of law in connection with a patent action likewise impressed him as 'the acme of scientific treatment' and a marvellous example of the 'methodical logical use of knowledge' which he termed scientific method. There is little in Armstrong's writings to suggest a direct debt to philosophy of science. Informally, when a student under Edward Frankland, 'an intuitive understanding of the method of discovery' had been acquired (Brock, 1973, p. 62). Yet this 'art of discovery' was deemed universally teachable because 'all we have to do is develop and train faculties which all possess to some extent' (Brock, 1973, p. 106). It was this 'common sense' view of science which underwrote his prescription for 'a liberal course of detective literature' in the curriculum for budding scientists (Armstrong, 1903, p. 381).

As for laboratory practice, it is significant that he preferred to describe the accommodation as a workshop and that he was severely critical of architects who produced for schools 'mere slavish copies' of laboratories provided in institutions of higher education. For much elementary work an ordinary classroom with a few tables would suffice; apparatus need not be elaborate or expensive; 'the simplest contrivances should be used, and the originality both of teacher and pupils developed by the construction of their own apparatus'. The only exception to this was the need to purchase a good balance (Gordon, 1893, p. ix). To the architects of school laboratories, Armstrong stressed the need to keep three S's in mind - Sense, Simplicity, Space (Armstrong, 1903, p. 454).

For Armstrong, then, laboratory work was primarily a training in scientific method, with the emphasis on *doing* in order to *understand*. Once the methods of experimental investigation had been acquired in the workshop, these would be used continually in subsequent science lessons to 'find out' the information which others were forced to acquire by rote and demonstration.

The theoretical foundations of heurism were cut away more by changes in the psychology of learning than in the philosophy of science (Brock, 1973, p. 39). It is true that the common sense model of science embodying a generalized scientific method became progressively discredited in philosophical terms in the early twentieth century (Jenkins, 1976). It is equally true that few contemporary debates about the aims of science education gave much weight to these philosophical considerations. As Elkana has pointed out, science teaching tended to lag behind developments in philosophy of science by 'at least forty years' (Elkana, 1970, p. 21). Certainly, those who argued for a new direction for science education, and specifically that the strongest claim of science for a central place in the school curriculum was its cultural value, placed most stress on psychological evidence to discount 'mental training' and much that Armstrong had stood for (British Association, 1918, pp. 133-4; Science Masters' Association, 1936, pp. 13-15).

When the prime aim of science education became awareness of the 'intellectual glory' of the scientific exploration of natural phenomena and of the contribution of science to our social heritage, laboratory practice focused more on the informational aspects of science. The role of experiment was to assist understanding of the laws and theories which made up 'the cathedral of science' and to exemplify the connections of these to the operations of everyday life. To these ends the demonstration experiment - for Armstrong 'a dogmatic, not a heuristic method' (Brock, 1973, p. 143) - enjoyed something of a revival in the inter-war years. George Fowles' classic *Lecture Experiments in Chemistry*, representative of a genre going back to Faraday's *Chemical Manipulations* (1827) and a text on which a generation of young chemistry teachers cut their teeth, was published in 1937. Other practical books delivered a similar message. 'Unless instructions are followed in every detail your experiments will not be successful, and will not teach you what they are intended to teach', readers of R. B. Grove's *A Science Course for Girls* (1934, p. 7) were warned. 'Practical work is the great enemy of sloppiness ... The student is up against the hard facts of nature. He must conquer them or he does not get results at all', a similar text advised (McKay, 1930, p. vii). Acolytes could scarcely be left unaware of the demands of rigour, precision, and meticulous attention to detail which the construction of the cathedral, and its further elaboration, entailed. In these years the manipulation and mastering of apparatus in the quest for an objective truth was a dominant concern of many science teachers (Layton, 1984, pp. 137-9).

The transition from what Elkana, over-generalizing somewhat, terms an inductivist-realistic philosophy of science to a positivistic-instrumentalist one represents another major change in science education in the twentieth century (Elkana, 1970, p. 21). Certainly, the reforms of the late 1950s and early 1960s emphasized the corrigibility of scientific theories. Attributions to specific philosophies of science were rarely made, but the instrumental role of theories, which 'may be no more than convenient and provisional fictions', was stressed (Science Masters' Association, 1961, p. 9). At the same time, the prescriptions for reform were not without inconsistencies; science was still 'a human quest for Truth', suggesting an ultimate goal of full correspondence to reality. Laboratory work, whilst designed so that the student could 'experience the same interactions between the experimental and intellectual aspects' of science as a research scientist (Chemical Bond Approach Project, 1963, p. 1), nevertheless implied that theory-free observation was the norm. Observation 'takes concentration, alertness to detail, ingenuity, and often just plain patience', students were informed (Chemical Education Material Study, 1963, p. 1), endorsing a view of it as an activity independent of conceptual levels of thought. Elsewhere, stress was placed on intuitive thinking, and the role of laboratory work 'in exploring a field before formal definitions or laws are introduced' (Bruner, 1962, pp. 55-68). Yet the same laboratory programme had as a prime objective that 'experiments should be true experiments' (Educational Services Incorporated, 1960, p. 4). The procedures of scientific enquiry were to become the method by which science was learnt. Of course, the frequent need for the investigation to take place within a conceptual framework which the learners could not possibly be expected to have constructed 'on their own' robbed the experience of any resemblance to 'a true experiment'. As others have argued (Driver, 1975) the curriculum reality for many learners became 'guided discovery', a strategy which deflected few students from a concern with 'what ought to happen' and projected 'an authoritarian, doctrinaire image of science that discovery methods were expressly introduced to counter' (Hodson, 1985, p. 40). The portrayal of the science community as 'the very paradigm of institutionalized rationality' (Newton-Smith, 1981, p. 1) remained largely unaffected.

## IMPLICATIONS FOR THE FUTURE

Viewed from the standpoint of philosophy of science, inconsistencies over ends and over-simplifications with regard to means characterized many of the science curriculum reforms of the 1960s (Elkana, 1970, pp. 21-3; Hodson, 1985). As far back as 1969, Koertge was arguing convincingly that the intended learning outcomes of a knowledge of scientific theory and an understanding of the process by which science grows needed to be addressed separately; there was no single, multi-purpose pedagogical solution through 'discovery learning' in the student laboratory (Koertge, 1969). Subsequently science educators encountered the writings of Popper and Kuhn (Smolicz and Nunan, 1975; Cawthron and Rowell, 1978; Donnelly, 1979), and more recently have attempted to explore the implications of a wider community of philosophers of science, including the 'new realists' such as Bhaskar (1975), Newton-Smith (1981) and Hacking (1983).

For some, such as Manicas and Secord (1983), there is to be found in these more recent works a 'new philosophy of science' which embodies a critique of the 'standard' logical-empiricist and the paradigmatic views. Its relationship to student laboratory practices is mediated by its implications for student learning. In contrast to a foundationist view according to which scientific propositions are based on sensory data 'out there', the realist theory assumes that 'our conceptual constructions of the world are what is "real" and it is through these that we interpret and re-interpret our experience'. Accordingly, student learning of science entails an appreciation of 'theories' as imaginative human constructions. Because pupils' own 'constructions' and 'ways of seeing' the natural world are not necessarily those which scientists have adopted, a function of laboratory work becomes the assisting of students to see the world as scientists have constructed it, without undermining confidence in the students' own abilities to make sense of experiences (Driver and Oldham, 1986). 'Disconfirming' or 'surprise' demonstrations, and laboratory counter-examples (Rowell and Dawson, 1983), may have a contribution to make to the restructuring of children's ideas, although conceptual conflict does not, of necessity, produce an alternative conceptual scheme. As Driver, Guesne and Tiberghien (1985) have pointed out, the role of experiment in relation to the ability to generalize needs careful re-examination.

Two points can be made about Manicas's and Secord's assumption of a 'new philosophy of science'. First, whilst there has been a critical reaction amongst philosophers of science to the non-rationalist positions of those like Kuhn and Feyerabend, it is by no means settled that 'realism is the truth and temperate rationalism the way' (Newton-Smith, 1981, p. 273). The issues are still energetically contested as indicated, for example, by Van Fraassen's (1980) spirited and elegant defence of constructive empiricism.

Second, it would seem that the preference of some psychologists for the 'new realism' is not unconnected with their prior adoption of a constructivist view of learning and a belief that a realist philosophy of science provides some warranty for this, or, at least, is compatible with it. If this is so, the situation differs little from that in the nineteenth century when Huxley drew on Mill's logician's view of scientific method and in the 1960s when curriculum reformers displayed a propensity for inductivism (Hodson, 1985, p. 35). Here again, philosophy of science is being raided for a version which appears to lend support to a view of learning previously adopted for reasons which have nothing to do with the nature of science. In passing, it is noteworthy that not all 'constructivists' incline to 'realism'; Pope and Gilbert (1983), for example, argue - unconvincingly as it happens - that 'if a science teacher takes a "realist" view of the nature of a subject being taught', then learning will be seen in terms of ' "facts" for transmission' (p. 253). This appears not to acknowledge the varieties of 'realism' and to assume a strict logical entailment between epistemology and method of teaching.

Of more direct significance for student laboratory practice than the debate over non-realist versus realist views about scientific knowledge is that over the nature of experimentation. Philosophers and sociologists of science have in recent years begun to re-examine this cardinal dimension of laboratory life. Hacking (1983), for example, provides strong evidence that experimentation often has a life independent of theory. 'To experiment', for him, 'is to create, refine and stabilize phenomena'. This involves a series of tasks - designing an experiment that might work, learning how to make it work and, perhaps the 'real knack', getting to know when the experiment is working (p. 230). The skill and judgemental elements in this process, what Ravetz calls 'the evolution of a fact' (1973, pp. 181-208), are rarely acknowledged in the student laboratory. The common experience that 'most experiments don't work most of the time' (Hacking, 1983, p. 230) is interpreted in terms of mechanical and technical difficulties, impurities in reagents, desiccation of specimens, manipulative deficiencies and the like. It is rarely that the 'standardized fact' is questioned. Expressed in different terms, the process of knowledge production in science involves 'artists' turning 'book-keepers' (Holton, 1952, p. 236); the private negotiations, tacit understandings, and 'feel' for apparatus remain unacknowledged and unexpressed in the public accounts of experimentation. In an interesting pilot study of teachers' views of 'difficult' experiments (in this case, experiments to demonstrate tropic effects in plant roots and shoots) Miller (1985) observed that anomalous results were frequently dismissed in favour of the Colodny-Went theory, the account favoured in many textbooks and examination syllabuses, despite the availability of published criticism of this theory by biological researchers. He concluded that 'the "meaning" of an experimental episode in school science is always open' and that 'the results of school experimental work are best seen as the raw material for various kinds of interaction between teacher and learner through which the acquisition of new scientific knowledge is negotiated'. This might counterbalance the prevailing situation in which student laboratory practice can promote a disposition to favour 'official knowledge' of experts over a student's personal observation and experience. An alternative view, advocated by Hodson (1985, pp. 44-5) is that if experimental work is to be used to support the acquisition of new knowledge then it is essential that a number of precise conditions are satisfied, including 'the experiment "works" successfully' and 'only one conclusion, or a limited number of conclusions, is possible from the experimental data'.

Perhaps the general verdict from this review of student laboratory practices from the standpoint of philosophy of science is that there is need for a critical reappraisal of the function of experimental work. As Hodson has argued, 'much practical work in school is aimless, trivial and badly planned. It is used unthinkingly without proper regard for its pedagogic and epistemological roles' (1985, p. 44). A clearer identification of the learning outcomes intended - a familiarity with materials and natural events, the enhancement of motivation to learn, the acquisition of laboratory skills, understanding of the processes by which scientific knowledge has been acquired, or understanding of an important scientific idea - needs to be accompanied by a critical review of available means so as to achieve a more appropriate match. Science education without some laboratory experience is unthinkable; but, equally, student laboratory practice is not a general panacea, the universal means to a multiplicity of ends.

## REFERENCES


Abney, W. (1904) Presidential address to Section L. *Report of the British Association for the Advancement of Science, 1903*. John Murray, London.

Adams, W. G. (1868) Letter dated March 1868, probably to the Council of King's College London. (Archives, King's College Library.)

Armstrong, H. E. (1903) *The teaching of scientific method*. Macmillan, London.

Ashford, C. E. (1902) 'The order in which subjects should be taught'. In: Conference of Public School Science Masters. Papers read at the Meeting held in the rooms of the University of London, on 19 January 1901 and on 18 January 1902.

Beveridge, W. I. B. (1961) *The art of scientific investigation*. Mercury Books, London.

Bhaskar, R. (1975) *A realist theory of science*. Leeds Books, Leeds.

British Association for the Advancement of Science (1889) *Report 1888*. John Murray, London.

_____ (1918) *Report 1917*. John Murray, London.

Brock, W. H. (ed.) (1973) *H. E. Armstrong and the teaching of science, 1880-1930*. Cambridge University Press, Cambridge.

Brown, F. O. (1968) Letter dated 16 September 1968. (Brown was demonstrator at the Clarendon Laboratory, Oxford during Clifton's later years. I am indebted to V. K. Chew of the Science Museum, London, for a copy of this letter.)

Bruner, J. (1962) *The process of education*. Harvard University Press, Cambridge, Massachusetts.

Bud, R. F. and Roberts, G. K. (1984) *Science versus practice. Chemistry in Victorian Britain*. Manchester University Press, Manchester.

Campbell Brown, J. (1914) *Essays and addresses*. J. and A. Churchill, London.

Cardwell, D. S. L. (1972) *The organisation of science in England*. Heinemann, London.

Carey Foster, G. (1874) Physics at the University of London. *Nature, 10*, 506-8.

Cawthron, E. R. and Rowell, J. A. (1978) Epistemology and science education. *Studies in Science Education, 5*, 31-59.

Chemical Bond Approach Project (1963) *Investigating chemical systems*. McGraw-Hill, New York.

Chemical Education Material Study (1963) *Laboratory manual for chemistry*. W. H. Freeman and Co., San Francisco, p.1.

Daniels, G. H. (1967) The pure-science ideal and democratic culture. *Science, 156* (3783), 1699-1705.

Davy, H. (1812) *Elements of chemical philosophy*. J. Johnson and Co.,

London.
Donnelly, J. (1979) The work of Popper and Kuhn on the nature of science. *The School Science Review, 60*, 489-500.
Driver, R. (1975) The name of the game. *The School Science Review, 56*, 800-5.
_____ Guesne, E. and Tiberghien, A. (eds.) (1985) *Children's ideas in science.* Open University Press, Milton Keynes.
_____ and Oldham, V. (1986) A constructivist approach to curriculum development in science. *Studies in Science Education, 13*, 105-22.
Educational Services Incorporated (1960) *Review of the secondary school physics program of the Physical Science Study Committee.* 1959 Progress Report. Massachusetts, USA.
Elkana, Y. (1970) Science, philosophy of science and science teaching. *Educational Philosophy and Theory, 2*, 15-35.
Fowles, G. (1937) *Lecture experiments in chemistry.* G. Bell, London.
Gardner, J. (1846) An address delivered in the Royal College of Chemistry, June 1846. *The Chemist, 4*, 294-301.
Glazebrook, R. T. and Shaw, W. N. (1885) *Practical physics.* Longmans, Green and Co., London.
Gordon, H. (1893) *Elementary course of practical science.* Macmillan, London.
Gore, G. (1882) *The scientific basis of national progress including that of morality.* Williams and Norgate, London and Edinburgh.
Griffin, R. (1844) *Griffin's catalogue of chemical apparatus, Part 3.* Richard Griffin and Co., Glasgow.
Grove, R. B. (1934) *A science course for girls, Book 1.* Nelson, London.
Guthrie, F. (1886) Cantor Lectures: Science Teaching. *Journal of the Society of Arts, 34*, 659-63.
Hacking, I. (1983) *Representing and intervening.* Cambridge University Press, Cambridge.
Hodson, D. (1985) Philosophy of science, science and science education. *Studies in Science Education, 12*, 25-57.
Holton, G. (1952) *Introduction to concepts and theories in physical science.* Addison-Wesley, Reading, Massachusetts.
Huggins, W. (1906) *The Royal Society or, science in the State and in the schools.* Methuen, London.
Huxley, T. H. (1893) On the educational value of the natural history sciences. In *Science and Education: Essays.* Macmillan, London.
Jenkins, E. W. (1976) H. E. Armstrong, heurism and the common sense of science. *Durham Research Review, 8*, 21-6.
Koertge, Noretta (1969) Towards an integration of content and method in the science curriculum. *Curriculum Theory Network, 4*, 26-44.
Landau, L. (1968) Theories of scientific method from Plato to Mach. *History of Science, 7*, 1-63.

Layton, D. (1973) *Science for the people.* Allen and Unwin, London.

_____ (1975) Huxley as educator: a reappraisal. *History of Education Quarterly, 15*, 219-25.

_____ (1984) *Interpreters of science.* John Murray/Association for Science Education, London.

Liebig, J. von (1863) Über Francis Bacon von Verulam und die Methode der Naturforschung. (Cited in Hacking, 1983, p. 153.)

Losee, J. (1972) *A historical introduction to the philosophy of science.* Oxford University Press, London.

Manicas, P.T. and Secord, P.F. (1983) Implications for psychology of the new philosophy of science. *American Psychologist, 38*, 399-413.

Martin, M. (1979) Connections between philosophy of science and science education. *Studies in Philosophy and Education, 9* (4), 329-32.

McKay, H. (1930) *Individual work in science, Part 4. Teacher's Book.* University of London Press, London.

McPherson, R. G. (1959) *Theory of higher education in nineteenth century England.* University of Georgia Press, Athens.

Mill, J.S. (1843) *A System of Logic, ratiocinative and inductive, being a connected view of the principles of evidence, and the methods of scientific investigation.* Parker, London.

Miller, R. (1985) Negotiating knowledge through experiment: teachers' views of 'difficult' experiments. Unpublished paper presented to the British Society for the History of Science conference on The Uses of Experiment, Bath.

Morrell, J. B. (1969) Thomas Thomson: professor of chemistry and university reformer. *British Journal for the History of Science, 4*, 245-65.

_____ (1972) The chemist breeder: the research schools of Liebig and Thomas Thomson. *Ambix, 19* (1), 1-46.

Newton-Smith, W. H. (1981) *The rationality of science.* Routledge and Kegan Paul, London.

Pickering, E. C. (1871) Physical laboratories. *Nature, 3*, 241.

Pope, M. L. and Gilbert, J. K. (1983) Explanation and metaphor: some empirical questions in science education. *European Journal of Science Education, 5*, 249-61.

Price, D. de Solla (1966) The science of science. In M. Goldsmith and A. Mackay (eds.) *The science of science.* Pelican Books, London.

Ravetz, J. R. (1973) *Scientific knowledge and its social problems.* Penguin Books, Harmondsworth.

Rintoul, D. (1898) *An introduction to practical physics for use in schools.* Macmillan, London.

Roscoe, H. E. (1906) *The life and experiences of Sir Henry Enfield Roscoe.* Macmillan, London.

Rowell, J. A. and Dawson, C. J. (1983) Laboratory counterexamples and the growth of understanding in science. *European Journal of Science Education, 5*, 203-15.

Science and Art Department (1866). *Thirteenth Report.* London, HMSO. Appendix B, 37.

Science Masters' Association (1936) *The teaching of general science.* John Murray, London.

_____ and Association of Women Science Teachers (1961) *Science education. A policy statement.* John Murray, London.

Shepherd, R. E. (1979) Individual practical work in the teaching of physics in England: a study of its origins and rationale. Unpublished M.Phil. thesis, University of Leeds.

Simpson, A. D. C. (Ed.) (1982) *Joseph Black 1782-1799. A commemorative symposium.* The Royal Scottish Museum, Edinburgh.

Smolicz, J. J. and Nunan, E. E. (1975) The philosophical and sociological foundations of science education: the demythologizing of school science. *Studies in Science Education, 2*, 101-43.

Strong, E. W. (1955) William Whewell and John Stuart Mill: their controversy about scientific knowledge. *Journal of the History of Ideas, 16*, 209-31.

Sviedrys, R. (1976) The rise of physics laboratories in Britain. *Historical Studies in the Physical Sciences, 7*, 405-36.

Todhunter, I. (1873) *The conflict of studies and other essays.* Macmillan, London.

Van Fraassen, B. C. (1980) *The scientific image.* Clarendon Press, Oxford.

Weinhold, A. F. (1875) *Introduction to experimental physics* (translated by Benjamin Loewy). Longmans, Green and Co., London.

Whewell, W. (1838) *On the principles of English university education*, 2nd edition. John W. Parker, London.

_____ (1858) *Novum Organon Renovatum.* John W. Parker, London.

Williamson, A. W. (1870) *A plea for pure science, being the inaugural lecture at the opening of the Faculty of Science in University College London.* Taylor and Francis, London.

Worthington, A. M. (1881) *An elementary course of practical physics.* Rivingtons, London.

_____ (1896) *A first course of physical laboratory practice.* Longmans, Green and Co., London.

# 2.2

# The Interplay of Theory and Methodology

**Joseph D. Novak**

## INTRODUCTION

There are no easy solutions for the improvement of science instruction. We are faced with two significant realities: (1) the amount of knowledge to be learned is overwhelming for students and teachers; and (2) new knowledge is being constructed at a rate that is equally overwhelming. Given these realities, what alternatives to the improvement of science education remain viable? It has been argued elsewhere (Novak and Gowin, 1984; Novak, 1985) that our best hope may be in new efforts to help students learn how to learn so they have a better chance to grasp the major ideas in science disciplines and also to help them learn how knowledge is constructed, partly to facilitate their learning and partly to help them understand the nature of science. The two strategies presented in this chapter are examples of what Baird (Chapter 4.2) describes as *metacognitive* strategies that facilitate students' ability to take control of their learning.

## VEE DIAGRAMMING

Much of our research over the past three decades has been done in science laboratory settings. We have attempted various strategies and syllabus revisions to help students understand the meaning of their laboratory work, and also the nature of scientific enquiry. In terms of the model in Chapter 1.2, we have modified the instructional activities over and over again, but we have continued to see a wide discrepancy between our intended learning outcomes and the actual learning outcomes. While some of the innovations showed modest positive gains in students' understanding of their laboratory work, the gains were usually small and some of these may have resulted from teacher enthusiasm and/or Hawthorn effect rather than from the merits of our innovations. We generally saw little or no gains in students' understanding of and appreciation of scientific enquiry.

What we continued to observe was that most students not only failed to link concepts, principles, and theories with the data they obtained or conclusions they made, they often were not clear about what events or objects they were observing and what questions they sought to answer from their observations. We observed the same failings in student reports on their reading of research papers. My colleague, Bob Gowin, drawing on his training and expertise in philosophy, first devised five questions to help sharpen students' thinking and/or observation: (1) What is the 'telling question'? (2) What are the key concepts? (3) What methods of inquiry (procedural commitments) are used? (4) What are the major knowledge claims? and (5) What are the value claims? These questions did serve as a useful basis for negotiating the meaning of laboratory work (or research papers) with students, but there continued to be great difficulty in helping students to see how various elements involved in the construction of knowledge interacted with each other. Since 1977, we have used a Vee heuristic, which relates ten elements involved in the construction of knowledge. This is described below and illustrated by example in Figure 2.2.1 on page 68.

The Vee shape was chosen to point to the events or objects studied. We might say that this is where any enquiry begins, but this is not necessarily the case. We could begin with a question such as 'How can we help students understand the meaning of a laboratory experience?' We might subsequently choose to create an education event where students are taught to apply the Vee heuristic in their laboratory work. Or we could be driven by our constructivist philosophy and choose to address a question like 'How can students be taught to understand that knowledge claims are constructed from observations and dependent upon our concepts, principles and theories?' Again, we might choose to construct an educational event that will use the Vee heuristic. In fact, any enquiry could begin with any element of the Vee. The traditional view was to begin with 'hypotheses', or what might be called 'anticipated knowledge claims'. But hypotheses do not exist in a conceptual/theoretical vacuum; they too are dependent on all other elements. As pointed out in Layton's discussion (Chapter 2.1), views about the nature of science have changed over the years. There is, however, an emerging consensus toward 'constructivist' views that focus on the meanings the researcher brings to an enquiry and that shape both the enquiry and the interpretation of the data. The Vee heuristic derives from an 'event-centred' constructivist epistemology.

We have found it useful to define each element in the Vee and to illustrate each with examples. The focus question(s) state(s) what we wish to know about the events or objects studied. A good focus question will also help us to decide what kind of records we choose to make and how we should table, graph, or otherwise transform our records. The events or objects we choose to observe must be considered carefully. Often we are not making direct observations of an event or object (e.g., photosynthesis or atomic decay) but rather we are observing some record-making device that we interpret to represent the event or object we are interested in. The construction of knowledge claims from the records we obtain requires not only theory, principles, and concepts that relate to that event or object but also a consideration of theory, principles and concepts that permit us to understand how the instrument (e.g., Warburg manometer or Wilson cloud chamber) functions to produce these records. Is it any wonder that university students have difficulty understanding their laboratory work, to say nothing about school students? These problems are illustrated in later chapters of this book.

Records of events or objects are what we define as facts when they are valid records. A faulty instrument will produce records, but they are artifacts and can lead to faulty claims. As John Dewey remarked often in his lectures, 'there is nothing on the face of a fact that tells you what it means'. Even when we obtain valid records (facts), we still must interpret what they mean. The Vee heuristic can be applied to assist in this interpretation.

Novak and Gowin (1984) have provided descriptions of techniques for teaching students how to do Vee diagramming in the laboratory.

## CONCEPT MAPPING

One of the major objectives of the new science curricula of the 1950s and 1960s was to encourage the teaching of science in a manner that would depart from the all too common practices of memorizing what Schwab and Brandwein (1962) called a 'rhetoric of conclusions'. Most science teaching had been (and unfortunately continues to be) a process where students memorize by rote the definitions of scientific terms and solve problems by relatively thoughtless substitution of numbers or symbols into formulas or equations. The solution was thought to be to emphasize that science derives from an enquiry process and that new knowledge is continually being discovered. Discovery learning, not rote memorization, was to be encouraged. The record shows (Stake and Easley, 1978; General Accounting Office, 1984) that the 'curriculum revolution' failed to achieve its goals. It has been argued earlier (Novak, 1969, 1977) that the curriculum reformers failed to distinguish between rote and meaningful learning as distinct from reception and discovery of instructional strategies. Reception instruction can be highly meaningful and discovery instruction can be largely rote. There were also significant fallacies in the epistemology underlying the push for instruction to illustrate the processes by which scientists 'discover' scientific laws, a shortcoming recognized by Elkana (1970) and works cited by Layton (Chapter 2.1).

Our research programme, based on Ausuabel's theory of meaningful learning since 1964, required a better way to represent what learners knew before and after instruction, and concept mapping was developed first as a research tool to ascertain what the learner knows. Concept maps provide a simple tool for this, a tool that can help both teachers and students recognize what the students know (and the misconceptions they have) that are relevant to the learning. As Edwards and Fraser (1983) have shown, concept maps are comparable to clinical interviews for revealing 'what the learner already knows'. The advantage is that concept maps can be used with classroom groups with little training, and they are comparatively easy to interpret or to score.

Concept mapping is a powerful tool to identify both valid and invalid concepts and propositions students hold prior to (or after) a segment of laboratory instruction. There is overwhelming evidence that most science instruction, including laboratory work, does little to alter students' misconceptions.

The result is that they 'do the experiments' but they do not see the meaning of the events or records they have studied. Their misconceptions are not modified. (See Helm and Novak, 1983.) We find that the use of concept maps not only can alert teachers to misconceptions students bring to the laboratory but also help them to recognize and modify their misconceptions (Novak, Gowin and Johansen, 1983; Robertson-Taylor, 1985; Heinze-Fry, 1987). Novak and Gowin (1984) have provided descriptions of techniques for teaching students how to do concept mapping.

## THE IMPORTANCE OF SHARED MEANINGS

We define education as: an activity that leads to a change in the meaning of experience. To educate, the teacher and student must seek to share their meanings in classroom and laboratory experiences. Learning is the responsibility of the students; it cannot be shared. Too often it is assumed that teaching causes learning; it does not. Teaching, even the best teaching, can only organize information and experience in a manner that permits or facilitates the grasping of meaning by the learner. Educating involves a give-and-take between teacher and student, a kind of cognitive pushing and pulling where teacher and student negotiate the meaning of the materials being studied (Gowin, 1981). Concept maps and Vee diagrams have proved to be enormously useful tools for helping teachers and students (or students with students) negotiate the meaning of the material being studied. When used effectively with students, they not only facilitate learning of the material but also help students learn how to learn (Novak and Gowin, 1984). Students cannot take command of their own learning when they lack the knowledge and skills to learn meaningfully. The result is that most students resort to essentially rote learning, and this does not lead to a change in the meaning of their experience. There is no lasting value in such learning.

Several research studies have been completed that show positive results for both cognitive change and attitudinal changes when concept maps and Vee diagrams are used with students. Robertson (1984) found that most students held negative attitudes toward laboratory work and did not understand the purposes of the work. However, when concept mapping and Vee diagramming were used with the same population of students in a later study, she found highly positive attitudes and high levels of understanding (Robertson-Taylor, 1985). Sex differences that were observed in her first study (no females were in the group that understood the laboratory sessions analysed) were absent in interview and questionnaire data obtained in her second study. Two of our PhD studies show that most students enter laboratory work with strongly 'positivistic' epistemologies and conventional 'enquiry-oriented' laboratory work did not move these students toward more 'constructivist' epistemologies. However, those students who made efforts to produce good concept maps and Vee diagrams made positive gains toward more constructivist views and also showed better understanding of laboratory experiments.

The male tendency to proceed more analytically than females in their work tends to favour males in typical 'cookbook' laboratory activities. The 'silencing' of the female described by Belenky *et al.* (1986) is clearly evidenced in laboratory routine. However, when students are given metacognitive tools that help them share meanings in laboratory, both males and females benefit, but the empowerment of females is particularly striking. Disadvantaged students and 'slow learners' are often denied laboratory opportunity since they tend to be 'lost' - and often misbehave. In our own work (e.g. Melby-Robb, 1982) and the work of the Purdue University group (e.g. Kahle, 1984) slow learners and minorities have been shown to be 'on task' and effective in the laboratory when metacognitive tools are used.

With the work of Kinigstein (1981) and Hallowell (1985), we have observed that school socialization has moved most students to predominantly rote mode learning by grades four or five. In consequence, much of school learning becomes a task to be dispensed with as readily as possible. When metacognitive strategies such as concept mapping are introduced, we find resistance on the part of most students, because this requires them 'to do more work' and 'to think hard'. By grade seven, this pattern is firmly established in some 80 to 90 percent of students (Novak, Gowin and Johansen, 1983) and is also predominantly evident in high school (Gurley, 1982; Kahle, 1984). Most of our recent work has shown the enormous constraints of the 'frame factors' (Figure 1.2.1) that operate in school settings and a teacher working in isolation, even using metacognitive tools, will find enormous resistance on the part of students and colleagues when efforts are made to move students toward more meaningful learning patterns. However, the laboratory provides a somewhat unique opportunity for this change in emphasis, since its event-centred context is a departure from most other school learning. In this setting, Gurley (1982) found that 90 percent of her students were 'on task' in the laboratories when Vee diagramming was used, but only 40 percent of her students were 'on task' in classes that did not use Vee diagramming. Typical of the comments made by students during interviews conducted by the school psychologist is the following:

> Vees are good. The left side tells you what relates to the experiment, and you want to know how you even got to asking the focus question in the first place! The left side helps tell you. I refer back to the left side to help explain the results for the knowledge claim. I skim over Vees when I study. If you were just doing a lab without asking a focus question, you couldn't tie it all together. I think you get more out of it [the Vee] than from lab book questions. The Vee makes you think more. And lab questions are just 'yes/no, draw a picture'. (Gurley, 1982, p. 161)

## THE USE OF VEE DIAGRAMS AND CONCEPT MAPS FOR IMPROVING LABORATORY INSTRUCTION

To illustrate how the Vee can help, we cite the work of Chen and Buchweitz. Figure 2.2.1 is a Vee diagram that Chen (1980) constructed from the laboratory guide for one of the exercises (on kinematics) in an introductory college physics course. Chen found that students in the course: (1) did not understand the concept of coefficient of friction; (2) did not understand why they had to reduce the separation between gates by $\Delta X/2$ to measure $\Delta t$ for the bar on the glider to pass the second gate; (3) could not distinguish between average velocity and instantaneous velocity; thought $\Delta X/\Delta t$ is $(X_2 - X_1)$ over $(t_2 - t_1)$ or $X/t\theta$; (4) got a slope close to 1 for the acceleration vs $\theta$ graph without realizing that it should be 980 $cm/S^2$ or close to 1000 $cm/S^2$. Chen's Vee analysis showed: (1) the key questions were not clearly given, hence students were not sure why they were making the specific observations; (2) the main objectives of the lab should have included study of the relationships between X, V, t and acceleration because these relationships were derived from graphing and are significant for comprehending the concept of constantly accelerated motion; (3) initial velocity equal to zero was given without any reason; (4) the concept coefficient of friction, although it was to be taught later, was needed in this lab; (5) the explanation of reducing the gate separation by X/2 was obscure, and the adjustment is trivial; (6) some terms in the instructions were vague (e.g., the term velocity).

From the Vee analysis, Chen found that some concepts needed were not included, or their relevance to the experiment was not made clear, in the original instructions. Furthermore, the instructions did not clearly explain how the data gathered and the data transformations made related to concepts and principles and to the key questions. Chen rewrote the laboratory instructions so as to reduce or eliminate the deficiencies. As a result, there was very substantial improvement in students' performance and the attitude toward the laboratory exercise. The physics course was taught in mastery mode, hence Chen used the number of test trials needed to achieve passing (criterion) level as an index of success (average of 1.6 test trials for 'new' version vs 2.7 trials for 'old'), and an attitude questionnaire.

Buchweitz (1981) developed concept maps and Vee diagrams for each of the six laboratory experiments used in a physics majors course on electricity and magnetism. He found that two of the six experiments had serious deficiencies either in terms of the lack of new conceptual linkages developed through the events observed and/or serious omissions in the conceptual/theoretical knowledge required by students to understand these events. Using data from 403 students enrolled, Buchweitz found that no significant physics achievement took place for the two laboratory sessions with serious deficiencies but highly significant gains were shown for the four laboratory sessions that were conceptually and methodologically interpretable by the students. His data were used to revise the laboratory programmme in future years.

**Conceptual**

**Theory:**
Newtonian mechanics

**Principles:**
Velocity increases when objects are accelerated
Accelerations of objects vary with slope of track

**Concepts:**
acceleration
slope
time
velocity

**Focus questions**

1. How can uniformly accelerated motion be described in terms of distance, velocity and time ? (For our purpose this will involve plotting *x* & *v* as functions of *t*.)

2. How can we express the relationships among distance, velocity, time interval and acceleration in uniformly accelerated motion ?

**Methodological**

**Value Claims:**
Completing the experiment and analysis of the experimental results will lead us to understand uniformly accelerated motion more fully and more meaningfully.

**Knowledge Claims:**
1. *x* vs. *t* graph for a uniformly accelerated motion with $V_o = \theta$ is

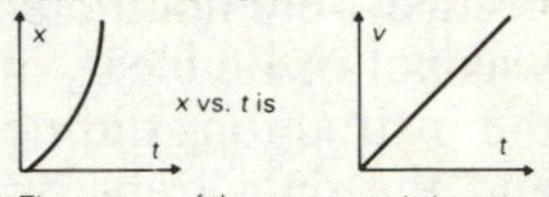


2. The tangent of the curve *x* vs. *t* at a particular time *t* is equal to the *v* at the same time.
3. The area under the curve *v* vs. *t* up to a particular *t* is equal to the value for *x* at the same *t*.
4. Acceleration = the slope of *v* vs. *t*
= twice the slope of *x* vs. $t^2$
= $1/2$ the slope of $v^2$ vs. *x*
5. g = cm$^{2}$/ = the slope of *a* vs. $\theta$

**Transformations:**
Graphs of data

Plot *x* vs. *t*, *v* vs. *t* with $\theta$ fixed, find the tangent to the former curve and the area under the curve up to a particular value of *t* of the latter curve.

Plot *x* vs. $t^2$, $v^2$ vs. *x*, find the slope of each graph, also the slope of *v* vs. *t*. for $\theta = 10, 15$ milliradian:

Find *a* directly from the following equations instead of finding the slope
$a = v/t$, $a = v^2/2x$, $a = 2x/t^2$ for all $\theta$'s:
Plot *a* vs. $\theta$ find slope.

**Records:** for $\theta$ to 5 milliradian:
1. $\frac{\text{width of the card}}{\text{the time for the card}} \frac{\Delta x}{} - \frac{v \text{ at } t}{\Delta t}$
, to pass the second gate

**Event:**
Moving a glider on an inclined air track with initial velocity equal to zero. (Digital timer and photocell gates are used for data recording).

**Figure 2.2.1:** Vee diagram of a kinematics exercise

The design of laboratory sequences can benefit by preparing Vees for each of the laboratory events planned for a term. If learning in successive laboratory sessions is to benefit from earlier studies, examining the laboratory sequence using a 'parade of Vees' can aid significantly for placement of laboratories into a sequence that will allow progressive gains in conceptual/theoretical understanding and methodological skills needed. The 6 to 20 pages are easily scanned for conceptual or methodological gaps or potential building blocks. Since concept maps drawn for each laboratory experiment represent the essential elements on the 'left side' of the Vees, these, too, can be helpful for better design of laboratory sequences.

## CONCLUSION

If laboratory programs are planned from concept maps and Vee diagrams, they will facilitate student construction of maps and diagrams for these laboratory events. Students will not only gain in appreciation of the fact that knowledge production in science requires wearing adequate 'conceptual goggles' but also a clear understanding of the relationship of concepts, principles, and theory to the interpretation of the meaning of the data. These tools can help students move from naive positivistic notions of epistemology toward more empowering constructivist views. While much more research is needed on the use of metacognitive tools to enhance laboratory learning, the research we have done to date is encouraging.

## REFERENCES

Belenky, M. F., Clinchy, B. M., Goldberger, N. R. and Tarule, J. M. (1986) *Women's ways of knowing*. New York, Basic Books.

Buchweitz, B. (1981) 'An epistemological analysis of curriculum and an assessment of concept learning in the physics laboratory'. Unpublished PhD thesis, Cornell University, Ithaca, New York.

Chen, H. (1980) 'Relevance of Gowin's structure of knowledge and Ausubel's learning theory to a method of improving physics laboratory instruction'. Unpublished MS thesis, Cornell University, Ithaca, New York.

Edwards, J. and Fraser, K. (1983) Concept maps as reflectors of conceptual understanding. *Research in Science Education*, *13*, 19-26.

Elkana, Y. (1970) Science, philosophy of science and science teaching.

*Educational Philosophy and Theory*, *2*, 15-35.

General Accounting Office (1984) *New directions for federal programs to aid mathematics and science teaching*. Gaithersburg, MD, US General Accounting Office, GAO/PEMP.

Gowin, D. B. (1981) *Educating*. Ithaca, NY, Cornell University Press.

Gurley, L. I. (1982) 'Use of Gowin's Vee and concept mapping strategies to teach responsibility for learning in high school biological sciences'. Unpublished PhD thesis, Cornell University, Ithaca, New York.

Hallowell, A. C. (1985) 'Learning theory applied to the development and evaluation of an education program about sea birds'. Unpublished MS thesis, Cornell University, Ithaca, New York.

Heinze-Fry, J. (1987) 'Evaluation of concept mapping as a tool for meaningful education of college biology students'. Unpublished PhD thesis, Cornell University, Ithaca, New York.

Helm, H. and Novak, J. D. (1983) 'Proceedings of the International Seminar Misconceptions in Science and Mathematics', June 20-23, 1983. Ithaca, NY, Cornell University, Department of Education.

Kahle, J. B. (1984) *An Investigation of Instructional Strategies which Enhance Biology Meaningful Learning*. Final report to the National Science Foundation. Lafayette, IN, Purdue University, Department of Education.

Kinigstein, J. B. (1981) 'A conceptual approach to planning an environmental education curriculum'. Unpublished MS thesis, Cornell University, Ithaca, New York.

Melby-Robb, S. J. (1982) 'An exploration of the uses of concept mapping with science students labeled low achievers'. Unpublished MS thesis, Cornell University, Ithaca, New York.

Novak, J. D. (1969) A case study of curriculum change - science since PSSC. *School Science and Mathematics*, *69*, 374-84.

____ (1977) *A theory of education*. Ithaca, NY, Cornell University Press.

____ (1985) The interplay of theory and methodology, in L. West and L. Pines (eds.), *Cognitive structure and conceptual change*. (In the Educational Psychology Series.) New York, Academic Press, 189-209.

____ and Gowin, D. B. (1984) *Learning how to learn*. New York, Cambridge University Press.

____ Gowin, D. B. and Johansen, G. T. (1983) The use of concept mapping and knowledge Vee mapping with junior high school science students. *Science Education*, *67*, 625-45.

Robertson, M. (1984) 'Use of videotape-stimulated recall interviews to study the thoughts and feelings in an introductory biology laboratory course'. Unpublished MS thesis, Cornell University, Ithaca, New York.

Robertson-Taylor, M. (1985) 'Changing the meaning of experience: Empowering learners through the use of concept maps, Vee diagrams, and principles of educating in a biology lab course'. Unpublished PhD thesis, Cornell University, Ithaca, New York.

Schwab, J. J. and Brandwein, P. F. (1962) *The teaching of science*. Cambridge, MA, Harvard University Press.

Stake, R.E. and Easley, J.A. (1978) *Case Studies in Science Education*, University of Illinois at Urbana-Champaign: Center for Instructional Research and Curriculum Evaluation and Committee on Culture and Cognition. Volumes I and II (prepared for National Science Foundation Directorate for Science Education; Office of Program Integration).

# Part 3

# INTENDED LEARNING OUTCOMES AND ACTIVITIES

# 3.1

# Learning Technical Skills in the Student Laboratory

**Elizabeth Hegarty-Hazel**

## INTRODUCTION

This description of a sobering hour spent at a school Science Society meeting conjures up visions of students engaged in what was billed as an Analysis Race. The idea was that students entered the room in turn and were timed as they set about the analysis of a three-radical compound.

Some of the more horrible sights included:

1. Benches littered with reagent bottles, some unstoppered
2. Large amounts of chemical used
3. Permanently roaring bunsen burners
4. Filter papers overfilled
5. Charcoal block tests which seemed to consist of largely blowing the unknown into the upper atmosphere
6. A boy who was about to wash out, with water from a fast tap, a test tube containing the unknown plus what seemed an enormous amount of superheated concentrated sulphuric acid. (Hannaford, 1958, p. 39)

Were these just the bad old days? The time was 1958, the country Australia. High class numbers (Hannaford quoted 60 students per teacher) and the lack of courses specifically on laboratory technique to precede sixth form chemistry in Australian schools were blamed for students' poor ability to perform technical skills. However, it is claimed (Newman,

1985) that even today poor performance of technical skills persists, and not only in Australia (despite reductions in class sizes to 20-30), but in other countries where class sizes are smaller again, so that technical errors and things going wrong are the norm rather than the exception.

The whole area of competence in technical skills has been significantly under-researched. A reason may be that techniques are somehow regarded as trivial by teachers and curriculum designers. It may be an entrenched view but it is odd nonetheless, because without technical skills, students' own activities would surely cease entirely. If students are incompetent at technical skills, then achievement of all other educational objectives for their laboratory classwork will surely be impaired.

Technical skills should not (and cannot without distortion) be considered in isolation. If a student uses a technique without a competent understanding of the equipment then it will be mere mechanical recipe-following and she or he will be let down as soon as there is the slightest variation in conditions. Further, the competent performance of a technique requires not only substantive knowledge of the equipment used but also of the scientific phenomena being investigated. In every branch of science there is a constant dynamic interaction between the syntactic and substantive structure (Schwab, 1962; Ford and Pugno, 1964) and Novak (Chapter 2.2) spells this out in a practical way for the laboratory.

Consider a student setting up microscopes to look at micro-organisms. Without a clear understanding of the procedural principles, she or he will not be able to deal with different in-built or external light sources, low or high power lenses, or with diaphragms in various parts of the equipment. Without an understanding of the structure of yeasts, the student might be satisfied with a microscope setting which showed the organisms as only tiny featureless dots - or might fail to have an appreciation of any structures which *were* revealed. An even more complex situation is when design of an investigation is the most important educational aim and microscopy the monitoring device or end product. Trying to juggle such a range of educational activities is an unwieldy task and we have separated the issues so that Atkinson discusses knowledge in Chapter 3.3 and Klopfer discusses scientific enquiry in Chapter 3.2. The present chapter is confined more to the performance and procedural knowledge of technical skills.

In terms of the model (Figure 1.2.1) technical skills are

intended learning outcomes for students. They are elements of a science discipline but their selection for student work has been influenced during the curriculum and instructional planning systems by the commonplaces (knowledge of students, teachers, milieu and subject matter), frame factors and constraints. The process by which the technical skills (inputs) with which students arrive are modified (outputs) depends on actual rather than intended outcomes. However, since knowledge of students is one of the commonplaces there are three prior stages in the model for influence - the processes of planning a curriculum and instruction and the process of instruction itself.

Throughout this chapter it is proposed to use the term technical skills to refer to all those laboratory skills involving technical manoeuvres and the manipulation of equipment (Hegarty-Hazel, 1986). The general term is rather broad and it is helpful to have some defined subdivisions to use in both research studies and practical assessments. Those developed by Englen and Kempa (1974) are helpful: methodical working, experimental technique, manual dexterity and orderliness.

## PROBLEMS AND PROSPECTS

The low priority allocated to technical skills was discussed in the introduction and this section will venture into some further explanation, despite awareness of a certain circularity of argument: if technical skills are low priority they may be badly done but if skills are badly done they will be given low priority. Certainly achievement of technical skills *is* often regarded as a low priority aim, by teaching staff, and little attention is often given to the development of teaching strategies and the monitoring of students' skills. At tertiary level, staff in some institutions definitely attribute very low priority of importance to technical skills when compared to other possible aims of laboratory classes (e.g. Boud, 1973, for physics; Lynch and Gerrans, 1977, for chemistry; Dunn, Kennedy and Boud, 1980, for physics, chemistry and biology; Hegarty, 1979, for microbiology). Some of the same research shows that university students attribute a higher priority of importance to technical skills, especially to those of obvious vocational relevance. The following quote (concerning skills courses in electron microscopy for college students) captures the emphasis:

> The pressures of the job market clearly indicate the value of training in specific skills within traditional scientific disciplines. Courses that emphasize instrumentation or techniques requiring complex skills will be attractive to the prospective student, balance theory with application in the laboratory, and increase the marketability of the degree in areas ranging from production-step responsibility to working in the rarified atmosphere of breakthrough research. (Wright, 1978, p. 39)

There is some evidènce which suggests that not only may the technical skills be important but that the technical training of teachers themselves may be problematic - at least at secondary level. For example, when secondary school chemistry, physics and other physical science teachers rated a list of possible teacher competencies using a scale ranging from 'Highly appropriate and badly needed by most physical science teachers' to 'Not needed or not useful', technical skills and use of equipment which were rated highly were use and care of simple electronic instruments such as voltmeters, pH meters and simple volumetric glassware (James and Schaaf, 1975). If such simple techniques are a problem for school teachers, is it not likely that they will prove a problem for their students? The same group of teachers (in Kansas, USA) also gave high ratings to the use and care of more complex equipment such as oscilloscopes, cathode ray tubes and radioactive sources. The teachers surveyed seemed acutely aware of deficiencies in their own technical skills. They also showed an interest in learning about the setting up and operation of laboratories with emphasis on what can go wrong and how this can be handled or avoided.

Yet, on the positive side, the laboratory *can* provide the best available setting for the learning of technical skills. Research has confirmed that conventional laboratory classes where students manipulate apparatus and perform techniques individually or in small groups are more effective than teacher demonstrations (Horton, 1930; Kruglak, 1952) or no laboratory (Kruglak, 1953; Yager, Engen and Snider, 1969; Ben-Zvi *et al,* 1976).

These research findings were unequivocal by comparison with the variability found in relation to other types of learning which may be acquired in a laboratory setting.

## CONDITIONS FOR LEARNING TECHNICAL SKILLS AND STRATEGIES FOR TEACHING THEM

It seems logical to turn to the literature on student learning to understand why there are problems with students' attaining competence in technical skills and the extent to which these reflect inappropriate teaching strategies.

Laboratory technical skills, like other psychomotor skills, involve the senses and the brain as well as the muscles. Performance requires coordination and smooth execution with standards of accuracy and sometimes speed being applied. Internal conditions of learning require an overview of the skills routine while external conditions require practice and feedback for improvement of accuracy, speed and the quality of part-skills (Gagné and Briggs, 1974). Once mastered, technical skills are well retained and continue to improve with practice (Fitts and Posner, 1967).

It is convenient to consider the competent learning of technical skills as a special case of a four step sequence: achievement, mastery, proficiency, competence (White, 1979). This requires accepting that the philosophy of mastery learning implies the provision of conditions for learning such that all or most students are able to reach mastery of the required skill.

Competence is not useful in a vacuum. Curriculum specialists and teachers must be able to identify those technical skills for which competence is expected and must also be able to define competence in terms of the milieu of society. Thus for one particular technical skill, competence will be a different matter for a chemist in a federal standards laboratory, a garage mechanic, an atomic physicist, or a biology teacher, and may be irrelevant for a typist or a stockbroker. This dependence on society for definition means that the need for competence on technical skills is relatively easy to establish in vocationally-oriented courses and relatively difficult for all other courses. Thus at junior levels of secondary and tertiary education, the relevant milieu is often accepted as being later levels in the educational setting rather than society at large. In this way junior secondary students may learn to use a microscope with a low power lens (10x) to observe largish microscopic organisms such as protozoans in pond water. Their minimal competence can be upgraded in the senior high school so that they are capable of using 40x lenses to examine smaller organisms such as yeasts. Use of a 100x lens with oil immersion with sufficient competence to observe stained bacteria or to use phase contrast to observe the fine structure of bacteria may become a realistic technical target during courses in biology or microbiology at the tertiary level.

It is almost axiomatic that teachers should be systematic and purposeful in matching a variety of teaching methods to clearly defined objectives. Teachers must know explicitly what is to be taught, students must know what is to be learned. The following is an application to the special case of technical skills of the instructional conditions proposed by Gagné and Briggs (1974) and by White (1979).

Teachers should provide for students a satisfying rationale (e.g. in terms of future learning or use in society), a specification of the technique and an overview of the steps in performance, ample individual access to the equipment involved, timely and detailed feedback on performance, opportunity for practice and improved performance and, if necessary, access to alternative learning modes. Teaching techniques fostering achievement of technical skills include various questioning strategies used by a teacher or included in written materials, demands for students to make diagrammatic representation of equipment, demands to interchange forms of representation, demands for note taking, record keeping or for maintenance of a laboratory work book or equipment roster. Grading for mastery of techniques should require criterion-referenced testing of actual accomplishments rather than any form of norm-referenced testing.

For students, there should be overt attention, participation and performance. There are also probably requirements for internal states such as understanding of relevant concepts and possession of a mental image of the equipment both at rest and at steps in its successful operation. Once proficiency is achieved, students are likely to retain it.

In view of the specifications above, consider the common situation where a student frequently encounters pieces of equipment once only during an entire course. It will be clear that there is not the faintest likelihood of the student attaining competence in even the most elementary skill. This common practice is likely to lead to feelings of incompetence and negative attitudes to science, but the literature shows little research on the relationship between teaching strategies, students' proficiency and students' feelings of competence. The issue of effects on students' attitudes is explored by Gardner and Gauld in Chapter 3.4. At present there is probably little alternative for teachers but recourse to trial and error. The shortcomings produced by insufficient practice will probably be obvious but it also needs to be remembered that prolonging practice too far can cause boredom (White, 1979 and see also Chapter 3.4). Proficiency will come with practice, but this is on an individual not class basis and classwork should be structured accordingly.

## RESEARCH ON IMPROVEMENT OF TECHNICAL SKILLS

Much research has yet to be done. There is little on modification of external conditions of learning - surprising since this is the area most immediately under the control of teachers and course designers. There has been some research on improving technical skills by attending to students' internal conditions of learning. This has chiefly focused on the provision of various types of overviews (e.g. summaries, flowcharts, checklists, diagrams, photographs, slides, movies or videotapes). Working with university chemistry undergraduates, Kempa and Palmer (1974) and Neerinck and Palmer (1977) used videotapes to provide overviews. Beasley's (1979) overviews were in the form of diagrams with brief accompanying instructions for the use of the analytical balances, pipettes and burettes in volumetric analysis by university chemistry students. Paying specific attention to internal conditions of learning, overviews were provided in different formats requiring either physical or mental practice. Mental and physical practice were found to be equally effective in improving the accuracy and precision of technical skills yet mental practice requires no equipment or expense but simply reading and rehearsal. A practical application of this finding would be to issue overviews for students to practise mentally before attending the laboratory or in the pre-lab sector of class time.

Research on external conditions has not specifically concentrated on the amount of practice required nor on the form of feedback. However, practice was the key element in some kits specially designed to improve the performance of university chemistry students (Runquist, 1979) and which had the desired effect in improving students' accuracy from 5% relative error to less than 1%. The kits contained a tape recorder and tape whose purpose was not clearly described by the author but which possibly improved students' grasp of the overview of the skills routine. The kits also contained chemicals, glassware and other materials needed for students to practise weighing, titrations, preparation of standard solutions and use of spectrophotometers. In a self-paced procedure, students were able to practise the techniques as they saw fit before completing an exercise where estimates of their error could be calculated. Information on errors should help students monitor their own performance. Exercises designed to do this are seldom explicitly taught in the laboratory but encouraging results have sometimes been reported. Some hour-long minilabs were specially designed for university physics students (Reif and St. John, 1979; St. John, 1980) and these required rough estimates of laboratory variables and ability to estimate errors. Students who had done the minilabs were found to out-perform students from an ordinary course in an examination incorporating technical skills outcome measures.

## EMPHASIS ON TECHNICAL SKILLS IN DIFFERENT CURRICULUM MATERIALS

The first part of this section will make use of content analysis techniques to determine the extent to which technical skills are required in some secondary and tertiary science courses. Content analysis reveals only frequency and one must always be mindful that frequency and form are not the same. To counterbalance any such impression which might arise from the content analysis studies, the later part of this section will be devoted to the issue of sequencing technical skills within the curriculum.

At high school level, a content analysis of twelve physics laboratory manuals commonly used at high school in the USA showed a range (0-10%), of experiments which specifically emphasized measurement and technique including uncertainty, error analysis, mensuration and the SI System (Kuehl *et al*, 1984). Half of the texts sometimes required use of unique apparatus while others differed in requiring it often or rarely. Similarly, analysis of five chemistry laboratory manuals showed that emphasis on performance categories varied considerably (Fuhrman, Lunetta and Novick, 1982). The requirement for carrying out qualitative observations was found in a minimum of 91.4% of exercises in any one manual (maximum 100%). Analysis using a laboratory task analysis scheme provided the following results for other performance categories: Carries out quantitative observation or measurement (50.7% to 82.8%). Manipulates apparatus, develops techniques (94.6% to 100%). Records results, describes observations (52.7% to 100%). Performs numerical calculations (26.1% to 65.7%). The two other headings which completed the performance category (interprets, works according to own design) stressed enquiry skills rather than technical skills.

At the tertiary level Hegarty (1978, 1979) analysed nine commercially available microbiology manuals according to student involvement in scientific enquiry. Exercises for development of technical skills (together with knowledge-oriented 'recipe' exercises) were represented at level zero, i.e. not involving experience of enquiry. It was found that an average of 38% of all exercises were set at this level (range 17.6% to 60%). At all levels, there were 92.95% of exercises requiring students to perform technical procedures.

These content analysis studies show that technical skills are required in almost every student exercise - underlining the obvious point that unless teacher demonstrations are used, technical skills are indispensable for student laboratory work. However, they also show that exercises, the sole purpose of which is to learn a skill, occur fairly seldom. *The potential therefore exists for students to encounter constant demands to use new skills and never become competent in any.* Content analysis cannot readily reveal opportunities for practice, or for transfer of skill use to a new situation.

Practice in the context of a well-judged curriculum sequence is illustrated in Figure 3.1.1. It is undertaken by final year science microbiology majors at the University of New South Wales over a seven week period (and is one strand only of coursework which altogether occupies some 16-18 hours of classwork per week). The technical skills of disc and dilution testing of antibiotics (steps 4 and 5) are standard procedures in diagnostic laboratories and are of direct vocational relevance. Previous teaching of procedural and substantive knowledge allows students to make some appropriate operator choices at even these steps. Continuing in this manner, the students have a sound basis for both the enquiry exercise (step 10) and the clinical exercise (step 12). They continue to use disc sensitivity testing throughout their course, taking responsibility for recognition of occasions when testing is warranted and for selection of appropriate antibiotics (step 13). Overall this

**Figure 3.1.1** Technical skills (antibiotic sensitivity testing) within a curriculum sequence for university microbiology majors

1. Mechanisms of action of antibiotics (lectures)
2. Effects of antibiotics on different genera of bacteria (lectures)
3. Principles of laboratory testing of antibiotics (demonstration and overview)
4. Choose range of antibiotics appropriate for specified organisms. Set up disc sensitivity test (laboratory)
5. Set up broth and agar dilution sensitivity tests. Correlate with disc test (laboratory)
6. Repeat techniques 4 and 5 to specified performance levels (laboratory)
7. Use of antibiotics in different clinical situations (lectures)
8. The laboratory recognition of sensitivity or resistance relationship between zone size and minimum inhibitory concentration (lecture)
9. Revision of practical and theoretical principles (test)
10. Design a method for monitoring an antibiotic (gentamycin) in patient's serum. Feedback on design. Comparison with current routine hospital procedures and recently developed 'high-tech' methods involving radio-isotopes (tutorial)
11. Carry out own method and routine method. Correlate results (laboratory)
12. Monitor patient's serum levels and determination of toxicity. Recommend changed dosage schedules as appropriate (simulation)
13. When culturing clinical specimens, decide those specimens or isolated organisms on which tests should be done and choose antibiotics (laboratory)
14. Perform competently, items 4, 5, 13 under examination conditions - closed book (practical examination).

sequence was designed to help students become competent, knowledgeable, enquiring and to enjoy their work on antibiotics - and so it proved in practice.

## ASSESSMENT OF TECHNICAL SKILLS

### Practical examinations and the practical mode

Common sense suggests that technical skills should be tested in a practical setting. Experience shows that despite the apparently obvious connection between student learning in the science laboratory and the testing of students' performance, the use of laboratory practical tests has a patchy history worldwide - although some countries (e.g. the United Kingdom) have a long tradition of practical examinations at both secondary and tertiary levels. Even in the UK, the use of practical examinations has varied depending on the value currently placed on testing of technical skills or observational skills, and the extent to which it has proved feasible or desirable to test higher level enquiry skills. It is a point of criticism that sometimes the testing of technical skills, procedural knowledge and enquiry skills are confounded so that poor performance on the former two results in low marks on the latter. In addition, there have been doubts about the validity, reliability and discriminating ability of practical tests. They are expensive, require extensive organization and supervision and constant renewal with attendant administrative difficulties.

The practices of assigning low proportions of marks to results of practical tests or of moderating students' marks on the basis of their performance in written tests are extremely dubious in view of studies which have demonstrated consistent low correlations between laboratory practical examinations and (knowledge-oriented) written tests (Abousief and Lee, 1965; Robinson, 1969; Tamir, 1972; Comber and Keeves, 1973; Kempa and Ward, 1975). Tamir (1972) introduced the term 'the practical mode' to signify this difference in performance (see also Chapter 5.2).

On this basis, the importance of practical examinations for assessment of practical skills seems well established (Kempa, 1986). The relationship within the practical skills to be assessed is a matter of priorities within a curriculum. The following will serve as an example. It is the schedule adopted for the assessment of chemistry laboratory work at sixth form level in the United Kingdom (University of London, 1977):

1. Manipulative skills (25-30%)
2. Skills in observation and the accurate recording of such (25-30%)
3. Ability to interpret observations (20-25%)
4. Ability to devise and plan experiments (10-15%)
5. Attitudes (10-20%)

Practical tests (and also written reports and interviews) have fairly high face validity because they are activities which are similar to those carried out by scientists in their normal work (Dunn, 1986). Paper-and-pencil tests have low face validity because they are not mirrored in scientific work. One would not wish to push this comparison too far - there is little or no possible counterpart for secondary or tertiary students. Another aspect of validity is the extent to which the assessment procedure measures the desired outcome, and on this criterion, a student's technical skill is best assessed by direct observation of the student using the skill. Assessment on the basis of results would have lower validity and again paper-and-pencil tests would rate lowest. In summary, direct practical tests although superficial in their relationships to the work of practising scientists, are the best tests available.

In practice, compromises are required and use of a range of testing methods may be advisable (Kempa, 1986). For example, direct observation of technical skills may be extremely valid, reliable, and capable of discriminating between students. However, it is extremely time-consuming and on some occasions may be replaced by measurement of outcomes with tolerable deficiencies. Use of criterion referenced testing is the only acceptable method of providing student feedback on performance of technical skills (Dunn, 1986). Teachers must decide what constitutes acceptable mastery of a skill. Care must be taken not to include in assessment for grading, mastery level skills where most students will top out on the test. Skills on which mastery is required are best separated in a compulsory prerequisite examination on a pass/fail basis.

Paper-and-pencil tests can be used to assess students' procedural knowledge or to complement practical tests of technical skills. (For an early example of such tests, see Persing (1929) and for a more recent one, see Lunetta, Hofstein and Giddings (1981)). However, what is or can be measured by such tests is uncertain. Use of written tests to replace practical tests is a dubious procedure. Kruglak (1955) showed only low to moderate correlations between laboratory performance tests of very specific skills and paper-and-pencil items which were intended to be exact analogues. Correlations for more complex aspects of laboratory performance were even lower. (See also Chapter 5.2.)

## Direct observation

The most direct method for assessment of students' technical skills is for suitably trained judges to observe them in action. This method has reasonable face validity in that it is most similar to the situation for which the student is being trained although it is far from clear what standards should apply (those of practising scientists or those of secondary or tertiary teachers?). Direct observation meets the criterion of validity where the assessment procedure measures the desired outcome. All other methods of assessment involve some steps of inference.

There are two major criticisms: one is pragmatic and may be the single most important reason for discontinuation of direct observation programmes or for their patchy use. This is the necessarily highly individualized and time-consuming nature of the method. The second criticism is more fundamental and concerns the distinction between performance and underlying competence (Wood and Power, 1987). Here there is always the danger of false negatives - students who perform poorly are judged to lack underlying competence whereas they may have a fair grasp of procedural knowledge and technical skills but not be able to apply them to the particular examination question. Direct observation should be accompanied by suitable 'elaborative procedures' (Wood and Power, 1987) whereby students would be interviewed or otherwise given the opportunity to explain why they did X or what they would do if conditions changed to Y. This may sound even more dauntingly time-consuming than direct observation is already but would probably be necessary only sometimes if performance criteria (described below) were made available to *students* at the time of learning a technique (as well as to staff at the time of an examination).

Different methods for direct assessment have been reviewed and illustrated (Dunn, 1986; Kempa, 1986). The research study reported by Englen and Kempa (1974) used three different methods of assessment by direct observation, subjective rating, checklists and intermediate mode checklists.

Subjective rating is threatened by idiosyncrasy and vagueness but checklists have a long and honourable history (Hendricks, 1950; Kempa, 1986, Chapter 5) and see Tamir's chapter, 5.2, on national evaluation processes in Israel). Intermediate mode checklists have a shorter history but have been successfully modified for use with assembly of equipment (Giddings and Hofstein, 1980).

Research findings for the three different methods of assessment (Englen and Kempa, 1974) were as follows: (1)

Subjective rating resulted in low mean grades by comparison with other methods and this was attributed to use of negative or penalizing performance characteristics. (2) Use of the checklist assessment method resulted in lower grade variances than either of the other methods but had advantages of validity. However, it was impractical for use in normal classes since it involved long spans of individualized attention to very few students. (3) The intermediate mode was found to have most of the advantages of the checklist (similar mean grades and variances) without its disadvantages of being too detailed and too consuming of time and energy for normal purposes. The basic mechanism appeared to be as for subjective rating but with use of both positive and negative performance features presumably due to specification of performance criteria.

## Measurements of outcomes, errors

There is a school of thought which says the outcome is everything - that in real life either you get it right or you don't. This would be arguable in most spheres of 'real life' but the more so in education. Kempa (1986, Chapter 6) argued for a balance between direct observation and measurement of outcomes in the assessment of technical skills at secondary school levels in the United Kingdom. Wood and Power (1987) warned against judgements of performance in the absence of elaborative procedures for tapping students' underlying competence, including procedural knowledge.

It is common practice to require students to submit the final product of their experiment either physically or in the form of laboratory reports or practical scripts and for teachers to make inferences on this basis about students' proficiency in technical skills. This may have a certain amount of face validity in that practising scientists often need to submit technical reports but at the student level critics have challenged the underlying assumption that a high correlation exists between proficiency on a practical task and the results it achieves.

Pickering and Monts (1979) challenged the common procedure of grading laboratory notebooks as a measure of technical and observational skills and found the two to be unrelated. Observational questions posed to freshmen chemistry students at Columbia University were concerned with colours of solutions and solids, relative viscosity and odour of liquids and relative sizes of crystals. Ratings of laboratory notebooks proved to be more influenced by completeness than accuracy of observations and were also correlated with neatness. The authors concluded that keeping a notebook encourages the habits of a good scientist in the laboratory (thorough record-keeping) and discourages cheating. However, they concluded that explicit grading is unwarranted and unrelated to the measure of technical or observational skills. In Chapter 5.2, Tamir is more accepting of this traditional method. There are pitfalls in assessing students' technical skills on the basis of their calculated results. In analytical chemistry, it is recognized that errors of measurement can occur due to the system, the instrument or the operator. Normally replicate measurements are taken and results can be assessed according to errors in individual measurements and expressed according to degree of confidence. Provided students operate under standardized conditions with carefully prepared materials, it is commonly accepted practice to assess their performance in terms of the accuracy and precision of their calculated results. The assumption is that there is a high correlation between correct results and good techniques such as pipetting and buretting. However, it is likely that such correlations under-represent students' technical skills because of common student errors in calculations. When checking calculations in quantitative volumetric experiments, Beasley (1980) noted many mathematical errors in the results of introductory college chemistry students - e.g. in the subtractions of mass or volume readings. More complex errors included lack of understanding of the logic of back titration required in the analysis of an antacid substance.

When he recalculated students' results from their raw data, Beasley found a correlation coefficient of only 0.2 between students' calculated results and his own. There was much higher correlation (0.7) between researcher-calculated results and the true result of the exercise. The following recommendation was made:

> If we are to make judgements about the proficiency of a laboratory performance in quantitative experiments, and subsequently decisions about the level of skill learning attained by students, it seems appropriate that the following criteria should be satisfied:
>
> (1) Students operate under standard conditions with carefully prepared materials.

(2) Comparisons of student performance be made in terms of the accuracy and precision of their determination.
(3) The accuracy and precision of a student determination to be based on the raw data collected rather than the mathematical computation of that data.
(4) After an initial determination of an unknown sample, students are given the opportunity to make the correct determination of their data through pre-programmed mathematical methods.
(5) Independent grading of laboratory performance and computational skill be recorded.

(Beasley, 1980, p. 71)

Englen and Kempa (1974) reported that for chemistry, the General Certificate of Education Boards (UK) accepted the view that it is the acquired laboratory skills which should be assessed and not the outcomes of sequences of operations. (This view has been incorporated into the GCE Boards' experimental schemes of teacher assessment.)

## CONCLUSION

This chapter, together with Chapters 3.2 (scientific enquiry), 3.3 (scientific knowledge) and 3.4 (scientific attitudes, attitudes to science), isolates for conceptual convenience part of the seamless role of intentions and events in the student laboratory. Chapter 2.1 warns that researchers, curriculum designers, teachers and students must be constantly aware of the interactions between all the dimensions. Recommendations ventured here are made in that context.

Technical skills have been undervalued, the subject of too little informed analysis and too little research. The appraisal in this chapter is therefore limited but suggests that use of technical skills by students in the laboratory should be restricted to skills of vocational importance (especially at the tertiary level) or skills which have been identified as essential prerequisites for later educational levels or for successful participation in scientific enquiries. Great care should then be taken to structure the learning environment so that teachers and students are clear about the rationale for learning technical skills and that opportunities are provided for overviews, practice and participation in meaningful *sequences* of curriculum. This way, students should develop both competence and positive attitudes. Their technical skills, procedural and conceptual knowledge may grow together. They may be able to participate in genuine scientific enquiry unhampered by the poor technique which has contributed to the failure of so many enquiry oriented programmes to materialize, and to the disasters in the programmes which have materialized.

All technical skills which fail to meet the criteria above should be considered candidates for elimination from a curriculum. It is surely better to do a few things well than many things badly. As every other chapter in this book shows, there are many competing demands for laboratory time and money.

Criterion-referenced practical measures are required to assess students' performance of technical skills. Checklists specifying performance criteria can be used in regular laboratory classes and more detailed procedures used in practical examinations. Techniques regarded as 'basic' should be assessed as for any other form of mastery learning with opportunities for feedback and remedial work until students achieve the specified level of performance. This assessment should not be permitted to inflate the results of assessment for grading and would be well kept on a compulsory pass-fail preliminary basis. Likewise, assessment of laboratory record books, whilst encouraging good scientific practice, has been shown to have little relationship to technical skills and would be best excluded from assessment for grading. Overall, practical assessments should be restricted to direct observation and judicious measurement of outcomes. Paper-and-pencil tests cannot be substitutes but could well be used as elaborative procedures to tap students' understanding of procedural knowledge.

## REFERENCES

Abousief, A. A. and Lee, D. M. (1965) The evaluation of certain science practical tests at the secondary school level. *British Journal of Educational Psychology*, 35, 41-9.

Beasley, W. F. (1979) The effect of physical and mental practice of psychomotor skills on chemistry students' laboratory performance. *Journal of Research in Science Teaching*, 16, 473-9.

_____ (1980) Analyzing laboratory performance - what do we really measure? *The Australian Science Teachers' Journal*, 26(1), 69-72.

Ben Zvi, R., Hofstein, A., Kempa, R. F. and Samuel, D. (1976) The effectiveness of filmed experiments in high school chemical education. *Journal of Chemical Education*, 53, 518-20.

Berger, C. F. (1982) Attitudes and science education, in Rowe, M. B. (ed.) *Education, in the 80's: Science*, Washington, D.C., National Education Association.

Boud, D. J. (1973) The laboratory aims questionnaire. A new method for course improvement? *Higher Education*, 2, 81-94.

Chalmers, R. K. and Stark, J. (1986) Continuous assessment of practical work in the Scottish HNC course in chemistry. *Education in Chemistry*, 5, 154-5, 160.

Comber, L. C. and Keeves, J. P. (1973) *Science Education in Nineteen Countries: International Studies in Evaluation*, New York, Wiley.

Dunn, J. (1986) Assessment of students, in Boud, D., Dunn, J. and Hegarty-Hazel, E. (1986), *Teaching in Laboratories*, Guildford, SRHE/NFER Nelson.

_____ Kennedy, T. and Boud, D. J. (1980) What skills do science graduates need? *Search*, 11, 239-42.

Englen, J. R. and Kempa, R. F. (1974) Assessing manipulative skills in practical chemistry. *School Science Review*, 56, 261-73.

Fitts, P. M. and Posner, M. I. (1967) *Human Performance*, Belmont, California, Wadsworth.

Ford, G. W. and Pugno, L. (eds.) (1964) *The Structure of Knowledge and the Curriculum*, Chicago, Rand McNally.

Fuhrman, M., Lunetta, V. N. and Novick, S. (1982) Do secondary school laboratory texts reflect the goals of the 'new' science curricula? *Journal of Chemical Education*, 59, 563-5.

Gagné, R. M. and Briggs, C. J. (1974) *Principles of Instructional Design*, New York, Holt, Rinehart and Winston.

Giddings, G. and Hofstein, A. (1980) Trends in the assessment of laboratory performance in high school science instruction. *The Australian Science Teachers' Journal*, 26(3), 57-64.

Hannaford, B. (1958) The need for a course in laboratory technique as preliminary to the study of elementary chemistry. *The Australian Science Teachers' Journal*, 4(1), 39-40.

Hegarty, E. (1978) Levels of scientific enquiry in university science laboratory classes: implications for curriculum deliberations. *Research in Science Education*, 8, 45-57.

______ (1979) The Role of Laboratory Work in Teaching Microbiology at University Level, unpublished Ph.D. thesis, University of New South Wales, Australia.

_____ (1986) Research, in Boud, D., Dunn, J. and Hegarty-Hazel, E. (1986), in *Teaching in Laboratories*, Guildford, SRHE/NFER Nelson.

Hendricks, B. C. (1950) Laboratory performance tests in chemistry. *Journal of Chemical Education*, 27, 309-11.

Horton, R. E. (1930) Measured outcomes of laboratory instruction. *Science Education*, 14, 311-19.

James, R.K. and Schaaf, J. (1975) Practitioners' ideas on laboratory skills, competencies needed for physical science teachers. *Science Education*, 59, 373-80.

Kempa, R. F. (1986) *Assessment in Science*, Cambridge, Cambridge University Press.

_____ and Palmer, C. R. (1974) The effectiveness of videotaped recorded demonstrations in the learning of manipulative skills in practical chemistry. *British Journal of Educational Technology*, 5, 62-71.

_____ and Ward, J. E. (1975) The effect of difference modes of taste orientations on observational attainment in practical chemistry. *Journal of Research in Science Teaching*, 12, 69-76.

Kruglak, H. (1952) Experimental outcomes of laboratory instruction in elementary college physics. *American Journal of Physics*, 20, 136-41.

_____ (1953) Achievement of physics students with and without laboratory work. *American Jounal of Physics*, 21, 14-16.

_____ (1955) Measurement of laboratory achievement. Part III. Paper-pencil analogs of laboratory performance tests. *American Journal of Physics*, 23, 82-7.

Kuehl, E. *et al.* (1984) An evaluation of high school physics laboratory manuals. *Physics Teacher*, 22(4), 222-35.

Lunetta, V. N., Hofstein, A. and Giddings, G. (1981) Evaluating science laboratory skills. *The Science Teacher*, 48(1), 22-5.

Lynch, P. P. and Gerrans, G. C. (1977) The aims of first year chemistry courses, the expectations of new students and subsequent course influences. *Research in Science Education*, 7, 173-80.

Neerinck, D. and Palmer, C. (1977) The effectivement of videotaped recorded demonstrations in the learning of manipulative skills in practical chemistry. Part II. *British Journal of Educational Technology*, 8(2), 124-31.

Newman, B. (1985) Realistic expectations for traditional laboratory work. *Research in Science Education*, 15, 8-12.

Persing, K. M. (1929) Testing Laboratory technic in high-school chemistry. *Journal of Chemical Education*, 6, 1321-7.

Pickering, M. and Monts, D. L. (1979) Grading observational skills - exam questions or notebooks? *Journal of College Science Teaching*, 8, 238-9.

Reif, F. and St. John, M. (1979) Teaching physicists thinking skills in the laboratory. *American Journal of Physics*, 47, 750-7.

Robinson, J. (1969) Evaluating laboratory work in high school biology. *American Biology Teacher*, 31, 236-40.

Runquist, O. (1979) Programmed independent study, laboratory technique course for general chemistry. *Journal of Chemical Education*, 56, 616-17.

Schwab, J. J. (1962) The concept of the structure of a discipline. *Educational Record*, 43, 197-205.

St. John, M. (1980) Thinking like a physicist in the laboratory. *The Physics Teacher*, September, 436-43.

Tamir, P. (1972) The practical mode - a distinct mode of performance in biology. *Journal of Biological Education*, 6, 175-82.

University of London (GCE) (1977) Notes of guidance and report form for the optional internal assessment of special studies, University of London. Cited by Giddings and Hofstein (1980).

White, R. T. (1979) Achievement, mastery, proficiency, competence. *Studies in Science Education*, 6, 1-22.

Wood, R. and Power, C. (1987) Aspects of the competence-performance distinction: educational, psychological and measurement issues. *Journal of Curriculum Studies*. In press.

Wright, R. D. (1978) Skills courses: electron microscopy modules. *Journal of College Science Teaching*, 8, 39-41.

Yager, R. E., Engen, H. B. and Snider, B. C. F. (1969) Effects of the laboratory and demonstration methods upon the outcomes of instruction in secondary biology. *Journal of Research in Science Teaching*, 6, 76-86.

# 3.2

# Learning Scientific Enquiry in the Student Laboratory

**Leopold E. Klopfer**

## INTRODUCTION

Virtually all science teachers recognize that empirical enquiry is the hallmark of the natural sciences. We believe that the sciences have succeeded in reducing human ignorance and building understanding largely because of their commitment to forms of enquiry which appeal to experimental and observable evidence to test ideas. One important challenge of science teaching at every educational level is to convey a firm sense of the nature and functions of empirical enquiry to our students. Engaging students in science laboratory activities and having them reflect on their work can contribute to meeting that challenge.

Scientists have used the processes of enquiry to produce imposing structures of knowledge about living things and energy and matter and other natural phenomena. Though the knowledge structures in the various sciences generally are well-integrated around a small number of central ideas, especially in the mature sciences, enquiry continually challenges the soundness of existing science knowledge. Through enquiry, scientists can correct the fund of knowledge in all the sciences, revise their ideas when necessary, and weave the structures of scientific knowledge into an ever-stronger fabric.

As science teachers, we have taken on the responsibility for teaching our students both the content of science - the body of structured knowledge about the natural world - and the processes of science - its methods of enquiry. Each is a formidable task. Teaching the content is formidable because the amount of science knowledge is vast and grows every day, and because most science knowledge has an integrated structure. Teaching the processes is formidable because the spirit of enquiry which pervades science is an exacting taskmaster, calling for sustained commitment and hard thought, checking and rechecking, questioning critically and re-examining, and, as Percy Bridgman put it, 'doing one's damnedest with one's mind - no holds barred' (Bridgman, 1957, p. 150). Nonetheless, with properly designed and sequenced instruction, both the content and processes of science can be accessible to almost every student who wants to learn. The task for science teachers is to improve our instructional strategies, or devise new ones, which will help students to become proficient in science content and the processes of scientific enquiry.

For many years, numerous educators, scientists and philosophers have advocated the incorporation of enquiry in science teaching (Champagne and Klopfer, 1981). We can trace the idea directly back to such thinkers as John Dewey (e.g. Dewey, 1910, 1916) and Thomas Huxley (1877), see also Chapter 2.1. Given this long history, it is hardly surprising that the term *enquiry* has different meanings for different people who write about science education. Compare, for example, how enquiry (or inquiry) is used by Connelly *et al.* (1977); Herron (1971); Kuslan and Stone (1968); Olson (1982); Schwab (1962); Sund and Trowbridge (1977); and Tamir (1983). For the purpose of this chapter, enquiry is taken to be a general process by which human beings seek information or understanding. Broadly conceived, enquiry is a way of thought. Scientific enquiry, a subset of general enquiry, is concerned with the natural world and is guided by certain beliefs and assumptions. Scientists' beliefs about scientific enquiry and the assumptions underlying it tend to change over time, and the examination of these beliefs, assumptions and changes is the province of the philosophy of science. There currently is considerable debate among philosophers of science about some basic assumptions of scientific enquiry (e.g. Feyerabend, 1976; Kuhn, 1970; Lakatos and Musgrave, 1974; Popper, 1972).

The domain of scientific enquiry with respect to science education may be divided into three main components: (1) students' science process skills, (2) general enquiry processes, and (3) the nature of scientific enquiry. The first component includes the full range of science processes, such as observing and measuring, seeing and seeking solutions to problems, interpreting data, generalizing, and building, testing, and revising theoretical models. These are detailed later in this chapter. Within 'general enquiry processes' are included strategies, such as problem-solving, uses of evidence, logical and analogical reasoning, clarification of values, decision-making, and safeguards and customs of enquiry. The latter include the agreed-upon procedures - the ethos - that individuals in all forms of rational enquiry are expected to follow. The third component, the 'nature of scientific enquiry' is essentially epistemological, reflecting its connections with the philosophy of science. For example, the structure of scientific knowledge is tentative - the product of human efforts - affected by the processes used in its construction and by the social and psychological context in which the enquiry occurs. Scientific knowledge is also affected by assumptions about the natural world, such as causality, non-capriciousness, and intelligibility. Elaboration of the latter two components of the domain of scientific enquiry with respect to science education lies beyond the scope of the present chapter (see Klopfer, 1975).

The model (Figure 1.2.1) which encompasses the contents of this book makes quite clear that the instructional strategies we employ in the laboratory and classroom are closely linked to the particular learning outcomes we intend students to achieve. The intended learning outcomes concerning science content knowledge that laboratory work can help students to achieve are discussed in Chapter 3.3, and the present chapter focuses on outcomes concerning the processes of scientific enquiry. The next sections of this chapter offer illustrations of the linkage between instructional strategies and intended learning outcomes with respect to scientific enquiry and an elaboration of the students' science process skills that constitute scientific enquiry.

## UNDERSTANDING SCIENTIFIC ENQUIRY

While most of this chapter explicates the laboratory-based outcomes pertaining to scientific enquiry processes, we should pause at the outset to recall that we teachers of science also have the responsibility of helping our students to understand the nature of scientific enquiry. To some extent, students will come to understand the nature of scientific enquiry by engaging in enquiries. However, for most students and particularly younger students, teachers need to make some specially directed effort to help them attain the goal of understanding scientific enquiry.

Judy K is a science teacher who wants her students, not only to become proficient in the processes of scientific enquiry and to understand important science concepts, principles and theories, but also to understand some aspects of the nature of scientific enquiry. That is why it wasn't altogether surprising that, one morning, the students in the day's first biology class found that all the chalkboards in Judy K's room at Samuel Langley High had been altered. Running down the centre of each board was a bright red painted stripe. Printed in red across the top of the left half of each board was the label, 'What We Know', while the right-hand side of each board proclaimed 'How We Know'.

To begin the class, Judy explained how she expected the new arrangement of the boards to help the students when they thought and talked about their laboratory experiments and other aspects of scientific work. 'When we do a laboratory activity', she said, 'you know that we follow some particular procedures to make our observations and measurements. The observations and measurements are our data, and we process the data in some particular way and interpret them to reach a conclusion for our experiment'. As she spoke, Judy wrote the six key terms across the middle of the board with while chalk: observations, measurements, procedures, data processing, data interpretation, conclusions.

'Now, these all should go under one of the headings, either "What We Know" or "How We Know". For example, observations goes under "What We Know" on the left side of the board, but information about the procedures we use to make the observation goes under the "How We Know" heading on the right side'. Judy erased the two terms from the centre of the board and wrote them under their respective headings. 'Two other kinds of information also go under the "What We Know" heading. Can anyone tell me what they are and why they belong here?'

After much discussion, the class finally divided the six kinds of information into the two designated categories. Under 'What We Know', they placed observations, measurements and conclusions. Under 'How We Know', they placed procedures, data processing and data interpretation. This exercise encouraged discussion about how the procedures used in an experiment affect the observations and measurement, and how data processing and data interpretation affect conclusions.

Judy then told her class, 'As you can see, the lines on the boards and the headings are permanent. From now on, whenever we discuss laboratory activities, I want you to list information about your observations and measurements on the left side of the board. List information about the procedures you used to observe and measure under the "How We Know" heading on the right side. Also list information on the right about how your data were processed and interpreted. And finally, your conclusions from the experiment go under the "What We Know" heading on the left side.'

'Now, how do you think we can use the division of the boards into "What We Know" and "How We Know" when we're discussing science principles and laws that we find out from experiments or read in a book?' Judy asked. It didn't take long for the class to decide that they would write the verbal or symbolic statement of a principle or law being discussed on the 'What We Know' side of the board. Under 'How We Know' on the right side, they would write the supporting evidence for the law or principle, such as observations and experimental results. One student pointed out that these kinds of information appeared on the 'What We Know' side of the board when the discussion was about experiments.

'That's right, Melissa,' beamed Judy. 'You're seeing how the science principles connect with the experimental phase of scientific work. It's important to understand this connection, and it's also important to be clear which phase of scientific work we're discussing at any one time. To help us keep track, I suggest that we write on each side of the board with white chalk when we're talking about the experimental phase. When we're discussing laws and principles, we'll write on each side of the board with yellow chalk.'

'We'll also use the same divisions on the boards to discuss scientific theories, and we'll write with red chalk for this phase. On the "What We Know" side of the board, we'll write the statement of the theory, and on the right side, we'll write the supporting evidence. This includes observations, experimental results and science principles which the theory explains. Do you see the advantage,' Judy concluded, 'of using this board arrangement as the framework for our discussions from now on?'

It seems quite evident that Judy K's board division scheme can help her students develop their understanding of scientific enquiry since it provides a consistent framework for thought and discussion. The use of three colours of chalk in the scheme also helps to make students aware of three identifiable phases of scientific enquiry. These phases appear again in the discussion of intended outcomes in the next section.

## SCIENTIFIC ENQUIRY OUTCOMES

The intended laboratory-based outcomes pertaining to scientific enquiry processes are stated in terms of skills and abilities which students can develop. Five outcomes, each with several component behaviours, are described in this section. Two of the outcomes (A and C) emphasize the 'hands-on' aspects of scientific enquiry and the kinds of skills which develop by engaging in laboratory activities and the manipulation of data. Contrasting with these are the other three outcomes which place more emphasis on the reflective aspects of scientific enquiry. It is not accidental that the outcomes which are primarily manipulative and the reflective outcomes alternate. The two kinds of activity which the outcomes represent frequently alternate in actual scientific enquiries. The five outcomes also traverse the three phases of scientific enquiry which Judy K asked her students to distinguish. Outcomes A, B and C plus parts of outcome D concern the experimental phase of enquiry. The latter part of outcome D deals with the formulation of scientific laws and principles, and outcome E with the theory-building phase.

### Outcome A: The skills to gather scientific information through laboratory work

This outcome can conveniently be described by the following set of student behaviours that are components of scientific information-gathering skills.

A.1 Observing objects and phenomena;
A.2 Describing observations using appropriate language;
A.3 Measuring objects and changes;
A.4 Selecting appropriate measuring instruments;
A.5 Processing experimental and observational data;
A.6 Developing skills in using common laboratory and field equipment;

A.7 Carrying out common laboratory techniques with care and safety.

To indicate the content and scope of these behaviours, some illustrations drawn chiefly from enquiries relating to heat phenomena may serve well. Representative examples of observing objects and phenomena, behaviour A.1, would find a student watching an ice cube placed in a glass of water in a warm room or another student noting changes in water being heated in a beaker on a hot plate. For either of these situations, several dozen discrete things can be observed in a few minutes, and the oral or written communication of these observables constitutes the next behaviour, A.2, describing observations using appropriate language. The emphasis here is on the effectiveness of the communication of the observations, rather than on the form of the language used, which could vary widely depending upon the level of sophistication attained by the student, but could still communicate accurately what was observed. 'The outside of the glass got wet' is an appropriate description by 8th or 9th grade students of an observation on the ice cube in water system as 'Moisture accumulated on the glass's outer surface' is for older students.

When observations go beyond being only descriptive and beyond simple counting and when any instrument is employed to make them, the student displays behaviour A.3, measuring objects and changes. In the ice cube in water system, the initial temperature of water was measured with a thermometer and found to be 22°C. The temperature of the water in the beaker being heated on a hot plate changed from 22°C to 24°C after one minute, to 27°C at the end of the second minute to 30°C at the end of the third minute. To obtain the desired data in any measurement, the student must select the appropriate measuring instrument (A.4), appropriate in the sense that the instrument is capable of measuring the desired quantity and appropriate in that it is operative over the range of the quantity to be measured. A stop watch is *not* the appropriate instrument for measuring the temperature of water in a beaker; a mercury-in-glass thermometer is *not* appropriate for measuring the temperature of the melt in a blast furnace. Data are obtained by students in the form of recorded observations and measurements, and they must usually process these data to yield values for the quantities under study. Behaviour A.5, processing of experimental and observational data, refers to the students' mathematical manipulation and adjustment of their observations and measurements. In a typical calorimetry experiment to determine the amount of energy (in joules) gained by a sample of lead when heated, the measurements recorded are the sample's mass (in grammes), its initial temperature (in degrees C), and its final temperature; processing of these data include subtracting the initial from the final temperature and multiplying the difference by the sample's mass and specific heat capacity to yield the number of joules gained.

The last two student behaviours are practically self-explanatory. They refer back to the technical skills already discussed in Chapter 3.1 in greater detail.

### Outcome B: The ability to ask appropriate scientific questions and to recognize what is involved in answering questions via laboratory experiments

Behind this outcome is the suggestion that students have penetrated beyond the surface skills of gathering and processing data to some degree of insight into the functions of these procedures in empirical enquiry. To decide whether or not students have this insight, some indicators of the insight in their behaviour must be examined. The following set of behaviours can be used as indicators for the insight and for the attainment of the outcome.

B.1 Recognizing a problem;
B.2 Formulating a working hypothesis;
B.3 Selecting suitable tests of an hypothesis;
B.4 Designing appropriate procedures for performing experimental tests.

To clarify the meanings of these behaviours, we can examine them in the context of an illustration, again using an enquiry on heat phenomena.

A beaker of water has been heated to 80°C on a hot plate. Leaving the thermometer in the water, the student removes the beaker from the hot plate and places it on the desk. After five minutes, the thermometer reads 72°C. Since the water has lost some energy without anything being done to it, the student recognizes that there is a problem. The student wishes to investigate heat phenomena in liquids and realizes that it will be difficult to contend with apparently spontaneous losses of energy from the liquid samples to the surrounding air. Such energy losses must be minimized to carry out the investigation, and the problem is how to accomplish this. What materials should be used for the containers that hold the liquid samples? Is energy loss through the walls of a container the same for all materials?

A student's behaviour in recognizing a problem (B.1) may pass through several stages, as the foregoing illustration suggests, from an awareness of the problem area to the identification of a specific problem that can be investigated experimentally. The problem could be at any level of science understanding (descriptive, qualitative, quantitative or symbolic) or it might encompass several of the levels. The last question in the preceding paragraph is a specific problem susceptible of experimental investigation, and it might quickly lead the student to the behaviour of formulating a working hypothesis (B.2) that would give direction to the investigation. The student might hypothesize, for example, that energy is transferred more readily through the walls of containers made of some materials than through the walls of containers made of other materials. An alternative, equally plausible hypothesis might be that the amount of heat energy transferred depends on the thickness of the walls of the container and not on the material of which the container is made. Both of these hypotheses are at the quantitative level of understanding. Hypotheses that represent the qualitative or descriptive level are possible also, if that level is appropriate for the problem being investigated. Whatever the hypothesis may be, the student next takes steps to determine whether it should be accepted or discarded.

Selecting suitable tests of an hypothesis, behaviour B.3, involves choosing a particular empirical approach or a series of experiments that logically can be used to check the hypothesis. This behaviour is concerned with the question of whether or not a proposed experiment constitutes a valid test of the hypothesis, and it is not concerned with an experiment's manipulative details or the construction of apparatus. These concerns are included under behaviour B.4, designing appropriate procedures for performing experimental tests. To obtain a valid test of the hypothesis that the heat energy transferred from a container depends on the thickness of its walls and not on the material of which the container is made, the student would have to employ a two-fold experimental approach. First, it is necessary to measure energy transfers in containers made of the same material but with different wall thicknesses. Second, the student needs to measure energy transfers in containers with exactly the same wall thickness but made of different materials. A suitable test of the alternative hypothesis, that heat energy is transferred more readily through the walls of containers made of some materials than through the walls of containers made of other materials, is more straightforward. The student simply would have to measure energy transfers in containers made of different materials.

Before performing the planned experiments, the student designs and devises appropriate procedures (B.4) for measuring energy transfers in containers made of different materials. One procedure could be: (1) Obtain or make containers of exactly the same size and shape but of different materials, e.g. metals, glass, ceramic, solid plastic, foam plastic, paper; (2) fill each container to the same level with boiling water; (3) stir the water with a thermometer and record the water temperature; (4) continue stirring and record the water temperature every 60 seconds for a period of 30 minutes. In this illustation the equipment and procedures used are quite simple, but this is not so in many experiments that students may carry out. A determination of the velocity of light or of other electromagnetic radiatons calls for complex apparatus and an elaborate protocol.

When students exhibit the behaviours recounted in this illustration, it would seem to be a pretty fair indication that they have some insight into the function of laboratory work in empirical enquiry. Attaining that insight is the central feature of outcome B.

## Outcome C: The ability to organize, communicate and interpret the data and observations obtained by experimentation

The focus of this outcome is the phase of empirical enquiry following information gathering and data processing. In this phase, students organize the processed data and apply their mathematical skills to analyse them. A workable set of component behaviours for this outcome is:

C.1 Organizing data and observations;
C.2 Presenting data in the form functional relationships;
C.3 Extrapolating, when warranted, of functional relationships beyond actual observations, and interpolating between

observed points;

C.4 Interpretation of data and observations.

Included in the first of these behaviours, organizing data and observations, are the various steps and processes involved in producing a display of information in a table or chart. The product of this work should be readily readable and should follow customary conventions of the scientific field in which the enquiry is being done. For example, when numerical data are displayed in a table, the units for all measurements should be included and the proper number of significant figures should be used.

The next two behaviours deal with the student's preparation and use of graphs. In an experiment to measure the volume of a sample of air at different temperatures but under constant pressure, a student found that the volume of the sample was 18.7 cc at a temperature of 100°C (or 373K), 14.6 cc at 20°C (or 293K), 13.7 cc at 0°C (or 273K), and 11.6 cc at -40°C (or 233K). To present these data in the form of a functional relationship, behaviour C.2, the student plots the data points on a sheet of graph paper with absolute temperature (K) on one axis and volume on the other axis. Since the points can be connected by a straight line through the origin, the graph shows the functional relationship between the two variables: volume of air is directly proportional to absolute temperature.

Had the relationship not been linear, the curve of the graph would have shown a different shape. By plotting points for the observed values of variables on suitably ruled graph paper, a student can make a presentation of any functional relationship. Extrapolation, when warranted, of functional relationships beyond actual observations and interpolating between observed points, behaviour C.3, can also be done from a graph. In the illustrative experiment, observations were made at 20°C and at 0°C, but the volume of air at 10°C (or 283K) was not measured. Interpolation on the graph of the relationship reveals that the volume of the sample of air at 283K was 14.2 cc. Similarly, extrapolating above the highest observed temperature and below the lowest observed temperature shows that the volume of the sample of air would be 21.2 cc at 425K and 8.6 cc at 173K. Both the interpolation and the two extrapolations are warranted here because there are no intervening conditions that alter the functional relationship between temperature and volume of air. An extrapolation to 73K would not be warranted, however, because the air would have changed from a gas to a liquid before that temperature was reached and the temperature-volume relationship does not take into account this intervening condition. Finally, interpreting experimental data and observations, behaviour C.4, is the first stage in the student's analysis of the results of an experiment.

## Outcome D: The ability to draw conclusions or make inferences from data, observations and experimentation

In an empirical enquiry, after a valid test of an hypothesis has been carried out and the enquiry has proceeded through the phases of data collection, processing, organization and interpretation, it is time for the student to check whether or not the findings support the hypothesis. In doing so, the student uses the data to make a logical inference about the acceptability of the hypothesis. Essentially, the student must answer the question, 'Is the evidence consistent with the hypothesis?' If the answer is in the affirmative, the student can draw the conclusion that the hypothesis is supported. For a negative answer, the conclusion is that the hypothesis is to be rejected. To illustrate, in an investigation of heat energy transfer from different containers, if the data show that the temperature of water drops more in metal containers than in plastic containers over the same period of time, this evidence is consistent with the hypothesis that heat energy is transferred more readily through the walls of containers made of some materials than through the walls of containers made of other materials. The hypothesis under test has been supported, and the student can draw the conclusion that the hypothesis can be accepted. The same logical process is involved in drawing conclusions about the acceptability of hypotheses whether the data obtained are quantitative, as in the illustration, or qualitative or descriptive. However, the logical process itself involves, not quantitative, but verbal reasoning skills.

A further phase of empirical enquiry goes beyond drawing conclusions about the acceptability of particular hypotheses to the generalizability of findings. Mathematical relationships frequently are central in this generalizing phase. To illustrate, in an enquiry into how the volume of the air changes at different temperatures, a student found that, at constant pressure, the volume of a sample of air increases linearly as its absolute temperature increases. Does this finding represent a general principle applicable to all samples of air? Is this an empirical law covering all gases, not only air? In the course of answering these questions, the student formulates appropriate generalizations, empirical laws or principles that are warranted by the relationships found. In addition, the student considers the results of experiments with other samples of air and carries out or checks the reports of other similar enquiries using different gases. If the original finding is corroborated, the student is justified in formulating an empirical generalization: at constant pressure, the volume of a gas is directly proportional to its absolute temperature. This phase of enquiry involves the student in making comparisons of the results of several enquiries and deriving from all the available evidence an abstract relation covering a range of related phenomena. The outcome of the student's reasoning, the generalization which is formulated, is a synthesis of the various results in a compact form.

To summarize this discussion, the outcome focused on in this section can be described by these component behaviours:

D.1 Evaluating an hypothesis under test in the light of observations and experimental data;
D.2 Formulating appropriate generalizations, empirical laws or principles that are warranted by the relationships found.

### Outcome E: The ability to recognize the role of laboratory experiments and observations in the development of scientific theories

This outcome concerns students' insight into the function of empirical procedures in another keystone of scientific enquiry, the development of theories. Once again, it is useful to have a set of behaviours which can serve as indicators of how much insight students have into the involved process of theory-building in empirical enquiry. The appropriate context for these indicative behaviours is the building, testing and revising of a theory. The following identifiable behaviours are involved:

E.1 Recognizing the need for a theory to relate different phenomena and empirical laws or principles;
E.2 Formulating a theory to accommodate known phenomena

and principles;

E.3 Specifying the phenomena and principles which are satisfied or explained by a theory;

E.4 Deducing new hypotheses from a theory to direct observations and experiments for testing it;

E.5 Interpreting and evaluating the results of the experiments to test a theory;

E.6 Formulating, when warranted by new observations or interpretations, a revised, refined, or extended theory.

Most of these behaviours engage students in rather sophisticated mental processes, as the following discussion of each behaviour illustrates.

Recognizing the need for a theory to relate different phenomena and empirical laws or principles (E.1) refers to the student's acceptance of theory-building as a legitimate part of scientific enquiry. Theory-building wasn't always considered legitimate in science. During the 19th century many chemists refused to give serious consideration to the atomic theory or any other hypothetical model of matter. They asserted that the only proper concern of the science of chemistry was macroscopic properties and changes that can be observed, and, eschewing speculative ideas, they based their science solely on various chemical laws and principles generalized from their laboratory experiences. Chemists today, on the other hand, like all scientists, recognize that empirical laws are not sufficient to organize and correlate all known phenomena, and they engage in the formulation of theories, which serve three major functions in science. First, a theory has a correlative function in that it ties together in a consistent, rational manner the various phenomena and generalizations in the area that it covers. In its explanatory function, a theory is used to account for or explain the observations and generalizations in its area. The heuristic function of a theory is to suggest new hypotheses, problems, and experiments that will give direction to further enquiries. The science student who is aware of these functions will more likely go beyond observations and empirical generalizations to the level of formulating and testing theories.

Formulating a theory to accommodate known phenomena and principles (E.2) identifies the first phase of the theory-building process. This phase, like the formulation of empirical generalizations (behaviour D.2), involves a synthesis of the student's knowledge to develop an abstract relationship, but now we are operating at a higher level of abstraction. The student tries to formulate a broad, general statement about the phenomena in an area of enquiry, and this statement will usually consist of a small set of postulates or assumptions. For example, after some time spent investigating heat phenomena, the student might propose that the various observations and generalizations which were made can be explained by conceiving of heat as a fluid substance. This theory of heat could be expressed in a set of postulates, like the following: (1) Heat is a colourless, odourless, invisible fluid substance. (2) Heat fluid occupies space and has mass, like other substances, but it has a very small mass. (3) Heat fluid flows spontaneously from regions of high concentration to regions of low concentration (from hot objects to cooler objects). (4) Heat fluid is always associated with matter and it increases disorder in the arrangement of particles of matter. (5) Heat fluid readily enters some gases, liquids, and solids, but it does not readily enter other gases, liquids, and solids. (6) When matter changes its state from solid to liquid or from liquid to gas, it absorbs heat fluid, and when matter changes its state from gas to liquid or from liquid to solid, it releases heat fluid. If this theory of heat has merit, the student can use it to account for or explain various heat phenomena. The specification of the phenomena and principles that can be explained in this way is behaviour E.3.

The analyses which the student makes under this behaviour, specifying phenomena and principles that are satisfied or explained by a theory, are quite similar to the analyses made in evaluating hypotheses (behaviour D.1), but here the student is operating across an additional level of abstraction. When evaluating hypotheses, the student analyses the relationship between an hypothesis and observational evidence, but here the analysis involves the relationship between a theory and both generalized evidence, expressed as empirical laws and principles, and discrete observations. Examples of some observations and empirical laws regarding heat that are satisfied by the theory given above are: metals are good conductors of heat but plastics are not - explained by postulate 5; when water at 60°C is added to water at 20°C, the resulting temperature of the water mixture is greater than 20°C - explained by postulate 3; the volume of a given quantity of any solid, liquid, or gas increases when it is heated - explained by postulates 2, 3 and 4; additional heating is required to change water at 100°C to steam at 100°C - explained by postulate 6; at constant pressure, the volume of a gas is directly proportional to its absolute temperature - partly explained by postulates 2 and 4. The greater the number of observations and principles that are encompassed by a theory, the more successful it is in fulfilling its correlative and explanatory functions. Being able to specify many phenomena which are satisfied by the theory the student has formulated will increase the student's confidence in its adequacy.

The heuristic function of a theoretical model, which is to suggest new hypotheses, problems and experiments that will direct further enquiry, is exemplified in behaviour E.4, deducing new hypotheses from a theory to direct observations and experiments for testing it. This phase of theory-building involves two identifiable mental operations. First, beginning with the statement of a theory, the student reasons from and in terms of it to certain deductions (hypotheses) that the theory logically suggests or implies. This mental process is not unlike the logical derivation by deduction of new propositions from a given set of theorems in geometry. Having deduced a new hypothesis, the student then proposes a plan of experiments and/or observations which will test the hypothesis. This mental process is also involved in the previously-mentioned behaviour B.3, selecting suitable tests of an hypothesis. The significant difference between behaviours E.4 and B.3 is that here the proposed plan of enquiry serves not only to test the acceptability of an hypothesis, but also to test the adequacy of the theory from which the student generated the hypothesis. To illustrate, postulate 2 of the heat theory given above states that heat fluid, like other substances, has mass, though its mass is very small. From this and from postulate 3, which asserts that a hot object contains more heat fluid than a cold object, a student could deduce the hypothesis that an object has a greater mass when it is hot than when it is cold. Since, by postulate 2, the mass of heat fluid is very small, the comparison of the mass of the hot with the cold object would have to be made over a large temperature difference, say 100°C or more, to test this hypothesis.

Another hypothesis a student might deduce from the theory concerning heat fluid is suggested by postulate 5. According to this postulate, heat fluid readily enters some substances but does not readily enter some others. The student might deduce from this that a characteristic of different substances, say different metals, is their differing capacity to increase their temperature when the same amount of heat is available. The hypothesis is that each kind of metal, for example, has a 'specific heat' which can be used to identify it. For either of these illustrative hypotheses, as well as for many others that could be deduced from the heat fluid theory, the student would next propose appropriate experiments and observations that would lead to a determination of whether or not the hypothesis could be supported. Such investigations would involve the processes of enquiry described under outcomes A to D. Clearly, new cycles of enquiry have thus been stimulated by a theory as it fulfils its heuristic function.

As in behaviour E.3, the student's mental processes involved in behaviour E.5, interpreting and evaluating the results of experiments to test a theory, include analyses of relationships. Here, the student seeks to analyse the relationships between the empirical evidence obtained and the hypothesis tested *and* between the empirical evidence and the theory from which the hypothesis was deduced. In addition, when these analyses are at hand, the student makes a judgement about the adequacy of the theory itself. Judgements of the theory's adequacy generally are based both on evidence of consistency and precision throughout the theoretical structure and on the degree to which it satisfies scientists' criteria for a 'good' theory. Scientists commonly base their evaluation of a theory on two kinds of criteria, viz., analytical criteria related to how well the theory fulfils its correlative, explanatory, and heuristic functions; and certain essentially aesthetic considerations about the theory's elegance and persuasiveness. In this phase of theory-building, represented by behaviour E.5, the science student has opportunities to join with others in discussions and even arguments about the value of a theory, since it is not unusual for controversies to ensue among scientists when competing theories are being evaluated.

Suppose that a student has the results of a large number of experiments with many different metals which show that the 'specific heat' of every metal tested differs from that of every other metal. These results confirm the student's hypothesis that each kind of metal has a 'specific heat' which can be used to identify it, and this confirmation engenders increased confidence in the heat fluid theory from which the hypothesis was deduced. Another student, however, has the results of many careful experiments repeatedly carried out to test the hypothesis that an object has a greater mass when it is hot than when it is cold. In none of the experiments was an increase detected in the mass of an object when its temperature was raised as much as 500°C. These results indicate that the student's hypothesis is not supported by the evidence, and the failure to confirm it suggests that postulate 2 of the heat fluid theory, which states that heat fluid has mass, is not correct. The student might now reason that the entire theory which conceives of heat as a fluid substance is thrown into question. If heat fluid has no mass, the student would say, it is inconsistent to assume that heat is a substance, since no other substance without mass is known. But the case against the theory is not decisive. As the first student, who has gained confidence in the heat fluid theory, could argue, the mass of heat fluid may be much smaller than originally anticipated and it may actually be so small that the addition of heat fluid mass to the mass of an object in a temperature increase of only 500°C cannot be detected with the instruments used in the experiments. From this point on, a lively discussion evidently can proceed, as each student marshals evidence, reasoned arguments, and judgements in the process of interpreting the results of experiments and evaluating a theory.

Through the accumulation of new observations, through the interpretation and reinterpretation of results of experiments, through discussions and debates, any theory in science becomes modified and sometimes is overthrown. The science student engaging in empirical enquiry will, before long, encounter the phase of theory-building when it becomes necessary to reformulate a theory previously espoused. At this juncture the student is engaging in behaviour E.6. Whatever route the reformulation takes, the student must make sure that the new theory fulfils the correlative, explanatory, and heuristic functions and satisfies the criteria of elegance and persuasiveness expected of every acceptable theory.

We have seen in the discussion and illustrations of the last few pages how observation and experimentation, as well as hefty doses of careful thought and human interactions, contribute to the empirical enquiry process of theory-building. But what working scientists find most satisfying when they build and successfully test a good theory - and science students can share the feeling when they develop scientific theories - is that it lets them cover and relate and explain a whole range of phenomena in a concise yet comprehensive way. Having travelled the repeated hypothesize-observe-interpret cycles that empirical enquiry demands, scientists will proudly display the theory that their work has wrought. They can point to the handful of verbal or mathematical statements that express the theory's powerful ideas, which explain a significant segment of the natural world. They are pleased that the ideas in the newly developed theory promote a deeper understanding of the theory's domain and have more interconnections with related ideas than explanations that were available before. These interconnections of ideas in networks become more and more important as the theory-building process continues. One result is that scientists in the various fields have developed certain theories whose ideas have a very large scope, ideas that tie together and help to explain so much of a science's domain that they can be thought of as fundamental.

## FROM OUTCOMES TO REALIZATION

The close connections between the formulating and testing processes discussed under the last two scientific enquiry outcomes and the resulting laws, principles and theories, which are the cornerstones of scientific knowledge, should remind us to beware of taking too seriously the accustomed distinction between science content knowledge and the process of scientific enquiry. After all, as students become proficient in the behaviours constituting the enquiry process outcomes that the preceding discussion focused on, they acquire knowledge of specific phenomena, of science concepts, of empirical laws and principles, and finally of interconnections among all these by means of theories. As students' skills and abilities in scientific enquiry grow, their science content knowledge grows also. Moreover, a full understanding of science principles and theories includes knowing about the scientific enquiry processes by which they were established. While making distinctions can be useful for expository purposes, science content knowledge and scientific enquiry processes are all of a piece.

As the model (Figure 1.2.1) reminds us, when we think about science instruction, we should beware of making sharp distinctions between learning outcomes to be expected from the science laboratory and outcomes from non-laboratory classroom work. The consideration of scientific enquiry outcomes in this chapter has played back and forth continually between laboratory activities and reflection on these in the classroom. When it comes to the attainment of scientific enquiry outcomes, the students' learning in the science laboratory and in the classroom is all of a piece.

Two kinds of questions probably will occur to science

teachers. The first concerns what teachers can do to help students attain the scientific enquiry outcomes which have been described. The second question is a concern about how likely it is that science students in elementary, secondary and even tertiary education courses will attain these outcomes.

For teachers wishing to help students attain the scientific enquiry outcomes, three suggestions should be considered. First, use the laboratory. Get your students and yourself involved in the continuing interactive cycles of investigative laboratory activities and reflection exemplified throughout this chapter. In a sense, much of the present book is an amplification of this suggestion, and you should find many specific hints in the following pages for using the laboratory to develop students' science process skills and their understanding of scientific inquiry. This general strategy can be recommended, not only for elementary and secondary school science courses, but also for tertiary level courses where a principal concern may be to develop the students' understanding of key aspects of the nature of scientific enquiry. Such understanding really cannot be developed in a deep way unless it is based on first-hand contact with the actual phenomena and the problems of investigating them that the laboratory provides, and unless students have opportunities to reflect on these experiences.

The second suggestion is to seek out some of the many specific instructional techniques for enhancing students' science process skills contained in discussions of science teaching methods and teachers' handbooks. Several of these resources which were mentioned previously are Connelly *et al.* (1977); Kuslan and Stone (1968); Sund and Trowbridge (1977) and Tamir (1983); and others include the works of Klinckman (1970); Simpson and Anderson (1981); and Woodburn and Obourn (1965). A third suggestion is to get involved with a particular course or curriculum which emphasizes enquiry and adapt its materials and approach to the situation in the school or tertiary institution where you teach. Enquiry was emphasized in many of the programmes and courses developed through the science curriculum reform efforts of the 1960s and 1970s. The most comprehensive worldwide listing of these efforts is in the ninth and tenth reports of the International Clearinghouse on Science and Mathematics Curricular Developments (Lockard, 1975, 1977), where there is also an indication of the intended emphasis on scientific enquiry in each project's summary description.

By following the foregoing suggestions or by other means, teachers can devote special attention to scientific enquiry in their instruction. Yet how likely is it that students will attain the desired scientific enquiry outcomes? For the United States, evidence bearing on this question has recently become available in the form of meta-analyses of science education research studies (Lott, 1983; Shymansky, Kyle and Alport, 1983) and in a Project Synthesis report (Welch *et al.* 1981), which collated findings from nationwide surveys and case studies. The findings from both these sources generally concur that many students become competent in the component behaviours described in this chapter under scientific enquiry outcome A, the skills to gather scientific information through laboratory work. Also, some students become competent in the component behaviours of our outcome C, the ability to organize, communicate and interpret the data and observations obtained by experimentation. However, in both the meta-analyses and the Project Synthesis report, only sparse evidence was found of students attaining competence in component behaviours of our outcomes B, D and E or of their gaining understanding of the nature of scientific enquiry. It is true that some researches and surveys incorporated in the meta-analyses and Project Synthesis review have methodological limitations. Nevertheless, we cannot say that very many US students attain competence in such scientific enquiry-related behaviours as formulating, testing and evaluating hypotheses, formulating empirical laws and principles, or building, testing, evaluating and revising science theories.

The Project Synthesis working group felt that the findings from its studies were persuasive enough to force a reconsideration of the common belief of science educators that all students should be expected to attain all the desired scientific enquiry outcomes equally. They proposed instead that the expected enquiry outcomes should be differentiated by taking into account both psychological considerations, such as known differences in students' intellectual ability, predilections and personal goals, and ecological considerations, such as the degree to which enquiry is viewed favourably in a particular community, school or academic institution. Their argument, in summary, is:

> We propose that science educators must attend seriously to the incontrovertible evidence about the uniqueness of the

> individual and the contextual differences in schools. We believe that this evidence is so compelling as to warrant the construction of plans to provide educative experiences which are tailored to the particular characteristics and goals of individuals. These individually tailored plans also must take account of the expectations and possibilities in the student's particular school environment. By this mechanism, societal goals with respect to science learning can become meshed with the individual and personal goals of the student and the community's expectations.
> ... Planning for this kind of matching in the inquiry domain is particularly apt, because there is a large scope here for legitimate variations in desirable inquiry-related outcomes for different students. Our stance is that all students should not be expected to attain competence in all inquiry-related outcomes which science educators ... have advocated in the past. For some students and in some school environments, it may not be appropriate to expect any inquiry-related outcomes at all. (Welch *et al.,* 1981, pp. 45-46)

Can the Project Synthesis group's argument be accepted? Does it pose a fatal threat to this chapter's intended learning outcomes with respect to scientific enquiry? The resolution of this dilemma will continue to occupy science educators, and in this book, is pursued further by Fensham (Chapter 5.4).

## REFERENCES

Bridgman, P.W. (1957). *The Logic of Modern Physics.* New York: Macmillan.

Champagne, A.B. and Klopfer, L.E. (1981). Problem solving as outcome and method in science teaching: insights from 60 years of experience. *School Science and Mathematics, 81*, 3-8.

Connelly, F.M., Wahlstrom, M.W., Feingold, M., and Elbaz, F. (1977). *Enquiry Teaching in Science: A Handbook for Secondary Teachers.* Toronto: Ontario Institute for Studies in Education.

Dewey, J. (1910). Science as subject-matter and as method. *Science, 31*, 127-129.

_____ (1916). Method in science teaching. *General Science Quarterly, 1*, 3-9.

Feyerabend, P. (1976). *Against Method.* London: New Left Books.

Herron, M.D. (1971). The nature of scientific enquiry. *School Review,*

*79*, 171-212.

Huxley, T.H. (1877). Lecture on the study of biology. In Huxley, T.H., *American Addresses*. New York: Appleton, pp. 129-164.

Klinckman, E. (ed.) (1970). *Biology Teachers' Handbook* (2nd ed.). New York: Wiley.

Klopfer, L.E. (1975). A structure for the affective domain in relation to science education. *Science Education, 60,* 299-312.

Kuhn, T.S. (1970). *The Structure of Scientific Revolutions* (2nd ed.). Chicago: University of Chicago Press.

Kuslan, L.I. and Stone, A.H. (1968). *Teaching Children Science: An Inquiry Approach.* Belmont, CA: Wadsworth.

Lakatos, I. and Musgrave, A. (eds.). (1974). *Criticism and the Growth of Knowledge*. London: Cambridge University Press.

Lockard, J.D. (ed.) (1975). *Science and Mathematics Curricular Developments Internationally, 1956-1974.* (Ninth report of the International Clearinghouse on Science and Mathematics Curricular Developments.) College Park, MD: University of Maryland.

_____ (ed.) (1977). *Twenty Years of Science and Mathematics Curriculum Development.* (Tenth report of the International Clearinghouse on Science and Mathematics Curricular Developments.) College Park, MD: University of Maryland.

Lott, G.W. (1983). The effect of inquiry teaching and advance organizers upon student outcomes in science education. *Journal of Research in Science Teaching, 20,* 437-451.

Olson, J. (1982). Dilemmas of inquiry teaching: how teachers cope. In Olson, J. (ed.), *Innovation in the Science Curriculum.* London: Croom Helm, pp. 140-178.

Popper, K. (1972). *The Logic of Scientific Discovery.* London: Hutchinson.

Schwab, J.J. (1962). The teaching of science as enquiry. In Schwab, J.J. and Brandwein, P.F. (eds.), *The Teaching of Science.* Cambridge, MA: Harvard University Press, pp. 1-103.

Shymansky, J.A., Kyle, W.C., and Alport, J.M. (1983). The effects of new science curricula on student performance. *Journal of Research in Science Teaching, 20,* 387-404.

Simpson, R.D. and Anderson, N.A. (1981). *Science, Students and Schools.* New York: Wiley.

Sund, R.B. and Trowbridge, L.W. (1977). *Teaching Science by Inquiry.* Columbus, OH: Merrill.

Tamir, P. (1983). Inquiry and the science teacher. *Science Education, 67,* 657-72.

Welch, W.W., Klopfer, L.E., Aikenhead, G.S., and Robinson, J.T. (1981). The role of inquiry in science education: analysis and recommendations. *Science Education, 65,* 33-50.

Woodburn, J.H. and Obourn, E.S. (1965). *Teaching the Pursuit of Science*. New York: Macmillan.

# 3.3

# Learning Scientific Knowledge in the Student Laboratory

Elaine P. Atkinson

## INTRODUCTION

Scientific knowledge is extensive and continually expanding. A person can know more and more about science but can never know everything about it.

There are a number of different types of scientific knowledge. One of these comprises particular symbols or facts. This type of knowledge includes, for example, N = $6.023 \times 10^{23}$; O represents an oxygen atom; oxygen is a gas at room temperature. Gagné and Briggs (1979) point out that learned facts are of obvious value to a student and can also be used for further learning. Learning facts that are likely to be used regularly enables a student to become a more efficient learner. A science teacher has to decide which facts are of such importance that they should be committed to memory. Devoting nearly all available time to the learning of particular facts or symbols can be detrimental to the learning of other types of science knowledge.

It is important that a teacher utilizes appropriate instructional strategies when the intended learning outcomes are symbols or facts. Gagné and Briggs, drawing on research, stress that new facts should be preceded or accompanied by a meaningful context and presented sequentially. However, the order of presentation of major sub-topics can vary.

Another type of scientific knowledge consists of science concepts. There are numerous science concepts, for example, heat, temperature, light, force, and photosynthesis. Much recent research on students' understanding of science concepts (Nussbaum and Novak, 1976; Nussbaum, 1979; Garnett and Hackling, 1984) has clearly identified that students' concepts develop over time. Novak (in Chapter 2.2) and other researchers (Gilbert, Osborne and Fensham, 1982; Osborne and Wittrock, 1983) stress the importance of exploring a student's prior understanding of a concept before deciding upon an appropriate instructional strategy to further develop the concept. Misconceptions may be discovered. Novak provides advice on how to assist concept learning. Gagné and Briggs consider that symbols or facts relevant to the learning or further development of a concept should be previously learned or presented at the beginning of the instructional stage. They further emphasize the importance of sequencing such that each new concept should be preceded by the learning of subordinate concepts.

Building on symbols or facts and concepts leads to a third type of scientific knowledge, empirical laws and principles. Further development leads to theories and generalizations. Challenging students with problem-solving situations contributes to the acquisition of these types of learning outcomes.

The above is one way, but by no means the only way, of categorizing scientific knowledge. What is important is the recognition that there are different types of scientific knowledge which require different instructional strategies for intended learning outcomes to be achieved.

Science educators typically place great importance on laboratory work, arguing that scientific knowledge cannot be effectively learned from books. They believe a personal involvement in practical work helps students acquire further scientific knowledge as well as developing technical skills. Until recently the importance of laboratory work in enhancing the learning of scientific knowledge has rarely been questioned. However, educators such as Kruglak (1952), Uricheck (1972), Osborne and Freyberg (1985) have queried whether the intended learning outcomes are actually achieved through the science laboratory.

## LABORATORY EXERCISES AND SCIENTIFIC KNOWLEDGE

A purpose of many science laboratory exercises is to promote the learning of particular symbols or facts and concepts. Such experiments, together with those focusing on the learning of technical skills, were recently described by Dunn (1986) as tending to be 'controlled exercises' - activities which are wholly devised by staff and which can be completed by the student in one or two laboratory periods. He noted that controlled exercises typically involve the student in three steps - an introduction to the exercise through some form of pre-laboratory activity, carrying out the exercise by following a well defined procedure (usually written) and writing up the exercise and submitting a report.

Controlled exercises are standard student fare at both secondary and tertiary levels. Investigating levels of scientific enquiry, Herron (1971) classified controlled exercises as level 1 (no enquiry) where the aim, materials and methods are provided by a teacher. An answer, more or less predetermined, is found as the result of students' experiments. Analysing some of the major enquiry oriented curriculum projects at secondary level, Herron found the majority of exercises set at level 1 or lower. Herron's analysis included the first edition of BSCS and when Tamir and Lunetta (1978) analysed the third edition of BSCS (Yellow Version) they found that although the curriculum materials had changed so there were more enquiry-oriented exercises, the majority of exercises were still at level 1. At the level of tertiary education, Hegarty (1978, 1979) found that up to 100% of exercises in commercially available microbiology laboratory manuals were at level 1 or lower. Of two locally designed courses which she analysed, one contained 25% of exercises at level 2 or above, where the aim is set by staff but students take some responsibility for design of materials and methods. The remaining 75% of exercises were at level 1 or below; still a high percentage.

Other chapters in this book suggest that the balance of different types of learning outcomes should be thoroughly examined and in some cases changed. But even accepting that controlled exercises are standard fare in student laboratories, there are questions to be asked. How well do students learn the facts or concepts embodied? Are there ways of improving learning and improving memory?

Dunn (1986) explored these questions in some detail and

suggested that there is plenty of room for improvement. Advantages were found in having learning aids laboratories where material is available on a self-service basis, allowing for remedial learning, pre-laboratory preparation, background material and for enrichment purposes. Advantages were also found for computers in aiding or replacing controlled exercises (see also Chapter 5.3). At undergraduate level in higher education, two organizational schemes, the Personalized System of Instruction (PSI) or the Keller Plan and the Audio-Tutorial method have been readily adopted as vehicles for controlled exercises. On the basis of review of research Dunkin and Barnes (1985) described PSI as the single most powerful innovation in higher education. The keys to successful use of both PSI and Audio-Tutorial for learning science facts and concepts are self-pacing and provision for mastery learning. Although neither is limited to learning scientific knowledge alone, it is the Audio-Tutorial method which has also been widely used (in Biology) for problem solving and other higher level activities.

In addition to questions of designing instruction so students learn scientific knowledge well, teachers have become increasingly interested in promoting storage and retention of knowledge in long term memory. Tulving (1972) hypothesized long term memory as comprising two divisions; semantic memory and episodic memory. Semantic memory is 'the organized knowledge a person possesses about words and other verbal systems, their meaning and referents, about relations among them, and about rules, formulae, and algorithms for the manipulation of these symbols, concepts and relations' (p. 386). Episodes are memories of events that an individual took part in either as a witness or an active participant. Tulving suggests that an event has to have time and place relations with other events or experiences for it to be designated an episode.

Some theorists use the term 'episode' to represent a different concept from that of Tulving. Gagné and White (1978) described an episode similar to that of Tulving, but not requiring time and place referents with other events or experiences. Gagné and White postulated that a student carrying out a practical experiment would be more likely to store in memory an episode relating to the experiment, than if they saw the experiment demonstrated by the teacher. They further postulated that this would enhance the learning and retention of the scientific knowledge associated with the experiment. Atkinson and White (1981) reported on a study which investigated this claim.

## LABORATORY EXPERIMENTS - ACTIVE STUDENT INVOLVEMENT, TEACHER DEMONSTRATION

One possible strategy for enhancing student learning and memory of scientific knowledge is by active involvement rather than watching a teacher demonstration. Prior to the latter part of the nineteenth century, science instruction was typically by a lecture-demonstration method. Students watched teachers demonstrating experiments in the laboratory but did not conduct experiments themselves. However, with an increasing desire for a greater emphasis on science processes and less emphasis on the learning of scientific knowledge, student laboratory work assumed greater importance.

Cunningham (1946) reviewed 37 studies and Bradley (1968) six studies comparing the influence on student learning of students carrying out experiments rather than seeing them demonstrated by a teacher. Findings were conflicting. Students carrying out practical work appeared to develop better self-reliance and manipulative skills, but were less attentive to the task at hand and required more time to cover the subject as fully. Garrett and Roberts (1982) critically reviewed twentieth-century studies that considered the effect on student learning of having science experiments demonstrated by teachers rather than students conducting them in small groups. They reported that research findings up to the early 1960s revealed virtually no difference in student learning between these two approaches, but stressed this was highly related to the types of learning that tended to be assessed.

Many comparative learning studies have been conducted but, as discussed in Chapter 1.1, most reviewers would agree with the sentiments expressed by Shulman and Tamir (1973) that there has been a 'disappointing harvest gleaned from the body of comparative research in science teaching ... comparative experiments contrasting "method x vs method y" do little more than demonstrate the obvious or add to our confusion'.

One of the reasons for conflicting findings was the erroneous assumption that there was only one student laboratory method and one teacher demonstration method.

Researchers need to describe more fully the practical experiment and the context in which it is carried out, before making comparisons between these approaches.

Garrett and Roberts (1982) emphasized the need to define what is meant by the terms student laboratory and teacher demonstration. They pointed out that the student laboratory, in most cases, involves two to five students working in a small group and that students in the same small group are likely to have quite different experiences. Garrett and Roberts see the basic difference between the two approaches as lying in the control frame - the teacher demonstration approach being strongly controlled by the teacher while in the small group laboratory the student has the greater control over the selection, organization and pacing of the knowledge transmitted and received during the teaching learning process.

Another reason for the conflicting findings was the assumption that, having carried out or seen an experiment, a student will refer to that when various learning outcomes are being assessed. Researchers have generally neglected to investigate the extent to which a student actually remembers details of science experiments during such an assessment. Besides considering the relevance of various models of instruction in understanding student performance, researchers need to explore changes in thought processes in memory, and then see how these changes in memory are related to differences in performance. An understanding of how memory works will contribute to the design of more effective instruction.

## THE INFLUENCE OF LABORATORY EXPERIMENTS ON MEMORY AND THE ACQUISITION OF SCIENTIFIC KNOWLEDGE

When planning instruction on the gas laws teachers usually incorporate practical experiments on Boyle's Law and Charles' Law. In a study carried out by the author (Atkinson, 1980) with 78 Year 10 science students in four classes in a co-educational city high school, a set of five programmed booklets was written for use in teaching the gas laws. Two laboratory experiments were included. Half of the students saw the two experiments demonstrated by the class teacher while the others carried out the experiments in groups of two to four students.

Two days (retention 1) and also six weeks (retention 2) after instruction students were individually given the same retention test comprising 18 scientific knowledge questions associated with the instructional programme. After completing each question an interviewer asked the student to verbalize the thoughts and pictures which flashed through the mind while deciding how to answer the question. Responses were tape recorded. Examples are provided for three of the students: Jane (consistently high scorer retentions 1, 2), Anthony (high scorer retention 1 but markedly less at retention 2) and Kim (consistently low scorer retentions 1, 2). Their responses to one of the questions are described.

*Q. What is the name given to the Gas Law which can be represented mathematically as* $p_1v_1 = p_2v_2$*?*

Jane only paused very briefly when given this question which she answered correctly on both occasions.

> I was trying to remember which Booklet it was in.
> (retention 1)

> Just deciding and trying to figure out if it was Boyle's Law or Charles' Law. (six weeks later)

Anthony, at retention 1, read the question and wrote the answer correctly. At retention 2 he readily said 'I don't know'.

> We were doing the test with Mr _____ and he explained how the formula worked when he was doing the first experiment. (retention 1)

> I remember there were three laws, but I couldn't remember which one it went to. (six weeks later)

Kim fiddled with her pen, paused, and indicated 'Can't do it' on both occasions.

> I was thinking about the Booklet that I did in science and how Mr _____ explained it on the board. But I couldn't remember what the name of the Gas Law was.
> (retention 1)

I was thinking about Mr_____. He did it all on the board [pause] name of the Gas Laws [pause] and I was trying to picture how it was set out in the pamphlets [Booklets]. It was too long ago for me. They had it all set out on about two pages. But I'm pretty dumb. (six weeks later)

After answering the 18 questions on the retention test, each student was shown a photograph of the apparatus used in each of the two experiments and asked to tell all that could be remembered about the experiment. In both interviews, Jan, Anthony and Kim recalled the two experiments. However, the episodes relating to the experiments differed. Three sets of the responses are presented.

With Boyle's Law experiment:

Anthony (small group practical)

I remembered that if the amount of the weights applied on the top doubles, the volume halves. It had to do with a law, but I don't know what it's called. (retention 1)

The plunger kept getting stuck inside the tube. I remember putting weights on the top [pause] it increases the pressure inside the cylinder. I can't remember anything else. (six weeks later)

Kim (teacher demonstration)

Yes, Mr_____ did it, twice. The first time I didn't really understand it, but he did it again. I still wasn't sure. I can remember he said something about less pressure put on the more the syringe stays up. (retention 1)

Yes, it was a pressure experiment. Mr _____ explained how to do it and everything. He was saying about certain weights being put on and more pressure forced on it and things like that. (six weeks later)

With Charles' Law experiment, Jane responded (small group practical):

We did that experiment. Some girls tried to copy our answers. One girl said 'Can we watch again?' We had to

> use an ice-cream container instead of a litre beaker. And looked at the temperature changes, pressure changes on the volume of the gas. (retention 1)

> I can remember that experiment because I was in a bit of a stink that day - I got really mad because everyone copied our answers. I can remember spilling water all over the bench. We did it in an ice-cream container because we didn't have beakers. I can remember drawing a graph about it - um, that's about it. (six weeks later)

Although they remembered the experiments, when answering the 18 questions only Anthony overtly reported linking with the experiments, and then only for two questions at retention 1. The examples highlight some of the different episodes reported. Overall many students recalled various steps in the experimental procedure; few students referred to the purpose of the experiments which was to show the relationships between the volume, pressure and temperature of a gas under different conditions.

Yet within the one small practical group episodes can vary.

> Yes, I can remember about the experiment [Boyle's Law]. We did it in the science room. I broke a syringe! I didn't do anything after that - I watched. Weight things were put on top of the syringe and I took the numbers down for whatever it was. That's about all. I can remember stupid things that happened - half the tin at the top came off. (Alison)

> Ah! Yes. We had this bin full of weights that was heavy. The girls were carrying them around. One of the girls broke a syringe. We were first and ahead of everyone else. It is good to be in front. We waited for ages to finish the last part. Someone had a weight for a long time. A girl I liked had it. We had a bit of fun. (Peter)

These protocols simply touch on the complex group interactions which influence student learning.

Across the 78 students, all students at retention 1 and all except two students six weeks later recalled various aspects of the Boyle's Law experiment. However, Charles' Law experiment was not recalled to such an extent. Of the students who saw this experiment demonstrated by the teacher less than three-quarters reported an episode relating to the experiment six weeks later.

Although episodes concerning the experiments were stored in memory, in only one of the 18 questions in the retention test did many students overtly link to the experiments. Three explanations could be suggested why the students did not call on the episodes when answering the retention test questions: they did not need them, they did not see the episodes as relevant, they did not have strong links between the questions and the episodes. These explanations are not mutually exclusive, and there was evidence in support of each one.

Irrespective of whether they saw the experiments demonstrated or carried them out in a small group, students scoring highly on the retention test generally recalled not only the steps involved in doing but also the theoretical underpinnings of the experiments. Few of the other students made reference to links between theory and practice. The influence of the kind of episode stored on the learning of scientific knowledge appears a more important area to investigate further than just whether an episode is stored or not.

In school science the laboratory appears not to be providing the link with theory which had been expected. Teachers must not assume that students are making the link between the theory and practice. Continually stressing the purpose of the practical work and the relationship with theory is one way teachers can enhance the learning of scientific knowledge in the laboratory.

## INSTRUCTIONAL STRATEGIES TO PROMOTE STUDENT LEARNING OF SCIENTIFIC KNOWLEDGE

Over the last decade science educators have been exploring more effective ways of encouraging the learning of scientific knowledge.

Duit (1983) described a study conducted with Years 7 and 10 students from Western Germany. A pre-questionnaire was administered exploring the meaning of words associated with the concept of energy and the making of predictions relating to three simple problems of mechanics. Interviews were held to expand on the questionnaire responses. During the instructional unit a traditional approach to teaching energy through work and incorporating laboratory experiments was used. After the unit was completed, the post-questionnaire and further interviews were held. Duit sees the above instructional process as providing a useful model for concept learning. He cautions, however, that it is very difficult to change everyday knowledge which has developed over many years. He sees that learning scientific knowledge is a long-term endeavour and that we should be pleased with little impacts of our teaching efforts.

A number of researchers (Duit, 1983; Smith and Lott, 1983) now recognize that learning of concepts does not simply involve the acquisition of new concepts, but rather the modification of existing concepts or their replacement. A general teaching strategy to bring about significant accommodation by creating conflict between students' everyday concepts and science concepts was formulated by Nussbaum and Novick (1982) and reported by Smith and Lott (1983):

1. Initial exposure of students' alternative conceptions through their responses to an 'exposing event';
2. Sharpening student awareness of their own and other students' alternative conceptions through discussion and debate;
3. Creating conceptual conflict by having students attempt to explain a discrepant event;
4. Encouraging and guiding cognitive accommodation and the invention of a new conceptual model consistent with the accepted scientific conception.

Laboratory experiments are incorporated within the process.

Erickson (1983) also pointed to the need for some suitable teaching strategies to assess students' concepts either before or during an instructional unit. A general strategy that he adopted was similar to that described above. A variation was adopting a metaphorical interviewing technique in which students were asked to select one or two metaphors from a particular list which best described their thoughts or ideas about a piece of scientific knowledge and indicate why these metaphors were selected. He conducted individual as well as large group interviews, with individuals responding on their own questionnaire.

Recent research described above provides science educators with challenging opportunities to develop innovative and manageable practices within the classroom to foster intended learning outcomes of scientific knowledge.

## REFERENCES

Atkinson, E.P. (1980) Instruction - Memory - Performances: The Influence of Practical Work in Science on Memory and Performance. Unpublished Ph.D. thesis, Monash University.

_____ and White, R.T. (1981) Influence of practical work on test performance, *Research in Science Education*, 11, 87-93.

Bradley, R.L. (1968) Is the science laboratory necessary for general education science courses? *Science Education*, 52, 58-66.

Cunningham, H.A. (1946) Lecture demonstration versus individual laboratory method in science teaching - a summary, *Science Education*, 30 (2), 70-81.

Duit, R. (1983) Energy conceptions held by students and consequences for science teachers. Paper presented at the Misconceptions in Science and Mathematics International Seminar, Cornell University, New York.

Dunkin, M.J. and Barnes, J. (1985) Review on teaching in higher education. In *Handbook of Research on Teaching* (ed. Wittrock, M.C.). Third edition, New York: Macmillan.

Dunn, J. (1986) Strategies for Course Design. Ch. 2 in Boud, D., Dunn., J. and Hegarty-Hazel, E. (eds), *Teaching in Laboratories*, London: SRHE/NFER Nelson.

Erickson, G.L. (1983) Student frameworks and classroom instruction. Paper presented at the Misconceptions in Science and Mathematics International Seminar, Cornell University, New York.

Gagné, R.M. and Briggs, L.J. (1979) *Principles of Instructional Design*. Second edition, New York: Holt, Rinehart and Winston.

_____ and White, R.T. (1978) Memory structures and learning outcomes, *Review of Educational Research*, 48, 187-222.

Garnett, P.G. and Hackling, M.W. (1984) Students' understanding and misconceptions concerning chemical equilibrium. Paper presented at the AARE Conference, Perth, Australia.

Garrett, R.M. and Roberts, T.F. (1982) Demonstration versus small group practical work in science education. A critical review of studies since 1900. *Studies in Science Education*, 9, 109-146.

Gilbert, J.K., Osborne, R.J. and Fensham, P.J. (1982) Children's science and its consequences for teaching, *Science Education*, 66 (4), 623-633.

Hegarty, E.H. (1978) Levels of scientific enquiry in university science laboratory classes: implications for curriculum deliberations, *Research in Science Education*, 8, 45-47.

_____ (1979) The Role of Laboratory Work in Teaching Microbiology at University Level. Unpublished Ph.D. thesis, University of New South Wales.

Herron, M.D. (1971) The nature of scientific enquiry, *School Review*, 79, 171-212.

Kruglak, H. (1952) A comparison of conventional and demonstration methods in the elementary college physics laboratory, *Journal of Experimental Education*, 20, 293-298.

Nussbaum, J. (1979) Children's conceptions of the earth as a cosmic body: A cross age study, *Science Education*, 63 (1), 83-93.

_____ and Novak, J.D. (1976) An assessment of children's concepts of the earth utilizing structured interviews, *Science Education*, 60 (4), 535-550.

_____ and Novick, S. (1982) A study of conceptual change in the classroom. Paper presented at the April 1982 meeting of the National Association for Research in Science Teaching.

Osborne, R.J. and Wittrock, M.C. (1983) Learning science: A generative process, *Science Education*, 67 (4), 489-508.

_____ and Freyberg, P. (1985) *Learning in Science: The implications of children's science*, Auckland, New Zealand: Heinemann.

Shulman, L.S. and Tamir, P. (1973) Research in teaching in the natural sciences. In R. M. W. Travers (ed.), *Second Handbook of Research on Teaching*, Chicago: Rand McNally.

Smith, E.L. and Lott, G.W. (1983) Teaching for conceptual change: Some ways of going wrong. Paper presented at the Misconceptions in Science and Mathematics International Seminar, Cornell University, New York.

Tamir, P. and Lunetta, V.N. (1978) Analysis of laboratory inquiries in the BSCS Yellow Version, *American Biology Teacher*, 40, 353-357.

Tulving, E. (1972) Episodic and semantic memory. In E. Tulving and W. Donaldson (eds.), *Organization of Memory*, New York: Academic Press.

Uricheck, M.J. (1972) Measuring teaching effectiveness in the chemistry laboratory, *Journal of Chemical Education*, 49 (4), 259-262.

# 3.4

# Labwork and Students' Attitudes

**Paul Gardner and Colin Gauld**

## INTRODUCTION

Laboratory work, its supporters claim, not only provides science students with experiences which foster cognitive development and psychomotor skill: it provides opportunities for enhancing their scientific attitudes and their enjoyment of science as well. The central theme of this chapter is the link between labwork and attitudes. What does research tell us about students' preferences for learning through labwork? What claims are made about the effects of labwork on students' attitudes? What evidence is available to support (or refute) such claims?

In this chapter, we will be concerned with two broad categories of attitudes. Consistent with the usage of an earlier paper (Gardner, 1975), we will use the term *attitudes to science* when we are referring to students' favourable or unfavourable reactions to some specified attitude object (e.g. 'learning physics', 'doing chemistry labwork'). This construct includes terms such as *interest*, *enjoyment* and *satisfaction*.

In the second category, *scientific attitudes*, there is no specific attitude object; this term is used to describe personality traits related to habitual styles of thinking (e.g. 'open-mindedness', 'honesty') which scientists are presumed to display. It is of course possible to find scientists who are closed-minded or dishonest, but they are likely to be regarded with disapproval. In other words, scientific attitudes are valued by the scientific community, even if they are not always displayed.

## EDUCATORS' VIEWS ON THE IMPORTANCE OF LABORATORY OBJECTIVES

Although the theme of this chapter is the link between labwork and *students'* attitudes, labwork obviously occurs as a result of the attitudes and decisions of *educators*. It is, after all, teachers and curriculum developers, and not students, who press for the building and equipping of laboratories and for the inclusion of labwork in science curricula. Research on perceptions of such people about the potential value of labwork can therefore provide us with a useful backdrop for the later discussion of students' attitudes.

The question of educators' views can be asked in a number of different ways:

1. Do educators believe labwork to be important?
2. What outcomes are believed to be promoted through labwork?
3. What is the relative importance of objectives for labwork?
4. Which objectives are best achieved through labwork?

### Importance

Reporting on some of his own research, Henry (1972) noted that 52 of a sample of 54 senior high school chemistry teachers in Victoria, Australia, agreed that labwork was vital. Of the two who disagreed, one thought demonstrations could suffice, but still favoured labwork; the other thought that although it was not vital to the current courses, then it ought to be. In a review of literature by psychologists and curriculum developers, Henry documented claims that labwork led to many different desirable outcomes in the cognitive, psychomotor and affective domains. Such outcomes included better understanding (Bruner, 1960, p. 30), familiarity with techniques and apparatus (McDonell, 1959) and manipulative skills (Kerr, 1964, p. 21). Labwork is also claimed to influence a variety of affective outcomes. These include scientific attitudes, e.g critical judgement and 'a sense of confidence in man's power to interpret nature' (Commonwealth Education Liaison Committee, 1964, p. 7) and 'attitudes to science', e.g. 'to arouse and maintain interest in the subject' (Kerr, 1964, p. 21).

As part of his own research, Henry asked secondary school chemistry teachers to justify their belief in the value of labwork. Attitudes to science were mentioned in relation to teaching methods and the assistance given to learning. Some teachers, for example, believed that the greater pupil involvement generated interest, and motivated students to learn more effectively. Labwork also suited teachers who preferred student-centred modes of instruction. Scientific attitudes were mentioned under an item where it was claimed students gained 'a better appreciation of what was involved in observing, hypothesizing and experimenting, of the nature of scientific methods, and how chemical knowledge has been accumulated'. Teachers also mentioned that labwork helped develop 'the habits of being observant, careful, patient and persistent'.

## Relative importance

A study conducted by Kerr (1964) in England and Wales permits some inferences to be made about the relative importance of various labwork objectives in the eyes of practising upper secondary science teachers. Cognitive objectives clearly dominated the list; psychomotor skills were relatively important; scientific attitudes, although not explicitly mentioned, were implied in the objectives relating to investigation, scientific methods of thought and problem-solving. Attitudes to science were considered of lesser importance.

In the Curriculum Diffusion Project[1] conducted in England in the early 1970s, teachers were asked to rate the importance of 19 possible student outcomes of any familiar science curriculum project (normally a Nuffield project). High mean ratings were given to objectives referring to experimental skills and problem-solving procedures. Similar ratings were obtained when teachers were asked to rate the perceived importance placed upon these objectives by the curriculum developers. The objective 'knowledge of the basic facts of science' was given a higher mean rating, but perceived to be regarded as less important by curriculum developers. 'Favourable attitudes to science' received lower mean ratings. As in Kerr's list, cognitive and psychomotor objectives predominate over affective ones.

Boud and his colleagues (1980) presented a list of 22 aims

for laboratory work in tertiary biology, chemistry and physics courses to more than 600 West Australian practising scientists who were then asked to rank them in order of importance. Responses were received from 307 scientists employed in research and development, management, quality control and analysis, and secondary and tertiary level teaching. The three most highly ranked aims were concerned with training students in observation, in basic practical skills and in interpreting experimental data. One aim related to scientific attitudes ('to foster critical awareness') was ranked fourth; one specifically related to labwork ('to teach the principles and attitudes of doing experimental work ...'), seventh; a general attitude aim ('to stimulate and maintain students' interest ...'), eleventh. Clearly, this group of scientists regarded the cognitive and psychomotor outcomes of labwork as more important than the affective outcomes.

## Efficacy

Another way of considering the question of importance is to identify those objectives regarded as best achieved through labwork (efficacy).

Henry (1972) tackled the efficacy question by developing a comprehensive list of 113 objectives which were potentially attainable through laboratory work in chemistry (26 involving attitudes). He had also devised a list of over a hundred brief lesson descriptions, half of which involved labwork, and teachers were asked to rate samples of these lesson descriptions in terms of their suitability for achieving each of the 113 objectives. He found that labwork lessons were most clearly favoured for objectives which related to the cognitive and psychomotor aspects of experimental procedure, and to social interactions in the class.

Olson and Reid (1982) presented eight British science teachers with a list of 20 typical science teaching events, with a request to group them on the basis of perceived common properties and then to verbalize the constructs which underlay their groupings. One important construct was related to how the teachers saw their influence in the classroom. Many teaching events were classified on the basis of whether the amount of teacher influence was high or low. Interestingly, high-influence roles were frequently regarded as more satisfying than low. Teaching roles involving 'uncertain and unclear technique and efficacy' tended to be perceived as low-influence. Situations in which the teacher was in control of the flow of events were seen as more manageable and satisfying for the teacher, although less stimulating for pupils.

Their report was not specifically concerned with labwork, but it is possible to make some inferences from it. Cook-book exercises - those in which pupils follow together a specified lab procedure - are high-influence situations. Open-ended, enquiry-based lessons would provide some scope for teacher influence, but are much more risky: unexpected, unintended (and hence uncontrollable) events are more likely to occur. Many teachers might prefer to avoid experiencing too many such events. It is therefore possible for a teacher to believe that open-ended labwork is valuable without translating such a belief into practice.

## STUDENT REACTIONS TO LABWORK

There is a common thread in the findings for students: many of them enjoy labwork, and prefer it to other modes of learning. This is not, of course, the universal reaction of all students at all times.

### Upper secondary level

Several studies point to the popularity of labwork in the upper high school years. In South Australia, Keightley and Best (1975) reported that practical skills were regarded as the most important of twelve objectives by Grade 11 biology students. As part of an evaluation of PSSC Physics in Victoria, Australia, Gardner (1972) measured attitudes at the beginning and end of Grade 11. One scale measured attitudes to non-authoritarian modes of learning. 'Favourable' items referred to labwork, open-ended problems and personal involvement, while unfavourable items reflected a preference for teacher-centred or textbook-based instruction and rote learning. Most students showed a moderately (but certainly not strongly) positive attitude on this scale; mean scores remained fairly stable during the year. Correlations of this scale with two student personality variables supported the commonsense view that labwork is more likely to appeal to students who like to think and who are willing to work hard in order to succeed.

In Israel, Hofstein, Ben Zvi and Samuel (1976) gathered data on over 500 chemistry students in Grades 10, 11 and 12, and reported highly favourable attitudes towards labwork, although there was a significant decline in Grade 12. Comparative studies (Ben Zvi, Hofstein and Samuel, 1976; Ben Zvi *et al.* 1976) indicated that labwork was regarded as more interesting than teacher demonstrations, lectures or filmed experiments. This finding held for both traditional and new chemistry curricula.

In an evaluation of Project Physics in NSW, Quinlan (1981) employed a scale called 'Emphasis on Experiments'. Most classes rated the emphasis upon experimental work in physics as moderate to low. However, stronger emphasis tended to be associated with higher class mean interest in physics.

## Junior secondary level

Lower down the school, the same positive reactions to labwork can be found. In South Australia, Dawson and Bennett (1981) asked junior high school students to rate their liking for 17 teaching methods. Methods employing student involvement (doing experiments, making models) or visual stimuli (visiting the zoo, watching film/TV) were popular. Listening to teachers talk and copying notes were - not surprisingly - unpopular.

In her review of the Czechoslovakian literature on student interests in physics, Zieleniecova (1984) cited several studies which indicate that experimental work carried out by students themselves, as well as demonstrations performed by the teacher, are the methods which are most frequently mentioned by students as favourably influencing their interest.

Although most junior secondary science students do appear to like laboratory work, Head (1982) reports that, while he was collecting examples of pupils' writing in science, 'a significant minority of pupils expressed a dislike for practical work, suggesting that the situation may not be quite so simple as it seems at first sight'.

## Comparisons of students' and teachers' views

Keightley and Best (1975) asked Grade 11 biology students in South Australia (N=2616) to rate their liking for twelve educational objectives. One objective was 'Practical skills (for example, learning to use a microscope; to dissect animals; to use laboratory apparatus)'. This was accorded the highest mean rating of all twelve objectives; 47% of students responded that they 'very much like to learn' this, and 32% that they 'like to learn'.

Interestingly, teachers (N=82) had a somewhat different view of the importance of laboratory work. Asked to rate the importance of the twelve objectives, the most valued objective was 'Understanding of the biologist's views of the world', regarded as a major aim by 94% of the teachers. Another item, 'Scientific ways of thinking (for example: learning to observe; to analyse experiments to see what conclusions are justified; to think logically)', was almost equally valued (87%). The objective 'Interest and Attitudes' was considered a major aim by 80% of the teachers, whereas the practical skills item was regarded as of major importance by only 24%.

The evidence in this study would suggest that both students and teachers value laboratory work, but that perhaps they do so for different reasons. Students apparently appreciate the 'hands-on' experience itself. Teachers seem to place less value on this; rather the laboratory work would seem to be regarded as psychomotor means towards more purely cognitive ends. There is some dissonance in values here: the 'Scientific ways of thinking' so highly prized by teachers was rated 'very much like to learn' by only 21% of the students. Another 47% said they 'like to learn' this, while 24% responded 'Don't care'. The different perceptions of students and teachers are clearly brought out by an analysis of whether an objective was included among a respondent's top three priorities. 'Scientific ways of thinking' was included by 55% of teachers, but only 14% of students; 'Practical skills' was given high priority by 38% of students, but only 2% of teachers. This specific comparison illustrates Keightley and Best's more general conclusion:

> The discrepancy between what teachers hope to teach and what the students hope to learn suggests that the students are more instrumental in their approach to biology than the teachers. The teachers, on the other hand, are more conceptual in their approach (p. 65).

As we have seen, students generally enjoy labwork. This generalization is, however, very sweeping: not all enjoy it equally, and even a student who usually enjoys it may find some aspects of it unsatisfying. The question becomes: what factors influence attitudes to labwork? What conditions promote or inhibit interest and satisfaction?

## PHYSICAL CONDITIONS

### Laboratory facilities

Science facilities in most Australian schools are generally good. Nevertheless, variations in the quality of facilities can have effects. Ainley (1978) developed an 'encouragement to explore' scale, and found a positive association of students' attitudes with school laboratory facilities. Also, attitudes to science were more favourable in schools where the science rooms were clustered together rather than dispersed (physical proximity may lead to better planning by teachers, greater use of resources, and use of a wider range of resources) (see also Chapter 5.1).

Such studies indicate that the availability and quality of lab facilities, together with the quality of the educational events that take place within the lab, do affect student interest in science.

## INSTRUCTIONAL CONDITIONS

Labwork is of course a broad term for an enormously diverse range of experiences. Labwork can vary in the time allowed for it (ample vs insufficient), in the goals which are emphasized (verification vs discovery), in the intellectual challenge offered (difficult vs easy), in the way it is fitted into a course (integrated vs separated) and in the professional competence of the teacher (organized vs disorganized). Let us examine these in turn.

### Time allowed

One factor that affects attitudes to labwork is the time allowed to complete it adequately. Gardner (1981) evaluated the attitudes towards their course of students taking Refrigeration and Air Conditioning in two technical colleges in Melbourne. The students were distributed among nine classes. In one class, students' attitudes, when compared to those of students in the other classes, were discrepantly unfavourable on a set of 19 items referring to issues such as clarity of goals, satisfaction with progress and utility of the course. The item displaying the greatest discrepancy was 'I'm not allowed enough time at school to become expert at a practical skill'. This would indicate that excessive time pressure might generate more widespread dissatisfaction.

A study of high school biology classrooms in the Australian Capital Territory was consistent with this view. As one student put it:

> You usually go halfway through a prac., and that's it ... the timing is all wrong in biology. If you are doing an experiment in biology and the bell goes you just leave it and that's it. Usually you do not get around to coming back to it, so it is left unfinished ... there have been a real lot of unfinished experiments that I would have liked to have done. (Fordham, 1980, p. 113)

In most of the classes observed, practical exercises were seldom completed as a result of restricted time.

## Variety

One obvious reason why labwork tends to be interesting is that it offers a change from other modes of learning, in both science and other subjects. In traditional classrooms, teachers talk a lot and students are relatively passive; the lab, in contrast, is a place where teachers talk less and students are active. There is evidence to support the claim that variety stimulates interest. In Norway, Lie and Sjöberg (1984) asked secondary students, 'What is important to you in order to like a school subject?' The four most frequent responses (endorsed by over 83% of the sample) involved cognitive and interpersonal characteristics of the teacher but the fifth (endorsed by 82%) was 'that there is variation in methods' (Sjöberg, 1984). The South Australian research studies cited earlier (Dawson and Bennett, 1981; Keightley and Best, 1975) are consistent with the view that teaching methods which offer an alternative to conventional teacher-talk and note-copying are preferred by students. The opinion of many students was reflected in the response to an interviewer's question asked in the early stages of the Learning in Science Project (Osborne, Freyberg and Tasker, 1979):

Pupil A: ... the part I liked was doing experiments ... dissecting animals and that, because I didn't know much about it ... and I wanted to find out more about it ... I liked the practical side of it rather than writing.

For other pupils it was the lack of constraint rather than the provision of an alternative context for learning which made practical work enjoyable.

Pupil B: ... we had fun, we just mucked around and made a mess.

Pupil C: ... we enjoyed the experiments but didn't learn much.

## Cognitive challenge

Labwork can vary along a continuum from pure 'cook-book' where students follow a given procedure (perhaps without even understanding why) to genuine open-ended research where no one knows the answers to the questions which guide the experimental design. This dimension, which might be labelled *cognitive challenge,* probably affects interest. Labwork which is routine, which fails to generate a sense of surprise, whose outcomes are known in advance, is hardly likely to enhance student interest. (At the other extreme, labwork which is so difficult that it cannot be understood or carried out is hardly likely to generate interest either.)

The concept of cognitive challenge can help us to understand some of the findings of studies reviewed in earlier sections. For example, Fordham (1980) obtained measures of high school biology students' intrinsic motivation (curiosity) and their prior knowledge, and evaluated their learning environments by observing and rating and by recording teacher-student interactions and students' comments. The biology course was an Australian adaptation of BSCS, i.e. it was supposedly laboratory-based and inquiry-oriented.

Some of Fordham's observations were made in laboratory settings. He found, for example, that in nearly all classes, students hardly ever used any reference material other than the textbook and laboratory manual. All the students within a class almost invariably performed the same laboratory exercises. It was during these exercises that students expressed their strongest dissatisfaction with the curriculum: the laboratory manual, they complained, 'failed to come up with any surprises'. In Fordham's view, the practical exercises were deficient in that they failed to arouse intrinsic motivation or generate cognitive conflict. Questions in the laboratory manual were tedious. One student said:

> You usually don't feel like doing them. ... If the teacher says we have to hand in a set of questions [exercises], [friend] and I whip it up in one night and hand it in the next day ... typically for experiments we answer the questions after about three weeks (of doing them).

One of the most important justifications for labwork is that it provides opportunities for developing scientific modes of thinking. One of the students in Fordham's study reminds us that this is certainly possible:

> ... [the teacher] wants us to develop our own ideas, to become good scientists, she likes us to construct hypotheses and see how they can be tested ... keen on scientific method ... with some experiments she wants to know how it could be changed, how the results would be different ... and often we try it ...

This teacher encouraged her students at all times to seek out their own solutions to problems raised in the laboratory exercises. However, such comments were exceedingly rare: in only two (out of 17) classes was there any variation from the path of standard, prescribed activities.

Getting the right answer quickly, rather than genuine investigation and conflict-resolution seemed to be more common. Fordham's findings would seem to suggest that interest would be stimulated by increased cognitive challenge. This view is supported by the earlier writings of Andersen and Koutnik (1972), who argue that labwork ought to be directed primarily towards the *essence of science*, i.e. enquiry and discovery:

> Teachers who assign 'cook-book' laboratory experiments assume that students enjoy following directions which prove that the teacher is right. Anyone studying adolescents soon discovers that they enjoy designing and carrying out experiments but that they avoid and often

abhor following directions and proving the teacher right. (p. 66)

Although insufficient cognitive challenge may result in a failure to generate interest, too much challenge may be equally undesirable. Palmer (1981) argued that undergraduate chemistry students engaged in quantitative analysis classes become so involved in the physical aspects of experiments that the logic and purpose of the procedures are lost. To alleviate this problem, Palmer developed a computer simulation exercise which students had to master before attempting the 'real' labwork. The simulation led to greater student interaction (with methods being discussed, rather than merely answers being exchanged), a positive evaluation by students of the contribution made to their understanding, and an increased interest in chemical analysis.

## Integration of theory with labwork

One factor which influences students' reactions to labwork is the extent to which it forms a coherent part of their science course: is it integrated or separate?

Zieleniecova (1984) cited Czechoslovakian research (Bednarik and Siroka, 1974) on students taking physics in the first year of secondary school; 60% of a sample (N=683) responded that they best understood new subject matter if experiments were done during the exposition; 16% preferred detailed exposition by the teacher without any experiments. The remainder either preferred labwork as prior independent study, or expressed the view that they would not understand physics under any circumstances. Some evidence for the benefits of integration also comes from the work of Boud *et al.* (1980) and Quinlan (1981).

Such studies seem to point to the conclusion that labwork should be integrated into a course, and not treated as a separate part of it. This rather vague generalization might, however, be more useful if we had a clearer notion of what integration might mean in practice. One teacher might teach a topic in class and ask students to perform a relevant laboratory exercise a week later; a second might reduce the time delay to a matter of minutes; a third might use the labwork to introduce the concepts to be discussed later in class. All might claim that they are integrating labwork with 'theory', yet they are doing very different things. The research does not tell us much about the most fruitful methods of achieving integration.

The phrase 'integration of theory with practical work' might imply that each of the parts is itself well-integrated, and all that is needed is to link the two together. But as Novak and Atkinson have shown in their earlier chapters (2.2 and 3.3) and Gunstone and Champagne demonstrate in their later one (4.1) students often come to a new experience with a collection of concepts, beliefs and explanations which may be quite different from the currently accepted scientific model; this collection is not necessarily coherent and well-integrated. New knowledge may be grafted on to this prior collection without changing it significantly. After performing practical work, students may recall unusual events and social interactions without learning the central concepts which the practical work was meant to develop. Students may adopt rote learning as a technique for assimilating their version of the knowledge acquired. It is therefore not surprising that many students fail to achieve mastery of a coherent, meaningful and integrated set of ideas. The issue of integration is much deeper than merely deciding *when* to offer labwork in the curriculum.

### Teacher's organizational ability

There is evidence that teachers who are cognitively well-organized, with good understanding of their subject matter and a capacity for effective planning, appear to enhance student interest in science. Gardner (1972) found a positive relationship between students' enjoyment in physics and the organizational ability of their teachers (an item specifically concerned with the lab was 'Practical work in this class is a shambles: far too often, apparatus is mixed up, incomplete, broken or can't be found.'). Agreement with this item was associated with low levels of teacher cognitive organization, i.e. if labwork was characterized in this way, the teacher tended also to be described as vague and disorganized in other aspects of teaching.

## SOCIAL CONDITIONS

Labwork occurs in a social setting and some of the dimensions of that setting may affect students' reactions to it. The major themes which emerge from the literature are that labwork offers opportunities for greater student autonomy, individual progression and social interaction.

## Autonomy

Although it is possible to become interested in an activity which someone else has selected, it is self-evident that the fullest expression of a person's interests is likely to occur under conditions of personal autonomy. The potential significance of this variable is seen in a study (Quinlan, 1981) which employed a 'student autonomy' scale as part of an evaluation of Project Physics in New South Wales. The scale assessed the amount of freedom that students perceived themselves to have in class. Most classes received a low rating, but the more autonomous classes were associated with higher class means on a measure of interest in physics. If autonomy is valued, and if labwork allows greater scope for it, we would have a partial explanation for its relative popularity.

Head (1982) pointed out that between the ages of about twelve and fifteen, many students change from being essentially conformist to demonstrating a greater awareness of their own individuality and a willingness to take personal initiatives. He also notes that, in the secondary school, labwork usually becomes more highly structured than it was at the primary level and allows students *less* opportunity to exercise their developing initiative. It is such a lack of freedom that Fordham (1980) and Bliss and Ogborn (1977) identify as a source of dissatisfaction at senior secondary and tertiary levels respectively. 'Good' laboratory stories reported by Bliss and Ogborn (1977) are often related to feelings of independence and freedom (p. 129). Students value being able to decide and act for themselves. According to Bliss and Ogborn,

> the importance of feelings of independence in 'good' laboratory stories, and in project stories, has, we think, to do with students finding it more meaningful - closer to real experimenting - to decide and act for themselves. (p.132)

## Individualization

One of the claims made for the laboratory is that it is a place where teachers can treat students as individuals. In the

Teacher's Resource Book for the PSSC course, for example, Haber-Schaim *et al.* (1976) argue that

> The laboratory provides an ideal setting where you can adjust your role as instructor to the individual needs of your students. Some require individual guidance, some are best left undisturbed, still others may enjoy an extra challenge. (p. 11)

One of the findings of the Keightley and Best (1975) study of the relative importance of twelve objectives in Biology was that teachers displayed much greater consensus over what they regarded as important than did students. For the six items endorsed by teachers as major aims of biology education, the percentages so endorsing the item ranged from 94% down to 48%. For the six items endorsed by students as very much liked, the percentages ranged from 47% down to 29%; moreover, there was a much more even rate of endorsement across the twelve objectives. These differences may be due partly to differences in enthusiasm (students are less willing than teachers to offer an extremely favourable response), partly to confusion (students' values may not be as clearly articulated as teachers'), but perhaps also partly to genuine individual differences. Students may come to study biology with different motivations in mind.

## Social interaction

When asked to conjure up a mental image of the laboratory, one might readily think of it in architectural terms. A laboratory is a room, usually large, fitted out with a variety of furniture, services, equipment and consumables. Alternatively one may think of it in instructional terms: it is a place for doing certain kinds of educational activities (learning how to connect circuits, analyse substances, prepare microscope slides ...).

Laboratories can, however, also be thought of in social terms. They are places which are associated with certain distinctive kinds of social interaction. Teachers may converse with students in ways which differ from their linguistic style in non-lab lessons. Students may talk to each other more freely, and may have somewhat greater control over the rate at which they work. All of these aspects of the social environment -

partly defined by the architectural environment and the nature of the work being done in it - may influence students' attitudes towards what they do there.

Henry (1972) in his study of chemistry teachers found support for a set of objectives reflecting such a dimension. Some teachers valued labwork because it enhanced the possibility of students associating enjoyably with other people and sharing their talents; it provided a social climate in which democratic group procedures could be fostered.

## LABWORK AND THE SCIENTIFIC ATTITUDE

### The traditional view of the scientific attitude

Prominent among the aims of science education has been the development of the scientific attitude or, as it has also been labelled, 'scientific-mindedness' (Gauld and Hukins, 1980). It has long been believed that if one wishes to encourage students to act in a scientific manner in the classroom, in the laboratory or in everyday life, it is not sufficient simply to teach knowledge about the various procedures that scientists use in their work or to develop the ability to use the variety of techniques and problem-solving skills which are thought to be central to scientific activity. Knowledge and skill must be accompanied by a willingness to put these into practice at the appropriate time and in the appropriate manner. It is this willingness that constitutes the scientific attitude.

The scientific attitude, as it is discussed in the science education literature, is not a uni-dimensional construct, but possesses a number of components. Central to the concept is the notion of 'critical-mindedness' or 'critical awareness', a group of attitudes related to the evaluation of ideas, information and evidence when engaged in solving problems or making decisions. These include objectivity, impartiality, openness, scepticism, intellectual honesty and a willingness to exercise caution when handling evidence. Alongside the attitude of critical-mindedness are general attitudes such as curiosity, humility and anti-authoritarianism and a commitment to general beliefs such as belief in the understandability of nature and in the existence of natural cause-and-effect relationships.

The special role which experimental work plays in science suggests that the laboratory can provide an environment for encouraging the development not only of scientific skills but also of an appropriate respect for empirical evidence and a willingness to respond by taking evidence seriously in making decisions about hypotheses.

## The effect of labwork on scientific attitudes

Hofstein and Lunetta (1982) indicated that little has been done to investigate the extent to which scientific attitudes are developed through labwork. While there are more studies than the single one they cite, much less work has been done in this area than in many others related to science education. In a brief review of the little research data available, Welch and his colleagues (1981) evaluated the state of primary and secondary science education in the USA, and concluded that, as far as scientific attitudes were concerned (or, in their terms, 'safeguards and customs of inquiry'), these were 'not generally part of the classroom scene' and were 'valued by only a small (percentage) of students, though most recognise their necessity for scientists'.

The effects of 'content-centred' and 'process-centred' laboratories on the scientific attitudes of college freshmen enrolled in a general biology course were compared in a study by Murphy (1968). After the one-semester course he found no difference between the two methods. (There was also no difference in the effects of the two methods on knowledge of or interest in biology.) The role of the 'laboratory-investigative' approach in changing an undifferentiated group of attitudes - curiosity, openness, responsibility, and satisfaction - was one part of a study by Raghubir (1979). Using this method, twelfth-grade biology students tackled a number of problems with minimal assistance from the teacher. For this group of students, the attitude gains were significantly greater than the gains of a group taught by a more traditional lecture-laboratory approach.

A physical science course at Jefferson College in Missouri was designed so that in lectures and laboratories the relationship between carefully chosen topics and everyday life was continually stressed (Daughton, 1973). A survey of two successive groups of students showed that there was a significant increase in the percentage of students who claimed that they would question advertising claims that a product had been scientifically tested and that they used 'a logical and systematic approach to the solution of daily problems'. However, we must note that self-reported claims that the students adopted more scientific attitudes in everyday life neither tell us that they did so in practice, nor inform us as to the ways in which these improved attitudes were displayed.

Kaplan (1963) investigated the relative effects of a problem-solving laboratory method which stressed scientific thinking and placed minimal reliance on the teacher, and a method concerned almost exclusively with the transmission of content. The outcome of interest was the extent to which the student depended on 'objective perception as a source of information' in everyday life and the author argued that, although the results were not conclusive, the problem-solving method appeared to have some advantage in developing this attitude. Coulter's (1966) findings on the effectiveness of inductive (problem-solving) laboratory, inductive demonstration, and deductive laboratory approaches were similar.

Although honesty in recording and interpreting data is supposedly a highly-prized scientific attitude, Fordham (1980) in his study of 17 senior biology classrooms notes that emphasis by the teacher on the 'correctness' of data and conclusions can inhibit rather than foster this trait. If their experiment was not 'successful' most students would copy information from others. In the words of one student:

> If the experiment doesn't work we go to somebody else and get their results ... you have to hand it up and it looks better when you get results that you are supposed to. When you read the aim of the experiment you get a good idea of the type of result you are expected to get. And if you don't get that result and you put it down, it's pretty obvious you won't get as good a mark as someone who got it to work. (Fordham, 1980, p. 114)

In the Fordham (1980) study, labwork failed to arouse the curiosity of students in almost all of the 17 classrooms. There was little attempt to use the motivational aspects of curiosity or to develop student curiosity about scientific problems. Instead, students were aware of the outcomes of activities before they commenced and found the questions in the laboratory manual 'tedious'. Teachers generally discouraged original thought by the students and tended to mark questions associated with labwork by comparing responses with the 'correct results'. It is unlikely that scientific attitudes would be fostered under such conditions.

This study, more than any other reviewed here, leads us to the core of the matter. *Merely being in the laboratory and doing labwork there do not, by themselves, foster scientific attitudes: it is the quality of the experiences that students have there that is crucial.*

## A role for the laboratory in developing a scientific attitude

The implications of the constructivists' research will have to be considered seriously. Teachers will need to be encouraged to develop ways of identifying the prior beliefs that their students hold. Only then can relevant experiences (e.g. labwork, demonstrations, class discussion) be designed that might have some impact upon those beliefs. Relevant experiences are those which directly engage and challenge students' prior conceptual frameworks, not in an attempt to demonstrate them to be silly or wrong, but for the purposes of showing how, in the scientific community, ideas (one's own and other people's) and new experiences interact.

Conflicting ideas, particularly when the conflict is a personal one for the student, can motivate some students to generate alternative views, discuss the conflict with peers, listen to the views of a credible teacher, or use other sources of information. If conflict can be resolved successfully one might expect that similar procedures are more likely to be used next time with further expectation of success.

The teacher in such a laboratory should be aware of the sorts of beliefs which students bring to a particular topic and should have given consideration to experiments and activities likely to challenge those beliefs. There should also be resources available which are likely to be of assistance in following up ideas on the way to resolution of the conflict. (For examples of such teaching strategies see Erickson, 1979; Nussbaum and Novick, 1982; Osborne and Wittrock, 1983; Rowell and Dawson, 1983; Osborne and Freyberg, 1985).

## CONCLUSION

We have seen that student attitudes to labwork are influenced by a wide variety of factors. The importance of the laboratory can be appreciated when one reads the testimonies of science educators in countries where lab facilities are often poor. Even in more highly developed countries, there is some evidence of a link between the quality of resources and students' attitudes.

Instructional variables influence students' reactions. Insufficient time to complete labwork properly can have negative effects. Labwork which is moderately challenging and open-ended rather than routine and predictable may help arouse and maintain student interest. On the other hand, excessive challenge may be counterproductive. Recent research suggests that prior computer simulation may help students towards increased understanding of the purpose and technique of a particular laboratory procedure, thus leading to a more effective subsequent learning experience in the lab.

Teachers make a difference, too, both through their own behaviour and through the way they influence the social climate of the class. Better student attitudes are likely to be found when the teacher is well-organized, when students have a degree of personal autonomy and scope for learning at individual rates, and when there is a climate of warmth and cooperation. The hopes that labwork might foster students' curiosity, openness and a willingness to solve everyday problems scientifically remain hopes: the available evidence does show weak relationships between labwork and these desired outcomes. While students generally enjoy hands-on experience and the opportunity to work individually or in small groups, we cannot conclude that such experiences will, by themselves, bring about major changes in styles of thinking.

The research literature does provide us with a few guidelines. Much more needs to be done to familiarize teachers with the constructivist perspective. More needs to be done to develop lab experiences which generate problems which are seen as significant by the student, rather than as tedious exercises leading to a 'right' answer. It is also important for the student to be given a substantial amount of control over the procedure for arriving at a solution. In addition, the outcome should provide satisfaction for the student and other opportunities should be provided for the student to act successfully in a similar fashion. One might expect that such a laboratory would not only help develop scientific attitudes but also stimulate enjoyment.

## NOTE

1. The project was directed by Professor P.J. Kelly, then at Chelsea College, University of London. The data are quoted in Gardner (1975).

## REFERENCES

Afolayan, M. A. (1978) An investigation of factors associated with science interests of secondary school students in Kadura State of Nigeria. Unpublished M.Ed. dissertation, Zaria, Ahmadu Bello University.

Ainley, J. G. (1978) *The Australian Science Facilities Program: a study of its influence on science education in Australian schools.* Hawthorn, Victoria, ACER.

Andersen, H. D. and Koutnik, P. G. (1972) *Toward more effective science instruction in secondary education.* New York, Macmillan.

Becker, H. J. (1983) Schools' uses of microcomputers: Report #1 from a National survey. *The Journal of Computers in Mathematics and Science Teaching*, 3(1), 29-33.

Bednarik, M. and Siroka, M. (1974) Pruzkum postoju a mineni zaku prvniho rocniku gymnasia k fyzice (A study of attitudes and opinions of pupils in the first year of general education secondary school concerning physics). *Acta Universitatis Palackianae Olomucensis, Facultae Rerum Naturalium*, 45, 193-215.

Ben Zvi, R., Hofstein, A. and Samuel, D. (1976) Students' evaluation of laboratory work in two chemistry curricula. *Studies in Science Education*, 2, 185-91.

_____ Hofstein, A., Samuel, D. and Kempa, R. F. (1976) The attitudes of high school students towards the use of filmed experiments. *Journal of Chemical Education*, 53, 575-77.

Bliss, J. and Ogborn, J. (1977) *Students' reactions to undergraduate science*, London, Heinemann.

Blum, A. (n.d.) Fact, theory and curricular evaluation in the affective domain. Unpublished Ph.D. dissertation, Hebrew University of Jerusalem.

Boud, D. J., Dunn, J., Kennedy, T. and Thorley, R. (1980) The aims of science laboratory courses: a survey of students, graduates and practising scientists. *European Journal of Science Education*, 2, 415-28.

Bougerra, M. L. (1984) *Tunisia.* National Report to the 12th IPN Symposium, University of Kiel, April.

Bruner, J.S. (1960) *The process of education.* New York, Vintage Books.

Commonwealth Education Liaison Committee (1964) *School science teaching.* Report of an Expert Conference held at the University of Ceylon (December 1963), London, HMSO.

Coulter, J. C. (1966) The effectiveness of inductive laboratory, inductive demonstration, and deductive laboratory in biology. *Journal of Research in Science Teaching*, 4, 185-6.

Daughton, W. J. (1973) An attitude approach to physical science. *Journal of College Science Teaching*, 3(1), 63-4.

Dawson, C. J. and Bennett, N. (1981) What do they say they want? Year 7 students' preferences in science. *Research in Science Education*, 11, 193-201.

Drottz, B.-M., Sjöberg, L. and Dahlstrand, V. (1985) Interests and achievement in engineering education. In Lehrke, M., Hoffman, L. and Gardner, P. (Eds.), *Interests in science and technology education*. Kiel, IPN/UNESCO.

Egbugara, U. O. (1980) An analysis of determinants of students' enrolments in School Certificate Physics in Imo State of Nigeria. Unpublished M.Ed. dissertation, Ibadan, University of Ibadan.

Erickson, G. L. (1979) Children's conceptions of heat and temperature. *Science Education*, 63, 221-30.

Fordham, A. (1980) Student intrinsic motivation, science teaching practices and student learning. *Research in Science Education*, 10, 108-17.

Gardner, P. L. (1972) Attitudes to physics. Unpublished Ph.D. thesis, Monash University.

_____ (1975) Attitudes to science: a review. *Studies in Science Education*, 2, 1-41.

_____ (1981) Refrigeration and air conditioning survey. Appendix A in *Evaluation of Basic Level Refrigeration and Air Conditioning Mechanics Course*. Melbourne, TAFE Services.

Garraway, H. (1985) Using microcomputer-based simulations to promote student interest and provide concrete learning experiences in physics, chemistry and biology, in Lehrke, M., Hoffman, L. and Gardner, P. (Eds.), *Interests in science and technology education*, Kiel, IPN/UNESCO.

Gauld, C. F. (1976) Scientific attitudes and personal development. *Australian Science Teachers' Journal*, 22(2), 73-82.

_____ (1982) The scientific attitude and science education: a critical reappraisal. *Science Education*, 66(1), 109-21.

_____ (1985) Empirical evidence and conceptual change, in Osborne, R. J. and Gilbert, J. (Eds.) *Some Issues of Theory in Science Education*, Hamilton, New Zealand: University of Waikato Science Education Research Unit, 66-80.

_____ and Hukins, A. A. (1980) Scientific attitudes: a review. *Studies in Science Education*, 7, 129-61.

Haber-Schaim, U., Cross, J. B., Dodge, J. H. and Walter, J. A. (1976) *PSSC Physics (4th Edn): Teacher's Resource Book*. Lexington,

Mass.: D. C. Heath & Co.

Harvey, H. W. (1951) An experimental study of the effect of field trips upon the development of scientific attitudes in a ninth grade general science class. *Science Education*, 35(5), 242-48.

Head, J. (1982) What can psychology contribute to science education? *School Science Review*, 63, 631-42.

_____ and Sutton, C. (1985) Language, understanding and commitment, in L. H. T. West and A. L. Pines (Eds.) *Cognitive structure and conceptual change.* New York, Academic Press, pp. 91-100.

Henry, N.W. (1972) An investigation of some educational objectives in Chemistry, judged to be achieved more appropriately by pupil laboratory work in the secondary school. Unpublished M.Ed. thesis, University of Melbourne.

Hofstein, A., Ben Zvi, R. and Samuel, D. (1976) The measurement of interest in and attitudes to laboratory work among Israeli high school chemistry students. *Science Education*, 60, 401-11.

_____ and Lunetta, V. N. (1982) The role of the laboratory in science teaching: neglected aspects of research. *Review of Educational Research*, 52(2), 201-17.

_____ and Scherz, Z. (1984) *Interest in science and technology: outline of the national studies - Israel.* National Report to the 12th IPN Symposium, University of Kiel, April.

Joyce, B. and Weil, M. (1986) *Models of teaching* (3rd ed.). Englewood Cliffs, N.J. Prentice-Hall.

Kaplan, E. G. (1963) Empiricism as a function of training in the scientific method. *Journal of Research in Science Teaching*, 1, 329-40.

Karmiloff-Smith, A. and Inhelder, B. (1974) If you want to get ahead, get a theory. *Cognition*, 3(3), 195-212.

Keightley, J. V. and Best, E. D. (1975) Student preferences for Year 11 biology classes in some South Australian schools. *Research in Science Education*, 5, 57-67.

Kerr, J. F. (1964) *Practical work in school science.* Leicester, Leicester University Press.

Kieffer, W. F. (1965) *Journal of Chemical Education*, 42(3), 119.

Klainin, S. (1984) The effects of an activity-based curriculum on student outcomes in chemistry in Thailand. Paper presented to 15th annual conference of Australian Science Education Research Association, Melbourne.

Lie, S. and Sjöberg, S. (1984) Jenter i fysikkfaget: Jentene trenger faget - faget trenger jentene (Girls in physics: the girls need the subject - the subject needs the girls). Oslo Universitetsforlaget.

Mason, J. M. (1952) An experimental study in the teaching of scientific thinking in biological science at the college level, *Science Education*, 36(5), 270-84.

McDonell, J. A. (1959) The teaching of physics. *Report of activities, Industries Summer School*, Science Teachers' Association of Victoria, 68-9.

Murphy, G. W. (1968) Content versus process centered biology laboratories, Part II: The development of knowledge, scientific attitudes, problem-solving ability, and interest in biology. *Science Education*, 52, 148-62.

Nussbaum, J. and Novick, S. (1982) Alternative frameworks, conceptual conflict and accommodation: towards a principled teaching strategy. *Instructional Science*, 11, 183-200.

Nwachuku, B. A. (1983) A study of some determinants of students' enrolment in School Certificate chemistry in some local government areas of Imo State. Unpublished M.Ed. thesis. Ibadan, University of Ibadan.

Obioma, G. O. and Obuche, R. O. (1985) The perceptions of Nigerian junior secondary school students of their interests in integrated science: a pilot study. In Lchrke, M., Hoffman, L. and Gardner, P. (Eds.), *Interests in science and technology education*. Kiel, IPN/UNESCO.

Odubunmi, E. O. (1983) The effect of socio-economic background and teaching strategy on learning outcomes in integrated science. Unpublished Ph.D. thesis, Ibadan, University of Ibadan.

Olson, J. K. and Reid, W. A. (1982) Studying innovation in science teaching: the use of repertory techniques in developing a research strategy. *European Journal of Science Education*, 4(2), 193-201.

Osborne, R. J., and Freyberg, P. (Eds.) (1985) *Learning in Science: the implications of children's science*. Auckland, Heinemann.

Osborne, R. J., Freyberg, P. and Tasker, R. (1979) Focus on experiments, *LISP Working Paper No. 2*, University of Waikato, Hamilton, N.Z.

Osbornc, R. J. and Wittrock, M. (1983) Learning science: a generative process. *Science Education*, 67(4), 489-508.

Palmer, G.E. (1981) Computer applications in the freshman laboratory. *Journal of Chemical Education*, 58(12), 995.

Pandit, S., Rajput, J. S. and Vaidya, N. (1978) Analysis of pupil feelings towards physics. *Educational Trends*, 12, 2.

Quinlan, C. (1981) Project Physics in N.S.W. *Research in Science Education*, 11, 86-87.

Raghubir, K. P. (1979) The laboratory-investigative approach to science instruction. *Journal of Research in Science Teaching*, 16(1), 13-17.

Rajput, J. S. (1985) Attitudes towards science subjects of teacher trainees in India. In Lehrke, M., Hoffman, L. and Gardner, P. (Eds.), *Interests in science and technology education*. Kiel, IPN/UNESCO.

Rowell, J. A. and Dawson, C. J. (1983) Laboratory counter-examples and the growth of understanding in science. *European Journal of Science Education*, 5(2), 203-15.

Sjöberg, S. (1984) *National study for Norway*. National Report to the 12th IPN Symposium, University of Kiel, April.

Stead, K., Freyberg, P., Osborne, R. and Tasker, R. (1979) Focus on attitudes. *LISP Working Paper No. 9*, University of Waikato, Hamilton, N.Z.

Suchman, J. R. (1962) *The elementary school training program in scientific enquiry*. Report to the US Office of Education. Urbana, University of Illinois.

Tamir, P. (1985) Interest in learning about plants and animals. In Lehrke, M., Hoffman, L. and Gardner, P. (Eds.), *Interests in science and technology education*. Kiel, IPN/UNESCO.

_____ Arzi, H. and Zloto, D. (1974) Attitudes of Israeli high school students towards physics. *Science Education*, 58, 75-86.

Welch, W. W., Klopfer, L. E., Aikenhead, G. S. and Robinson, J. T. (1981) The role of inquiry in science education - analysis and recommendations. *Science Education*, 65(1), 33-50.

Zieleniecova, P. (1984) *Interests in science and technology in Czechoslovakia*. National Report to the 12th IPN Symposium, University of Kiel, April.

# Part 4

# COMMONPLACES: LEARNERS, TEACHERS AND SUBJECT MATTTER

# 4.1

# Promoting Conceptual Change in the Laboratory

**Richard F. Gunstone and Audrey B. Champagne**

## INTRODUCTION

Throughout the 1984 school year, one of us taught a Year 7 science class at a high school in Melbourne, Australia. In that class, the concepts of solution, suspension and solubility were introduced via a rather routine laboratory exercise. Students were instructed to take a small quantity of each of five chemicals, add these to individual test tubes of water, shake, and record what they saw. The five chemicals were contained in the normal reagent bottles in which they had been supplied to the school. In order to provide a form of identification for the students, labels reading 'salt 1' to 'salt 5' were placed on the reagent bottles. The label on the bottle containing 'salt 1' was, quite by accident, placed in such a way that the supplier's reagent label was still clearly visible. Hence students could see that this bottle contained sodium chloride. During the course of the lesson the teacher asked one student (Darren) a rather vague question about the progress of his investigations just after Darren had added 'salt 1' to water. The following brief interchange resulted:

Darren: 'I knew that [i.e. the sodium chloride] would dissolve.'
Teacher: 'How did you know that?'
Darren: 'Because it's got chlorine in it and chlorine dissolves in swimming pools.'

There was a time when this statement would have been dismissed as a student howler, useful only for therapeutic staff room conversations. However, it can more usefully serve as a very specific example of the results of an important research thrust of recent times. This research thrust offers some very significant perspectives on student learning.

In the first half of this chapter these perspectives are reviewed. This review will show the widespread existence among students of beliefs about, explanations for, and conceptions of natural phenomena which are at odds with the tenets of science. The remainder of the chapter considers some implications for the nature and practice of laboratory work which arise from these perspectives. Specifically, we will consider some ways in which the modification of students' beliefs and explanations may be assisted by particular uses of the laboratory. Hence the conceptual change thrust in the chapter is concerned with the modification of concepts and beliefs which are held by students before formal science teaching commences.

In terms of the model described in Chapter 1.2 (see Figure 1.2.1), this chapter is concerned with particular forms of student inputs, ways in which these inputs can influence actual learning outcomes, and some consequent issues concerning laboratory work. Given that these particular forms of student inputs involve knowledge and beliefs of students before formal learning, aspects of this chapter relate to some earlier chapters on intended outcomes - 3.2 scientific enquiry, 3.3 scientific knowledge and 3.4 scientific attitudes and attitudes to science.

## CONSTRUCTIVIST PERSPECTIVES ON LEARNING SCIENCE

### Views of natural phenomena - an example

A sheet of clean, smooth glass lies on a horizontal table. Sitting on the glass is an air puck (a very smooth metal disc, on top of which is a small air compressor that forces air out of a hole in the disc, resulting in a miniature hovercraft). Beside the table is a vacuum cleaner with the hose connected so that air blows out from the hose. Consider how you would use the air blowing from the vacuum cleaner to cause the puck to move at a steady speed across the glass. (This question is taken from Lawson, 1984.)

We have used this task, with the apparatus we have described, on a substantial number of occasions. Groups to whom it has been given include senior high school physics students, physics majors in a one year post-graduate course of teacher training, and physics and science teachers in both in-service and higher degree courses. On each occasion, one person has initially been asked to demonstrate their answer by actually using the vacuum cleaner to move the puck. Without exception, the demonstrated answer has involved pointing the hose at the puck - for the whole distance that the puck accelerates across the glass. It is not uncommon for the unexpected acceleration to produce mild expletives, and it universally produces considerable surprise.

In thinking about this observation it is important to remember that all the groups of participants described would claim knowledge of Newton's Laws. Both the anecdote about Darren given in the introduction to this chapter, and the informal data about views of force and motion described above, contain the same general message. That is that students develop ways of interpreting the world around them which are not the result of formal school science learning. Further, school experience can combine with these personal views to produce new and unexpected explanations, as illustrated by the case of Darren's statements about chlorine and swimming pools. Such explanatory views of phenomena are often at odds with the tenets of science. Most significantly, it is clear from examples such as the force and motion one above that real-world events can often be interpreted via a view at odds with science - even when the science interpretation has been learned and used to pass examinations. But when faced with this particular 'real-world' problem, all those who have demonstrated an answer have done so by using a conception that 'motion requires a force' rather than the science conception that 'change of motion requires a force'. The use of a 'motion requires a force' interpretation of the world for this problem is not surprising. It is a view which has been frequently reported in the literature.

## Views of natural phenomena - some selected research

Investigations of the nature and extent of students' views about the world, of their personal understanding of science concepts, have ranged over a wide variety of content areas in recent years. In order to elaborate some of the issues of importance in this work, a few examples are considered.

In a study of views about physical changes involving water, Osborne and Cosgrove (1983) conducted individual interviews with 43 students aged 8 to 17 years. A number of common phenomena were shown: boiling, condensation of steam, evaporation, condensation on a cold surface, melting. Students were questioned about their interpretations of each event, e.g. 'What are the bubbles [in the boiling water] made of?' The results of the interviews guided the construction of a pencil and paper instrument used to survey the prevalence of views in a much larger sample. We will consider here the views about the nature of bubbles in boiling water. The interviews revealed four commonly held views about the nature of the bubbles - they comprise heat, air, oxygen and/or hydrogen, or steam.

In a larger survey, it was found that all of these views were reported by significant proportions of students. For the 589 students aged 12-15 years, 'steam' was the least commonly offered response (10%-15% at each age level). At age 12 'heat' and 'air' were each given by about 30% of respondents, with 'heat' still being offered by about 20% of 15 year olds. The only response to increase in frequency from age 12 to age 15 was 'oxygen and/or hydrogen'. For the 136 students aged 16 and 17, 'heat' was given by only a very small proportion. However, the sample for these ages was drawn from those who had elected to undertake a study of chemistry in senior high school. That is, the decline in frequency of 'heat' may be due to the nature of the sample rather than the effect of science learning. Even in this select population of 16 and 17 year olds, 'oxygen and/or hydrogen' was a marginally more frequent response than 'steam' (both around 35%), and 'air' was given by more than 20%.

Of these four views about the nature of bubbles in boiling, 'heat' seems clearly to be an interpretation of the phenomenon which owes nothing to school learning. This perhaps is an explanation formed solely by independent reasoning on the part of children: these bubbles only occur in boiling water; boiling only takes place when 'heat is added', etc. At the other extreme, 'oxygen and/or hydrogen' seems to arise from unexpected interpretations of school learning. In considering the oxygen and hydrogen view, Osborne and Cosgrove (1983) suggest that students giving this response have 'quite reasonably in our view, integrated their knowledge about water being made of oxygen and hydrogen, and air consisting of oxygen and other gases, into their earlier view that bubbles were made of air produced by the water' (pp. 827-8). This research uncovered other instances of unexpected integration of school learning with interpretations of phenomena acquired outside school: for example the view that water drops on the outside of a jar containing ice are formed by the passage of water through the jar, this water passage being an example of diffusion. The 'oxygen and/or hydrogen' view of bubbles was extended by some students into a relatively comprehensive explanatory scheme. Steam condensing on a cool surface was explained by the recombination of oxygen and hydrogen; gradual evaporation of water from a wet surface resulted from water changing into oxygen and hydrogen in the air.

Individual understandings of concepts of a classificatory nature have also been probed. Bell (1981; Bell and Barker, 1982) has explored views of the concept of animal. This work also used interviews, in this case based on drawings of instances and non-instances of the scientific concept of animal ('Is the — an animal? What tells you that?'). Results of the interviews again guided the development of larger sample surveys. Children were often found to use a definition of animal different from that taught in science. Criteria used to judge animalness by junior high school students included the number of legs, membership of another named group (e.g. 'I still think spiders are insects, not animals'), ability to move, habitat, and size. Amch and Gunstone (1985) have used a similar probe of views of animal with university science graduates. The use of non-scientific criteria, as found by Bell, was more limited with these graduates but was still present (e.g. an insect is not an animal; snail, worm, lizard are each not sufficiently complex an organism to be an animal; for one graduate, even habitat as a criterion to decide a fish is not an animal).

## Views of natural phenomena - after science instruction

There have been a number of systematic investigations of beliefs about and understandings of natural phenomena among university science students. Brumby (1981, 1984) has found, with both British and Australian university biology students, that phenomena involving adaptation and evolution are commonly interpreted via a conception much nearer to Lamarck than to Darwin. And this with students who achieved well in senior high school biology exams which included questions on Darwinian evolution. Champagne, Klopfer and Anderson (1980) found among 110 first year physics students at the University of Pittsburgh a number of widely held beliefs: that heavier objects fall faster; that for two objects suspended at rest on a pulley, being lower implies being heavier; that constant force implies constant velocity motion. They also reported that the frequency of these views was little different between those who had studied physics at high school (about 70% of the sample) and those who had not studied physics at high school. Gunstone and White (1981) subsequently undertook a similar study with all 470 first year physics students at Monash University, all of whom had studied physics for two years in senior high school. Results were broadly similar. Interpretations of real events frequently showed the existence of the same beliefs. A number of investigations of understanding of basic mechanics concepts among beginning university students have been conducted by McDermott and others at the University of Washington (e.g. Trowbridge and McDermott, 1980, 1981). These investigations showed that students who could acceptably define velocity and acceleration commonly could not determine relative velocities and accelerations in a real situation, that they often confused speed and position, and that these problems were frequently not affected by instruction.

The significant finding in such studies is that views commonly found among students before instruction are often still used to interpret the real world *after* instruction - even with students who have demonstrated an ability to use the science view to answer examination questions. It is as if students can simultaneously hold two competing conceptions of the world in their head. One of these is used for performance tests, the other remains the one used to make sense of the world. Under such circumstances it is clear that there can be little understanding of the concepts involved. Many would argue that a fundamental purpose of teaching should be to seek to have students move to a single conception which is used on all occasions.

## Views of natural phenomena - a summary

Although these general approaches to investigations of student beliefs and understandings seem relatively recent, there are already a number of interpretative reviews of the area. Some of these are concerned with a specific content area such as mechanics (e.g. Champagne, Klopfer and Gunstone, 1982; McDermott, 1984), while some encompass a wider range of science (e.g. Driver, Guesne and Tiberghien, 1985; Osborne and Freyberg, 1985).

There is in these reviews a broad consensus: students do have beliefs about explanations for natural phenomena before any formal science study of the phenomena; these beliefs and explanations are often at odds with the beliefs and explanations of science, and are often sufficiently robust to be unaffected by science instruction; apparently successful students of science can continue to interpret the world outside school via the beliefs and explanations they held before learning science, even when they have used a science explanation in conflict with their own to answer examination questions; while science seeks explanations which encompass a wide variety of phenomena (i.e. parsimony is important), students are commonly not concerned about this issue (i.e. opposing beliefs can be used to interpret situations seen by science as explainable by a single view).

## The origins of these views

Given our current understanding, any statements beyond the obvious would be most conjectural. However, the obvious is worth stating. In a very general sense, it can be said that these views arise from everyday interactions with the world. Such interactions involve both out-of-school and in-school experiences, peers, parents, media, and so on. In some cases it may even involve a researcher interviewing a student - the researcher's question may give rise to the formation of an explanation which the student had not previously conceptualized. However, this seems unlikely with views which show considerable robustness. It seems reasonable to argue such robustness derives from successful use of the view (in terms of making necessary sense of the world) for some time prior to being exposed to an alternative view in science classes. It is this previous use in the real world characteristic of student views which is one of the important issues arising from consideration of the origins of views, and we shall return to it in the second half of this chapter. The other important issue arising here is that individuals generate their own views, come to their own understandings, as a result of interpreting their experiences. This issue also is considered subsequently.

## The labelling of views

One issue on which there has clearly not been agreement among researchers is the appropriate generic name to use to describe the views of individuals. Names which have been used include 'alternative frameworks', 'alternative conceptions', 'naive theories', 'children's science'. Although in general this issue is not of significance in the present context, two points are relevant. Firstly, there is considerable, but not universal, agreement in the literature that the label 'misconception' is not helpful. Such a label carries a connotation of clear and unambiguous error and can hence cloud more important issues such as the origin and frequent resilience to change of the views. Secondly, some work has clearly labelled student views in terms of the historical evolution of scientific ideas (e.g. McCloskey, 1983 describes student views in areas of mechanics as showing impetus theory in the sense that the views paralleled those held by advocates of the impetus theory such as Buridan). Such labelling has been argued against (e.g. Lythcott, 1983) on the grounds that student views do not show any of the features of comprehensiveness that are to be found in the writing of, for example, Aristotle or Buridan. Di Sessa (1983) goes further to argue that the underlying intuitions that students use to interpret phenomena are not theories at all, at least in the sense implied by McCloskey (1983).

Some form of objection seems possible for most labels which have been used. It is for this reason that the more neutral terms 'views', 'beliefs', 'explanations' have been used in this chapter.

## A broader context briefly - constructivist perspectives on learning

Data of the nature discussed above are not new. Over twenty years ago, Atkin and Karplus (1962) drew attention to the out-of-school experiences of children, and the 'invention' (p. 47) by students of interpretations of these experiences.

Further, they suggested that most of these experiences and inventions 'reveal a type of natural philosophy - a "commonsense" orientation popular at a given point in history' (p. 47). Understandably, given the knowledge of these issues at the time, the authors go on to suggest that if children do not '*invent* the modern scientific concepts, it is necessary for the teacher to *introduce* the modern scientific concepts' (p. 47). Today it seems not quite so simple. However, the more important point is that this perceptive insight caused no ripples in the common research perspectives at the time.

Seventeen years earlier again, Oakes (1945) explored adults' predictions about, observations of, and explanations of the falling of two balls of different weight. His approach and his raw data are remarkably similar to part of the work reported by us at a much later time (Champagne, Klopfer and Anderson, 1980; Gunstone and White, 1981). Unfortunately we were unaware of Oakes' work at the time of our research. While it is true that Oakes came to his research with a somewhat different orientation (as evidenced by his concern with animistic and teleological aspects of explanations in later work, Oakes, 1957), it is still of interest that his 1945 paper seems to have had no impact on thinking about learning in science.

Why this should have been is clearly a question which cannot be answered absolutely. One important variable is the general direction of learning research at the time of the investigation. The more recent work on students' views has taken place in a broader climate of concern with issues such as understanding in learning, the role of context and prior knowledge in the idiosyncratic meaning which individuals extract from experience, and, most importantly, the growing acceptance of the view that individuals actively generate their own personal meanings. This latter issue leads to positions such as: the learner is an active processor, understanding cannot be taught but can be created by the learner from experiences directed by the teacher, the linking of new ideas with existing memory is fundamental to understanding, successful learning involves a personal restructuring of the way one thinks about the world. This constructivist view of learning has been well elaborated in the context of students' views of the world by Osborne and Wittrock (1983). Links between the persceptives on learning science reviewed here and developments in cognitive psychology have been discussed by Champagne and Klopfer (1984).

What is particularly significant, especially in terms of the remainder of this chapter, is that learning - both informally via day-by-day experiences and formally in science classes - can validly be seen as a personal act of construction. It is then hardly surprising that this individual construction may often lead to understandings of an unexpected form. Nor is it surprising that changing a student's view should involve active reconstruction on the part of the student. As a consequence, the following discussion on some implications for laboratory work which arise from this understanding of learning focuses on ways of contributing to conceptual change. That is, ways of increasing the chances of individuals attempting to reconstruct their view of the world so as to encompass the views of science.

## SOME IMPLICATIONS FOR LABORATORY WORK

General considerations of the implications for science education of the research discussed above have taken a number of directions. One of these is reflected by a number of attempts to change student views to those of science through particular forms of instruction (e.g. Champagne, Gunstone and Klopfer, 1985; Gunstone, Champagne and Klopfer, 1981; Nussbaum and Novick, 1982; Osborne, 1983; Rowell and Dawson, 1983). Possible directions for laboratory work are to be found in many of these attempts to produce conceptual change, as well as in the results of probing students' views. A number of these directions are now considered individually, and some more general implications are raised in the conclusion to the chapter.

### Links from the laboratory to the everyday world

Three points of significance here were argued in the first half of the chapter: that in attempting to make sense of the world around them students often form views which are at odds with those of science; that students do not see the virtues of parsimony of explanation; that student views are often not affected by instruction.

One rather obvious direction for laboratory work can be derived from these points. That is to increase the extent to which laboratory tasks and equipment use common phenomena and materials. For example, one of the most everyday of experiences involving chemical change is cooking. Yet these chemical changes rarely find a place in laboratory investigations of this topic. Further, it is uncommon for students to even see cooking as being an example of chemical change. This leads to a closely related aspect of this direction. One approach to the problem of parsimony is to have students explore a variety of contexts related to the scientific principle being taught. If we wish students to see the universality of a scientific explanation then we should consider having them explore a variety of situations which can be analysed by use of the explanation.

There is a *caveat* to these suggestions, again an obvious point. It is well illustrated by experiences in the teaching of the Year 7 class mentioned in the introduction to this chapter. In that class, a consistent thrust involved attempting to use everyday materials in laboratory work and to relate what was being learned to everyday experiences. This thrust was sufficiently strong to influence content in some parts of the course. While it had positive effects, the thrust also generated some occasional dissatisfaction. The students were in their first year of high school. As a consequence they had expectations of science which included using the unfamiliar (equipment, chemicals, etc.). The thrust towards everyday phenomena and equipment caused some students to periodically feel they were missing something.

This of course suggests a consideration of degree of use of everyday things rather than its inappropriateness. In the extreme, it is hard to find any reason to support a teacher telling a class in its first science laboratory that, in the context of discussing things dangerous, 'we've got gas taps here ... those of you who've been in a kitchen do not need to be told this' (Young, 1977, p. 259). Imagine a teacher believing that 11 year olds may never have been in a kitchen.

More generally, there does seem reason to believe that many students in science classes come to feel that what is taught in these classes has no use in understanding the everyday world. Experiences such as that described by Young (above) can only strongly reinforce that feeling.

These concerns with everyday issues cast doubt on the wisdom of one of the basic arguments invoked in the 1960s to support science curriculum reform – or, more fairly, on the common mode of expression of this argument. The then director of the US Commission on College Physics (cited by Lerch, 1973) expressed the argument as one of two basic axioms: 'It is in the lab and only in the lab, that the student can experience physics as it actually is' (p. 154). In fact for many students 'physics as it actually is' is in their terms somewhat the antithesis of many laboratory encounters. The sentiment of Fowler's axiom might be better expressed as 'It is only through direct experience with the world that students can experience physics/science as it actually is', tautology notwithstanding. There can be significant differences between a laboratory approach to teaching science and an interaction with real phenomena approach.

The comment in Fowler's axiom leads to another issue arising from this concern with relationships between the laboratory and everyday experience. This is the use of laboratories to 'verify' some forms of scientific principle. It is considered in a separate section below.

## Deriving content and sequence from students' views

An area of science in which there has been considerable investigation of children's views is that of electricity in direct current circuits (e.g. Osborne, 1983; Shipstone, 1984). A review of this work is contained in Driver, Guesne and Tiberghien (1985). Such studies show that there are a small number of views of electricity used by students in interpreting these circuits.

It is important to note that the term 'electricity' has been deliberately used here, rather than 'electric current'. One very important outcome of these investigations is that, before instruction, students very rarely distinguish the concepts 'current' and 'energy'. This is true even when students use these terms, as the terms are used as synonyms for 'electricity' (Gunstone and Shipstone, 1984). This point alone raises useful suggestions for the content and sequence of laboratory work with DC circuits. These include using introductory laboratory exercises whose purpose is to aid discrimination between current and energy, and questioning the very common practice of using energy transformers (usually light globes) as crude current measurers.

Of the common student views about circuits, one has it that there are two different sorts of electricity/current flowing in opposite directions from the battery, along the two wires and meeting in the device being powered (e.g. light globe). Another view is that there is a single direction of flow, but a gradual decline in electricity/current around the circuit as the electricity/current is used up. The nature of laboratory work which might challenge the first of these views seems clear - for example, in a simple circuit involving a cell and a single light globe, predicting and investigating the result of adding ammeters in each connecting wire or reversing the connections to the cell or adding one or more extra globes in series (including both identical and different globes). The same laboratory experiences can assist the challenging of the second view of electricity/current (diminishing around the circuit) - provided there is discrimination between the concepts of current and energy. The effectiveness of such laboratory work, as part of a teaching sequence designed to have students accept the scientists' view of DC circuits, has been investigated (Osborne 1983; Shipstone and Gunstone, 1984; Steinberg, 1983). These sequences have partial, but only partial, success. The limited nature of this success is perhaps due to the failure of many students to adequately discriminate between current and energy. Gauld (1986) also suggests a range of possible explanations for this limited success, including idiosyncratic reconstruction of memories of laboratory experiences, experience outside laboratories, and so on.

More absolute directions for instruction have been given for the concept of normal reaction (Gunstone and Shipstone, 1984). Many students do not believe that a rigid surface such as a table could possibly exert an upward force on an object lying on it. This has been described as a student view in conflict with the science view. However more careful consideration leads to the conclusion that such student beliefs are quite reasonable, if the table is indeed rigid. But of course it is not. The conflict arises from this confusing and incorrect assumption made in presenting the science view of the situation. Experiences which lead to an acceptance of distortion in the table (see Minstrell, 1982) largely resolve the issue. These experiences include considering whether or not an upwards force exists on one book placed on a hand, or several books placed on a hand, on a book hung from a spring, etc. The distortion resulting from a heavy object placed on a table can then be observed by shining a slide projector on to a mirror placed on the table and seeing movement in the reflected light.

Many of the commonly reported student views about phenomena lead to suggestions for the content and sequence of laboratory work. One most helpful source of these views is the reviews of this work cited earlier in the chapter. The other obvious source is students themselves. This leads to a quite familiar form of laboratory work with a somewhat different twist - what has often been termed laboratory for theory testing.

The use of school laboratories for the testing of hypotheses/theories is discussed in Chapter 3.2. As reflected in school texts and books on science teaching methodology, this approach involves the posing of a problem, the generation by students or teacher of hypotheses, and the experimental investigation of hypotheses. Generally the student-generated hypotheses are seen as a form of guess (informed, inspired or otherwise) rather than a belief. The research perspective considered here suggests that belief is frequently a more appropriate description. That is, for many problems, students will advance possible explanations which are in fact their already existing personal understandings of the situation. This means that a different light is thrown on the exercise. In particular, it cannot be seen as having students choose the accepted explanation on the basis of objective investigation between competing possibilities. Rather, it is most likely that observations may not be uniform throughout a group and that the interpretations of observations will be even more varied. (This issue is elaborated in the next section.) Active reconstruction of personal views is likely to be required of students. This is considerably different to the notion of adding knowledge to an empty mind which underlies previous perspectives on this form of laboratory work. Consequently moving from data to an accepted conclusion is likely to be difficult, time-consuming, and to demand considerable effort on the part of the teacher. Driver (1983) gives a very useful discussion of these issues.

Laboratories for theory testing can be based on existing knowledge of the views of students. Thus the teacher would be able to be prepared for at least most of the ideas advanced by students for investigation. The work on DC circuits described above is one example of this. Similar work in other areas has been reported (e.g. Champagne, Gunstone and Klopfer, 1985; Rowell and Dawson, 1983). Alternatively the approach may be used without knowledge of the beliefs which will emerge. An example of this (taken from Radman, 1985) involved a teacher beginning a sequence of lessons by asking her group of 11 and 12 year olds how hot water could get when heated. Discussion of the data produced in investigating that question led to the emergence of a number of student beliefs. One was the view that the volume of water present affected the maximum possible temperature. This had not been anticipated by the teacher, but was experimentally investigated and discussed. This resulted in an apparent reconstruction of understanding by at least some who initially held the volume belief.

## Observation

This is a crucial component of laboratory work. Some very important issues surrounding observation have been raised by the research focus being considered here. None of these issues is new, but this research context has given somewhat different perspectives on the nature of the issues in terms of school laboratories and learning.

Firstly, we consider the proposition that observation is frequently theory-dependent. That is, 'seeing is believing, if I hadn't believed it I wouldn't have seen it' (Brilliant, 1979, p. 40). We illustrate the importance of this, in at least some contexts, by considering some of the specific tasks used by Champage, Klopfer and Anderson (1980) and Gunstone and White (1981) to probe student beliefs.

Other examples of such theory-based observations were found in this same investigation. For example, an 18 litre bucket containing sand and a large piece of wood were attached by a cord and suspended at rest over a pulley. The object used as a pulley was a bicycle wheel mounted so that it could rotate as it would on a bicycle. Subjects were asked to predict, observe, explain for the case of a small teaspoon of sand added to the bucket. Some predicted that the bucket would move a small distance and then come to rest again ('at a new equilibrium position'). Of these, some reported observing a small movement (even though no one else did) and others suggested there must be a movement so slight it could not be seen.

Even when the observation is unambiguous, beliefs can cause the observation to be disregarded. Another of the tasks used by Gunstone and White (1981) involved the same piece of wood and a different bucket of sand hanging at rest on the bicycle wheel. In this case, the two objects were placed on the same level. The wood was then pulled down about half a metre and held in its new position. Subjects were asked to predict, observe, explain what would happen when the wood was released. About half the 470 in the sample predicted the wood would move. This figure does not include those who showed little understanding of reality by predicting movement because there was now more cord on one side. The substantial majority of these subjects reconciled the conflict between this prediction and the universal observation of no movement by arguing the observation to be wrong. Most commonly this was achieved by suggesting that friction in the bicycle wheel prevented motion, or that the wood was held for too long at its new position (so that it 'got used to a new equilibrium position'). Other reasons advanced were even more extreme, even including the distortion in the plastic bucket resulting from the weight of sand it contained.

What is of particular interest in this example is the widespread extent to which the observation did not affect the original belief about the situation. Rather, the belief affected the observation. This gives rise to two obvious implications for laboratory work. Firstly, there are many contexts for which it is unreasonable to assume that identical observations of phenomena have been made by all in a laboratory class. This is so even if it seems clear to the teacher that all students have observed the same phenomena. Secondly, when an observation is uniform throughout a class, neither the legitimacy nor the instructional purpose of the observation can be assumed to have been accepted by all participants. In both these cases the teacher has an important role to play. In essence, this role involves determining any conflict in observations, resolving the conflict, and considering the implications of the observation. All of these will usually involve time and interaction, points to which we return in the conclusion.

In addition to the theory-based nature of observation, there are other aspects of this skill which are highlighted by the research on students' views of phenomena. Despite the central importance of the observation skill to laboratory work, explicit development of an understanding of the observational skills is rarely a focus of science education. Consider, again, a task from the work of Champagne, Klopfer and Anderson (1980) and Gunstone and White (1981). This involved the dropping of single objects from a height of about 2 metres. The task involved prediction, observation, explanation of the relative speeds of the object at markers half way down and just above the ground. In both studies some remarkable observation was reported, e.g. speed at ground is $\sqrt{2}$ times speed half way down. In addition to being theory-related, such 'observations' reveal a lack of understanding of what in fact an observation is.

Distinguishing between observation and inference is a commonly difficult task but a crucial one if laboratories are to be validly used for learning from observation.

Another important issue arises from responses to the dropping - single - object task. Not one respondent made any comment on the difficulty of the task. We would argue that a response to the request to observe speeds at the two points which read 'it is too hard - you can't tell' or 'without better apparatus than my eyes I cannot say' would be most reasonable. The expectation of a correct observation may play a powerful role here. One further aspect of observing is suggested by Osborne and Cosgrove (1983). They report that the extensiveness and detail of observations of changes of state of water declined with increase in age of subjects.

All of these issues point to a need to explicitly develop through teaching an understanding of the skill and purpose of observing, just as is commonly done for other laboratory skills such as recording and analysing data. In this context, Driver (1983, Chapter 2) raises some important aspects of learning to observe. The basic issue she addresses is that of appropriate ways of focusing the observation task, and helping students decide what aspects of a situation are relevant and what are irrelevant in terms of the purpose of the task. Strategies suggested include making the observation, considering all results in a class, and then re-observing with focus on deciding which of the alternative results initially produced is appropriate; making the observation, then seeing a teacher demonstration/explanation of the phenomena, then making observations of a closely related phenomenon; direct instruction in the conventions and constructs used in some science observations (e.g. field lines in iron filings around magnets). A sound understanding by the teacher of the nature of observation is an important antecedent to the teaching of observing. Martin (1972, pp. 104-32) provides a useful introduction.

## Inappropriate uses of observation to verify scientific principles

It is not uncommon for quite unreasonably extravagant claims to be made for the use of the laboratory for experimental (observational) verification of scientific principles. Consider again the example of dropping objects (shot put and rubber ball). We inferred previously that most, but not all, people observe these two balls to arrive at the same time when dropped simultaneously from about 2 metres. The same form of predict-observe-explain task has been used with a number of groups of physics graduates undertaking teacher training, except that the balls have been released from about 10 metres above the ground. The universal prediction has been that the two will take the same to time fall. The universal observation has been that the shot arrives first. Estimates of the observed magnitude of the separation of the two balls on reaching the ground range from 0.3 to 1 metre. This range is much more a function of the observer than of the particular drop. The observation is always a very surprising one for physics graduates.

A clue to the important point in this observation is to be found in the justification given by some who predict that the heavier ball will arrive first when dropped from 2 metres. This justification is of the form 'I've tried dropping heavy and light things off [e.g. a bridge] and the heavier one always gets there first'. That is, their prediction is based on previous observation of the same phenomenon. The point is that the assertion that all objects have the same acceleration when dropped is an idealized and conditional statement. It is simply not true for events taking place in the earth's atmosphere. It can only be observationally 'verified' for a small range of objects over a very small range of distances - and then only because of the inability of the instrument being used (the eye) to detect very small differences. This particular example of inappropriate 'verification' is also of interest because of the widespread view that Galileo used the Leaning Tower of Pisa observationally to show uniform acceleration of falling bodies. Not only is this view questionable on historical grounds, there is strong reason to believe that any such experiment would show that the objects Galileo is supposed to have dropped would in fact arrive at different times (see, for example, Franklin, 1978).

Many scientific principles are idealized statements in the sense that they cannot be observationally verified. It is philosophically unreasonable to try to 'discover' or 'verify' such principles directly from observation. More importantly from the students' perspective, conflicts between a supposed verification of a principle and out-of-class observations are most confusing. In such cases it would be very surprising to find a student reconstructing their views to incorporate the science principle.

## Qualitative laboratory exercises

Clement (1982) briefly describes some qualitative laboratory exercises used with students. These were developed as one instructional response to his findings about the nature of student beliefs in mechanics. He argues that such laboratory exercises, with the absence of formulas to use for number stuffing, 'are an effective way of getting students to think about their own preconceptions. In general, when qualitative misconceptions arise, it is necessary for students to express them and to actively work out their implications in order to see the advantages of the Newtonian point of view' (Clement, 1982, p. 70).

Champagne, Klopfer and Gunstone (1981, 1982) also argue for the use of qualitative laboratory tasks. In this case the argument is based on analyses of the differences between uninstructed students, novice physics students, and expert physicists in both problem-solving approaches and in relevant explanatory views. These differences lead to suggested laboratory tasks, for use with beginning physics students. The laboratory tasks have two important characteristics. Not only are the tasks specifically qualitative, they are also much smaller and more focused than usual laboratory exercises. Each is in essence a single task. This is argued in order to allow much greater emphasis on the meaningful incorporation of observation and deduction from observation into the explanatory view of the student. Hence this approach to the use of the laboratory reduces the 'hands on' time for an experiment and implies considerable increases in time for discussion and reflection.

An example of such an approach is briefly described in Champagne, Gunstone and Klopfer (1985). This laboratory investigation considered falling objects. Four observation tasks were used: two equal volume and different mass balls were dropped from about 2 metres and about 10 metres; two equal volume and different mass cubes were dropped about 0.25 metres through water and through oil. In each case students predicted, observed, explained. It is clear that the actual performance of these tasks took relatively little time, even allowing for the common repetition of a task as students tried to move to an agreed common observation. The four tasks were not performed in close sequence. Both before and after each single task a considerable time was spent in discussion, to clarify individual bases for prediction, the nature of the observation, and most particularly the interpretation of the observation. It is of interest that the small group involved in this exercise subsequently demonstrated substantial understanding of the issues involved in the four tasks. More importantly, members of the group demonstrated considerably enhanced awareness of the nature and origin of their own understanding.

What is being argued here is that there are times when laboratory work with a quantitative focus can, because of its emphasis on measurement/formula/numerical answer, discourage a consideration of conceptual relationships. The then implied qualitative laboratory focus can be enhanced by using smaller and hence more focused tasks. In such a case, the active processing and interpreting of observations can be more readily fostered by the teacher. This fostering depends on considerably more time than is usual being given to intellectual interaction with observations and their consequences. (See Driver, 1983, Chapter 2 for further discussion of the problems which arise when observation is inadequately focused.)

## CONCLUSION

The issues relating to the purposes and uses of laboratory work introduced above have much in common. It may well be helpful to see them more as a whole than as discrete ideas. All are underpinned by a view of learning as an act of construction by the individual. In this view learning science is a disciplined creative process. Even when a teacher gives a pupil an explanation for how and why something behaves as it does, the pupil must still actively create meaning from that explanation. Teaching involves helping pupils to generate appropriate meanings from incoming information, to link these meanings to other ideas in memory, and to evaluate both newly constructed ideas and the way old ideas are related in memory (Osborne and Wittrock, 1983, p. 505).

Clearly then, the generation of links by the learners is argued to be of vital importance to the acquisition of understanding. This includes both the linking of new ideas with existing knowledge and experience, and the restructuring of links between ideas already in memory. The first of these (linking new with old) will be most challenging and difficult when the new and old are in conflict, a circumstance shown earlier to be a common one. When one explanation is already believed by the learner, the provision via instruction of a conflicting explanation produces a situation not readily resolved. The generation of real linkages here will require considerable effort on the part of the learner. In such cases the other type of linkage described above (restructuring of links between ideas already in memory) will also be necessary.

The perspectives on laboratory work which have been discussed in this chapter all have concern with the generation of links - links between new and existing ideas, and so on. Of course the links generated may not always be those anticipated by the teacher. Examples of this were given in the first half of the chapter. As a consequence, the laboratory should be used in a way which does more than provide some possibility of some links being generated. Teaching strategies must, at least in part, aim to help students understand the nature and appropriateness of the links being generated.

As we have already suggested on a number of occasions, this involves time for interaction and reflection. The nature of laboratory exercises is then changed in important ways. Less time is spent in interacting with apparatus, worksheets, instructions, and (in extreme cases) recipes. More time is spent interacting with ideas. The proposition that the development of concepts takes time is much older than the research perspective in focus here (see, for example, Brownell and Hendrickson, 1950, p. 113). However, the focus gives clear reasons for the need for time, and thus gives indications of profitable ways of spending this time.

To argue for more time in the laboratory to be spent on interaction and reflection is to argue for the use of discussion. This is a teaching strategy which has been widely under-used in laboratories. In a stereotypical laboratory exercise, a form of discussion may well occur after students have completed recording data. In these cases however the discussion is invariably very closed. The teacher is using the strategy to gradually build up the expected conclusion from the laboratory exercise without regard to students' interpretations of and beliefs about the exercise. Little, if any, conceptual change results. Perspectives on learning and laboratory work considered in this chapter provide ways of using discussion which will allow students to evaluate their own beliefs, observations, interpretations. This may promote appropriate consideration of the nature of their beliefs, observations, interpretations, and ways in which these link with other ideas. We would argue this process to be a necessary, if not sufficient, condition for conceptual change.

What is being described here is, of course, the promotion of students' understanding and control of their own learning: that is, the promotion of *metacognition*. The links between these ideas and the issues discussed in two other chapters of this book are strong. In Chapter 2.2 two strategies for developing students' reflection about their beliefs, observations and interpretations are presented. The concept of metacognition is elaborated in Chapter 4.2 and some approaches to the development of metacognitive perspectives in students are outlined. Both of these chapters are centrally concerned with the themes of personal reflection on and judgement about ideas and beliefs which run through the arguments above.

## REFERENCES

Ameh, C. O. and Gunstone, R. F. (1985) Teachers' concepts in science. *Research in Science Education* 15, 151-7.

Atkin, J. M. and Karplus, R. (1962) Discovery or invention? *The Science Teacher* 29(5), 45-51.

Bell, B. F. (1981) When is an animal, not an animal? *Journal of Biological Education* 15, 213-28.

_____ and Barker, M. (1982) Towards a scientific concept of 'animal'. *Journal of Biological Education* 16, 197-200.

Brilliant, A. (1979) *I may not be totally perfect, but parts of me are excellent*, Santa Barbara, CA, Woodbridge Press.

Brownell, W. A. and Hendrickson, G. (1950) How children learn information, concepts, and generalizations, in N. B. Henry (Ed.) *Learning and instruction* (49th yearbook of the National Society for the Study of Education), Chicago, University of Chicago Press, pp. 92-128.

Brumby, M. N. (1981) The use of problem solving in meaningful learning in biology. *Research in Science Education* 11, 103-10.

_____ (1984) Misconceptions about the concept of natural selection by medical biology students. *Science Education* 68, 493-503.

Champagne, A. B. and Klopfer, L. E. (1984) Research in science education: The cognitive psychology perspective, in D. Holdzkom and P. B. Lutz (Eds.) *Research within reach: Science education*, Charlestown, WV, Appalachia Educational Laboratory, pp. 171-89.

Champagne, A. B., Gunstone, R. F. and Klopfer, L. E. (1985) Effecting changes in cognitive structures among physics students, in L. H. T. West and A. L. Pines (Eds.) *Cognitive structure and conceptual change*, Orlando, FL, Academic Press, pp. 163-87.

Champagne, A. B., Klopfer, L. E. and Anderson, J. H. (1980) Factors influencing the learning of classical mechanics. *American Journal of Physics* 48, 1074-9.

Champagne, A. B., Klopfer, L. E. and Gunstone, R. F. (1981) *A model of adolescents' understanding of physical phenomena and its application to instruction*. Paper given at the meeting of the American Educational Research Association, Los Angeles, April, ED202670.

_____ (1982) Cognitive research and the design of science instruction. *Educational Psychologist* 17, 31-53.

Clement, J. (1982) Students' preconceptions in introductory mechanics. *American Journal of Physics* 50, 66-71.

Di Sessa, A. A. (1983) Phenomenology and the evolution of intuition, in D. Gentner and A. L. Stevens (Eds.) *Mental models*, Hillsdale, NJ, Erlbaum, pp. 15-33.

Driver, R. (1983) *The pupil as scientist?* Milton Keynes, Open University Press.

_____ Guesne, E. and Tiberghien, A. (Eds.) (1985) *Children's ideas in science*, Milton Keynes, Open University Press.

Franklin, A. (1978) Galileo and the leaning tower: An Aristotelian interpretation. *Physics Education* 14, 60-3.

Gauld, C. (1986) Models, meters and memory. *Research in Science Education* 16, 49-54.

Gunstone, R. F., Champagne, A. B. and Klopfer, L. E. (1981) Instruction for understanding: A case study. *Australian Science Teachers' Journal* 27(3), 27-32.

Gunstone, R. F. and Shipstone, D. M. (1984) Students' alternative frameworks in science: A perspective on some implications for instruction. *Conference papers of Australian Association for Research in Education*, 287-94.

Gunstone, R. F. and White, R.T. (19;81) Understanding of gravity. *Science Education* 60, 291-9.

Lawson, R. A. (1984) Student understanding of single particle dynamics. Unpublished Ph.D. dissertation, University of Washington.

Lerch, R. D. (1973) An evaluation of the divergent physics laboratory. *Science Education* 57, 153-60.

Lythcott, J. (1983) 'Aristotelian' was given as the answer, but what was the question? in H. Helm and J. D. Novak (Eds.) *Proceedings of the international seminar: Misconceptions in science and mathematics*, Ithaca, NY, Dept. of Education, Cornell University, pp. 276-85.

Martin, M. (1972) *Concepts of science education: A philosophical analysis*, Glenview, IL, Scott Foresman.

McCloskey, N. (1983) Intuitive physics. *Scientific American* 248(4),

114-22.
McDermott, L. C. (1984) Research on conceptual understanding in mechanics. *Physics Today* 37(7), 24-32.
Minstrell, J. (1982) Explaining the 'At rest' condition of an object. *The Physics Teacher* 20, 10-14.
Nussbaum, J. and Novick, S. (1982) Alternative frameworks, conceptual conflict and accommodation: Toward a principled teaching strategy. *Instructional Science* 11, 183-200.
Oakes, M. E. (1945) Explanations of natural phenomena by adults. *Science Education* 29, 137-42, 190-201.
_____ (1957) Explanations by college students. *Science Education* 41, 425-8.
Osborne, R. J. (1983) Towards modifying children's ideas about electric current. *Research in Science and Technological Education* 1, 73-82.
_____ and Cosgrove, M. M. (1983) Children's conceptions of the changes of state of water. *Journal of Research in Science Teaching* 20, 825-38.
_____ and Freyberg, P. (Eds.) (1985) *Learning in science: The implications of children's science*, Auckland, Heinemann.
_____ and Wittrock, M. C. (1983) Learning science: A generative process. *Science Education* 67, 489-508.
Radman, M. (1985) Unpublished B.Ed. assignment, Monash University.
Rowell, J. A. and Dawson, C. J. (1983) Laboratory counterexamples and the growth of understanding in science. *European Journal of Science Education* 5, 203-15.
Shipstone, D. M. (1984) A study of children's understanding of electricity in simple D.C. circuits. *European Journal of Science Education* 6, 185-98.
Steinberg, M. S. (1983) Reinventing electricity, in H. Helm and J. D. Novak (Eds.) *Proceedings of the international seminar: Misconceptions in science and mathematics*, Ithaca, NY, Dept. of Education, Cornell University, pp. 406-19.
Trowbridge, D. E. and McDermott, L. C. (1980) Investigation of student understanding of the concept of velocity in one dimension. *American Journal of Physics* 48, 1020-8.
_____ (1981) Investigation of student understanding of acceler- ation in one dimension. *American Journal of Physics* 49, 242-53.
Young, M. F. D. (1977) School science - Innovations or alienation?, in P. Woods and M. Hammersley (Eds.) *School experience: Explorations in the sociology of education*, London, Croom Helm, pp. 250-62.

# 4.2

# Metacognition, Purposeful Enquiry and Conceptual Change

**John R. Baird**

## INTRODUCTION

Effective independent learning has long been held as a major goal of education. Researchers, educationists, teachers, and politicians have advocated student autonomy in various ways, all of which reiterate Dewey's (1916) cogent comment:

> Since in reality there is nothing to which growth is relative save more growth, there is nothing to which education is subordinate save more education ... the purpose of school education is to insure the continuance of education by organizing the powers that insure growth. The inclination to learn from life itself and to make the conditions of life such that all will learn in the process of living is the finest product of schooling. (Dewey, 1916, p. 51)

Independence in learning requires that the learner accepts responsibility for, and can assume control of, the learning process. Effective learning proceeds by purposeful enquiry, whereby the learner monitors and evaluates the nature and progress of learning, and makes productive decisions based on these evaluations. Purposeful enquiry does not arise spontaneously - it must be learned. The school science laboratory can provide an important context for this learning. But does it? Is the laboratory more than a physical location for teaching content and content-related skills? Is it also a mental laboratory, a 'learning gymnasium', where students develop their capacity to learn? If students engaged on laboratory work were asked 'What are you doing?', 'Why are you doing it?' and 'What has what you are doing got to do with science or your everyday life?', what responses might you expect? Answers given to these questions have indicated that most students have only a limited idea of what they are doing, and few can explain why (Baird, 1984). At both secondary (Sirotnik, 1983; Tisher, 1984) and tertiary (Baird and White, 1982b; Moreira, 1980) levels, many students have been found to be uninformed, non-purposeful, dependent and dissatisfied. These students are not in a position to take control of their learning.

Given this situation, what is limiting learning development, and what can be done about it? In this chapter, some factors which limit purposeful enquiry are discussed, and an argument is advanced for a method of pushing back these limits. The argument centres on the assertion that the laboratory should train for enhanced student metacognition.

Metacognition refers to the knowledge, awareness and control of one's own learning. There is a burgeoning literature on metacognition and its components (e.g. Brown, 1980; Flavell, 1981). This is not the place for detailed theory about metacognition which, anyway, is largely presumptive at this time. However, a brief mention of each component will be given.

*Metacognitive knowledge* is stored in memory. It includes a person's knowledge about learning (its nature and process), effective learning strategies (what they are, when to use them) and personal learning characteristics (strengths and weaknesses, processing habits). It is reasonable to suppose that the nature or extent of a person's metacognitive knowledge can influence the extent to which he or she can take responsibility and control over learning.

*Metacognitive awareness* and *control* are learning outcomes associated with certain actions taken consciously by the learner during a specific learning episode. For example, a learner may be obtaining, graphing and interpreting values for an acid-base titration. Numerous cognitive processes are involved with such a procedure. These would include recognition of which is the acid and which is the base, retrieval from memory of features related to the action of acids and bases, and application of intellectual skills related to plotting of curves and interpretation of their shape. The level of *metacognitive awareness* derives from the learner consciously asking various evaluative questions regarding these cognitive processes, and determining answers to these questions, once asked. Such questions may include the following: 'What is the significance of the change in the curve at this point?'; 'How many more values should I take?'; 'Do I have enough information in order to be confident of the answer?'; 'Are the values I am getting reasonable ones?' Associated with awareness, *metacognitive control* involves the conscious decisions made by the learner regarding the approach to, progress through, and completion of, the task.

Interpreted this way, each of the three components of metacognition is inextricably linked with purposeful enquiry and effective independent learning. To return to the sorts of evaluative question asked at the start of this chapter, desirable performance outcomes would be expected for a student who thought to use, used appropriately, and obtained answers to such questions as 'What am I doing?', 'Why am I doing it?', and 'What has what I am doing got to do with science or everyday life?', and made decisions accordingly.

Figure 4.2.1 is a conception of the variables involved in learning. The figure is presented for three reasons. First, it serves to highlight the importance of learner-based variables on learning outcomes. By involving variables which are features of, or which bear on, the individual learner, it elaborates, from one perspective, the 'student inputs —> actual learning activities and actual learning outcomes' aspect of the general model (Figure 1.2.1). Second, it will form a basis for interpretation of the learning episodes below. Third, it reiterates the role of metacognition in learning. According to Figure 4.2.1, learner-based variables, directly or by interaction with instruction-based variables, affect the learning processes which occur. Learning processes occur at different levels of consciousness and control. Some processes are automatic and subconscious. These processes effect manipulation of information during processing and exchange of information between the site of processing and memory. Processing is then evaluated, consciously or subsconsciously, by the learner. Based on this evaluation, the learner makes conscious or subconscious decisions regarding further processing. Learning outcomes of learning processes include metacognitive awareness and control, which derive from conscious evaluation and decision-making respectively. Outcomes then determine *performance*. The standards of learning processes and their associated learning outcomes characterize purposeful enquiry which, in turn, is a necessary prerequisite for effective independent learning. In the next section, some variables are considered which, by constraining learning processes that lead to adequate metacognition, limit performance.

## LIMITING FACTORS

Five episodes involving laboratory work in science are presented below. These episodes occurred in a research study of two classes of Year 9 students at a Melbourne, Australia metropolitan high school (Baird, 1984, 1986). They illustrate interactions among the elements of Figure 4.2.1. The first three episodes centre on the uninformed nature of learning during directed work. Inadequate perceptions of class work are demonstrated (episode 1); some reasons for these inadequate perceptions are advanced (episodes 2 and 3). Episodes 4 and 5 involve students' inability to pursue independent enquiry because of conceptual and skill inadequacies.

*Episode 1.* In two successive lessons, students examined the invisible changes that take place when a match burns (Australian Science Education Project (ASEP) unit 'How Many People', p. 6). In the first lesson, students followed through a series of experiments. In the second lesson, the teacher led the class in a discussion about the effects of burning matches on the general living and non-living environment. As part of the research study, the teacher and all students wrote evaluations of each lesson. These evaluations included their perception of the topic of the lesson and the major task expected of the students, expressed as what they did and why they did it. Some typical responses for these two lessons are given in Figure 4.2.1.

While many students' perceptions of lesson *topic* were similar to that of the teacher, this was not the case for lesson *task*. The students' responses for task were limited to what they had to do, and the teacher's intentions as to why they were doing it seemed not to be appreciated by them. Subsequent assessment revealed that students' understanding of the topic was poor.

*Episode 2.* In this unit (ASEP 'Skin and Clothes'), students were to examine microscopically a vertical section of human skin, and compare this with a three dimensional 'block'

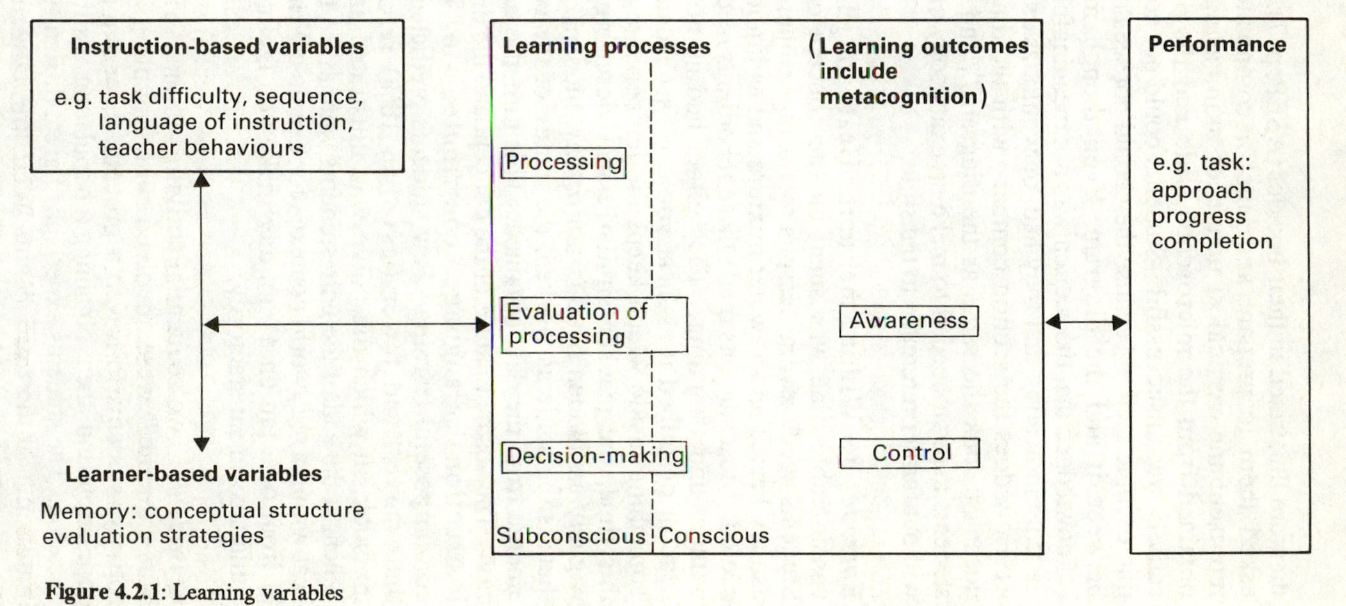


Figure 4.2.1: Learning variables

diagram illustrated in their booklet (ASEP, p. 7). The teachers asked them to prepare and label a composite diagram to represent the vertical structure of skin, using information obtained from the microscope slide and the diagram. The activity was unsuccessful. Students could not comprehend the microscopic section because they could not see any relationship between it and the diagram. Even though many students acknowledged that the section was not meaningful to them, they gave no indication that they had, or would, ask such questions as 'How does the section compare with the diagram?', 'Why doesn't it look the same as the diagram?', and so on. Most students then proceeded to make an exact copy of the diagram, without further reference to the slide.

*Episode 3*. Within the unit 'Heat and Temperature', considerable time was spent on the concept of latent heat. Students spent several lessons recording temperature changes as they heated an ice-water mixture and as liquid naphthalene cooled. Most students made these recordings mechanically and without understanding. They either had no notion of latent heat, or described it as 'heat you can't see', 'delayed heat', or 'something inside which keeps the temperature constant'. At this time, the teacher similarly was describing latent heat vaguely as 'hidden heat' (from *latens*, hidden). He believed that, as the students worked, the nature of the concept would emerge from the data obtained. This turned out not to be the case. The abstract nature of the concept and the difficulty of the instructional language contributed to uninformed, non-purposeful learning. Few students predicted the shape of their curves, and fewer (less than 10%) referred to their textbook, in which the curve was illustrated and explained. When, at the end of the unit, students were asked to apply latent heat within a real-world context - for example 'Why does sitting in front of a fan on a hot day make you feel cooler?' - few could answer satisfactorily.

*Episode 4*. Two sequential activities in the ASEP unit 'How Many People?' were involved with sampling. In the first activity, students followed a procedure for sampling a given population and then obtaining a population estimate by multiplication. The students experienced great difficulty with simple 0ratios, e.g. 'if there are 39 ants in 80 $cm^2$, how many are there in 120 $cm^2$?' Because of this, most students could not incorporate the notion of representativeness into their concept of

| Lesson | Teacher | Student 1 | Student 2 |
|---|---|---|---|
| (1) | | | |
| Topic | Burning matches. | What is given off by a burning match ? | A burning match. |
| Task | Develop in pupils' minds the consequences of a simple action | To conduct experiments to find if moisture given out by a burning match | Copy off the board |
| (2) | | | |
| Topic | Burning matches. | How many people? | Waste products |
| Task | Develop ideas on interaction of activities. | To do a concept map on does a burning match use materials ? | To complete the flow—chart type thing |

**Figure 4.2.2**: Typical topic and task responses for episode 1

the term. Consequently, when students had to estimate the size of a population of their choice by sampling, their selection of sample size often did not reflect the contrasting considerations of convenience and accuracy. Students were aware of their lack of proficiency with ratio transformations and this contributed to a reticence in attempting independent investigation.

*Episode 5*. Following a topic on 'Measurement and Errors' the teacher set one class the pendulum task - a test of Piagetian formal operational thinking, for which students were to 'discover and verify factors that may be involved in affecting the period of a pendulum' (teacher's notes). The teacher's intentions for this task were two-fold: to 'develop understanding of the reasons for errors in measurements, and how easy it is to make these errors' (teacher's notes); to have students design their own experiments and engage in independent investigation.

Students spent seven lessons on the task. Throughout this time, they demonstrated various metacognitive deficiencies which led to inadequate performance. As for episodes 3 and 4, their comprehension of the task was limited by inadequate understanding of key concepts. In this case, the concepts were 'factor' and 'period'. The term factor was unclear to many students, with the result that they considered many inappropriate variables, such as 'wind' and 'knocking the table'. One reason why they considered such transitory variables was confusion regarding the purpose of the activity, e.g. '[to find] the different kinds of errors that can be involved that will affect the period of a pendulum' (a student's notes). A second reason concerned the nature and meaning of period. Students were unable to distinguish the term from other experimental variables such as oscillation, the number of oscillations, the time over which measurements were taken, or rate of swing. In turn, this inability was due either to their lack of proficiency with ratios and proportional reasoning (discussed above), or confusion regarding the intuitively unusual feature of period - that the slower the speed of oscillation, the greater the period. As a consequence, many students ignored the term and reduced the task to 'see which factors affect the pendulum', or 'what affects the movement of the pendulum'. Some students spent up to five lessons invest- igating inappropriate variables.

The students' approach to, and interpretation of, the activity were also influenced by their real-world preconceptions and expectations. Intuitive knowledge interfered with their ability to control variables and to attribute small differences in results to errors:

> Student O: All four times, we got 15 [oscillations] exactly.
> Researcher R to Student C: What does that mean?
> Student C: That it doesn't matter what the weight is.
> R: Would you have expected that to happen?
> Student C: No.
> Student O: We are going to do the same again now, only using shorter lengths of string.
> R: Why do you need to use the four weights again?
> Student O: Because it might make a difference with a shorter string.
> R: Haven't you just shown that weight doesn't matter?
> Student C: If it's shorter, it will go quicker ... if it was really heavy, it would be pulling down on the string ... the heavy would go quicker.

These results support the discussion in Chapter 4.1 of this volume regarding the important influence on learning of the learner's existing conceptions. Much of the work was done in a non-purposeful, non-confident manner. The rate of progress was very slow - it was only after approximately six lessons that most groups had demonstrated successfully that length, but not mass, affected the period. Even at the end of the activity, many students still did not understand such crucial aspects as the meaning of 'period'. However, students valued and enjoyed the activity:

> I thought that it was a very good exercise in seeing how self reliant we are, whether we can cope with a set task or whether we have to be 'led by the nose'. I quite enjoyed the task, doing experiments that weren't pre-set for us, using my own initiative.

Poor performance on the pendulum activity may be interpreted in two ways. It may be that inability to complete the activity reflects general intellectual incapacity - such tasks as defining and controlling variables are intellectually inappropriate for the students' stage of cognitive development. An alternative explanation focuses on the metacognitive requirements for success. The nature of the activity is such that adequate performance is dependent on the prior achievement of a level of metacognition sufficient for the student to have resolved basic elements of task nature and purpose. However, as indicated above, many students had not passed the first hurdle - they were unclear about what they were doing and why they were doing it. In some ways this result is not surprising. Even if their particular intellectual skill deficiencies are set aside, it was obvious that their previous laboratory experiences had not prepared them for the novel metacognitive demands placed on them.

The five episodes above demonstrate how various learner and instruction based variables (see Figure 4.2.1) compromise learning outcomes by limiting purposeful enquiry. Students invoked a *processing style* labelled surface-level processing (Marton and Säljö, 1976) or content-prescribed analysis (Baird and White, 1982a). This style is characterized by a step-by-step approach, with little attempt to integrate aspects of the task or relate it to personal experience. Accordingly, application of *strategies* for *evaluation of processing*, and *decision-making*, are largely curtailed. Application of this style was influenced by various factors. In episodes 1-3, *teacher behaviours* forced students into a subordinate, receptive role. The teacher spoke most of the words, asked most of the questions, and did most of the relating and integrating. When students applied this style in episodes 4 and 5, where adequate metacognition was necessary for task completion, its limitations were highlighted. The students' *conceptual structure* sometimes militated against purposeful enquiry. Either existing conceptual structure was inadequate or conflicting (episode 5), or various factors combined to preclude generation of a meaningful concept (episodes 3, 4 and 5) or its integration with existing concepts (episode 2). Two factors which prevented generation of meaning were inadequacies in *intellectual skills* (episodes 4 and 5) and in the *language of instruction* (episodes 2-5). Additionally, as mentioned in episode 5, it may have been that the *task difficulty* involved certain cognitive abilities which were not possessed by some students. However, the degree of importance of this factor remains problematic.

Evidence from the above study and others (e.g. Baird, 1986; Baird and Mitchell, 1986; Baird and White, 1982a,b; Osborne and Wittrock, 1983; Sirotnik, 1983) indicates that the bases for these episodes generalize across year level, content area and school. In the next section each of the above factors will be set within the framework of three approaches for improving learning performance.

## OVERCOMING LIMITING FACTORS

If it is accepted that informed, purposeful enquiry is an important feature of the science laboratory, but that currently the standard of such enquiry is limited by inadequacies in students' perceptions, conceptions and skills, what can be done to overcome these limits? Three approaches are outlined below. The objective of the first approach is to improve the comprehensibility of instruction by taking account of students' existing conceptions, the objective of the second approach is to train in required intellectual skills, and the objective of the third approach is to train for enhanced metacognition.

*Approach 1: Improve the comprehensibility of instruction.* Episodes 3-5 demonstrated the difficulty students experienced in generating understanding key terms. Improving the potential comprehensibility of the information appears to have significant effects on performance. For the pendulum activity, if the task were changed to make it more relatable to the students' prior knowledge and experience, their perceptions of task nature and purpose, and their performance, improved significantly. Two examples of such a change would be to have students consider a clock and attempt to alter its 'tick-tock' rate, or consider a swing and attempt to alter the rate at which it rocks back and forth. Neither of these tasks requires the use of such terms as period, oscillation, or factor. Nor does the change necessarily affect the inherent nature of the task (Piaget, 1972). Other research has also shown that familiar materials and context improve performance by pre-school children (Brainerd, 1977), older children (Kuhn and Brannock, 1977), and adults (Sinnot, 1975). Chapter 4.1 argues similarly for the inclusion of everyday contexts and experiences within the curriculum of the science laboratory. The epistemology of *contextualism* (Zimmerman, 1983) takes, as a central tenet, the dependence of cognitive functioning on context. It has been argued that teacher language affects students' understanding and approach to learning. Apart from increasing the difficulty of task comprehension, the use of specialist, abstract language reinforces the students' conceptions of learning as something imposed from outside, rather than something generated from within. Consequently, it reduces students' confidence in and responsibility for the process of learning, and preserves teacher direction and control.

*Approach 2: Train in deficient intellectual skills.* Intellectual skill deficiencies severely limit a student's ability to achieve adequate metacognition (episodes 4 and 5). It cannot be assumed that deficiencies in basic skills are uncommon - there is widespread evidence that many students leave school without ever acquiring them (e.g. Capon and Kuhn, 1979). These findings have important implications for science curricula. Teachers must recognize that successful completion of significant portions of the curriculum may require explicit preliminary training in prerequisite skills.

*Approach 3: Train for enhanced metacognition.* Consideration of this approach requires that purposeful enquiry be distinguished clearly from what might be called 'directed' enquiry - a type of enquiry observed commonly in school laboratories. In directed enquiry, students follow relatively prescribed instructions: learners are subordinate and, while exhorted to 'be enquiring', they often act as question-answerers rather than question-askers. Little training is given in asking questions which direct and monitor the nature of the enquiry. Enquiry in science laboratories has often been linked with the 'scientific method', based on formulating and testing hypotheses and interpreting results. This training is little different in nature from other task-related 'learning skills' training, such as the SQ3R method (Robinson, 1946) and its many variants. Both are general in scope, but vague in instructions, and do not adapt to the requirements of individual students and specific learning situations. The relatively inflexible rules-for-action do not train for student-directed monitoring and control.

The study from which the above episodes were taken (Baird, 1984, 1986) was directed to enhancing all of the three components of metacognition (knowledge, awareness, control). As part of this study, another Year 9 class attempted the pendulum task. However, this class worked through the task in a manner designed to facilitate the achievement of adequate metacognition. This manner involved some structuring of approach, progress and completion. At three stages of the activity, groups of students considered, and wrote answers to various evaluative questions they had been practising asking during lessons (e.g. Baird, 1986, pp. 270-71). This procedure assisted students in overcoming conceptual confusion, and allowed them to approach their experiments more perceptively. As an example, three groups came independently to an insight on the problem that teachers hardly ever discuss, or perhaps even recognize. That is, that the timing of multiple swings to find the average for one *assumes* that amplitude does not affect period. The experimental design of these students called for the timing of individual swings. This result indicates that many Year 9 students are capable of sophisticated abstract reasoning *once* they have achieved a satisfactory level of metacognition. Other evidence from the study also indicated that students were thinking about their learning:

> *Do I fully understand this?'* No. I do not fully understand why I got the results I did in the last experiment. Have I read the stopwatch wrongly??? *Do I understand enough to justify stopping*? I think so. I have all results except that last one. *What do I have to do to achieve full understanding*? Redo the last experiment, and if I get the same result, ask! *Is it worth doing*? Yes. I want to see whether I made an error, and if not, why did I get that result? (A student's written response to some questions)

Even though the students ended up taking less time on the activity than the first class, their progress and performance were assessed more highly by the teacher.

Overall, the research study led to enhanced student attitudes to, and control of, learning. It established a framework for future attempts to train for improved learning.

## TRAINING FOR IMPROVED LEARNING

There is no simple answer to the question of how to train for improved classroom learning. However, two related points regarding teaching and learning should be made clearly at the outset. The first point is that the often assumed causal dependence of learning outcomes on teaching behaviours must be rejected. Fenstermacher (1986) argues eloquently that teaching should be directed to 'improving the student's abilities and capacities to be a student ... the concept of *studenting*' (Fenstermacher, 1986, p. 39, emphasis added). The quality of learning which then occurs is the responsibility of the student. Thus, training must be to make studenting as effective as possible through practice in the effective discharge of this responsibility. The second point is that, in order that it be effective, training must make allowance for the constructivist nature of the learning undertaken by the student. These two points serve to establish a perspective for developing a rationale for effective training. Further, while there may be no simple answer regarding training, it may be that by changing a relatively small number of underlying factors, we can achieve significant improvements. In a synthesis of 'thousands of research studies', Walberg (1984) posits the view that learning may be envisaged as involving relatively few, widely generalizable, factors. Bloom (1984) has extended Walberg's results in his search for ways to improve the quality of group based learning. By incorporating the three approaches above into a general programme, training bears on nine of Bloom's eleven most effective variables (e.g. 'feedback-corrective', 'student classroom participation', 'improved reading study skills', 'initial cognitive prerequisites'). Training will only succeed if, to borrow Hewson's (1981) terms, it is intelligible, plausible and fruitful for the participants. Thus, for both students and teacher, training must be understandable and meaningful, it must provide a plausible alternative to existing conceptual frameworks, attitudes and behaviours, and it must reap the benefits of improved learning outcomes.

The assertion here is that training should be based on enhancing each of the three components of metacognition. Improvement in metacognitive knowledge may involve a fundamental reconception of the nature of learning, and particularly, the learner's role in learning. Consider a student who believes that 'Right answers for everything exist in the Absolute, known to Authority whose role is to mediate (teach) them. Knowledge ... (is) to be collected by hard work and obedience' (Perry, 1970, p. 9). This student will be unable to accept responsibility for, and control of, learning until this conception changes. The same can be said of a teacher's ability to foster change if this is his or her view of teaching and learning. Similarly, enhancing students' metacognitive awareness and control requires that students and teacher expend effort. Learners must ask questions and have the knowledge, desire and opportunity to apply procedures for obtaining answers. They must monitor and evaluate the outcomes of enquiry, and make appropriate decisions. The teachers must foster a learning environment which is supportive of informed student contribution. This support may necessitate re-evaluation of teacher-student behaviours, curricula, instructional methods and materials, and assessment. Many of these considerations have been discussed more fully elsewhere

(Baird, 1986; Baird and Mitchell, 1986). Accordingly, only three aspects of training for enhanced metacognition will be mentioned further here.

The first aspect relates to the association among questioning techniques and cognitive and metacognitive outcomes. As described earlier, purposeful enquiry is centred on asking evaluative questions. Students can be trained to ask such questions, but training must extend to providing the students with techniques for gaining answers (Baird, 1984, 1986). The nature of these techniques may be as varied as the questions. For example, answering a question about the relative solubilities of two salts may be answered by experiment, about their chemical formulae by consulting a textbook, about 'How much do I already know about salts and solubility?' by constructing a concept map (see Chapter 2.2) or a Venn diagram, about 'Where am I up to, and what should I do next?' by checking the steps that have already been taken, or about 'What is the nature of this activity?' by constructing a Vee map (see Chapter 2.2). Answers to these questions will generate cognitive, metacognitive knowledge, and metacognitive awareness outcomes which, in turn, will influence metacognitive control through decisions the learner makes.

The second aspect refers to the important role of the science laboratory in metacognition training. If devised intelligently, laboratory work provides a potentially challenging range of contexts and situations which can allow the students freedom to pursue productive enquiry. In order that this enquiry be effective, the laboratory curriculum needs to allow students adequate levels of 'control, involvement and interaction' (White and Tisher, 1986, p. 881); it must therefore build on students' existing conceptions, intellectual skills and metacognitive competencies.

Finally, success of training depends on developing the participants' conception of the importance of *process* in learning. In addition to process as it relates to self-evaluation and decision-making during learning, and to an individual's intellectual development, process extends to what is taught. In science education, schools traditionally have taught the content of science without emphasizing the process of science - the events and methods which have led individuals to discover what is now content. By undervaluing process in science, schools have limited its relevance to students and have distorted its challenge to such an extent that active personal enquiry and interest have been replaced by detached rote memorization and application of procedures (Passmore, 1975). When conception develops, the challenge becomes clear:

> I am still resisting with all my might the 'meaninglessness' of the events of the world and trying to replace it by 'incomprehensibility'. But how difficult it is to carry through this point of view. (Max Planck quoted in McCormmach, 1982, p. 218)

Only when a person' s conception of learning is such that incomprehensibility replaces meaninglessness, and only when he or she has the desire and skills necessary to generate comprehensibility for the events of the world, can formal education truly be said to have succeeded. The science curriculum and the school laboratory can contribute to this success by training for enhanced metacognition.

## REFERENCES

Baird, J. R. (1984) Improving learning through enhanced metacognition. Unpublished Ph.D. thesis, Monash University.

_____ (1986) Improving learning through enhanced metacognition: a classroom study. *European Journal of Science Education*, 8(3), 263-82.

_____ and Mitchell, I. J. (Eds.) (1986) *Improving the quality of teaching and learning: An Australian case study - the PEEL project*. Melbourne, Monash University Printery.

_____ and White, R. T. (1982a) A case study of learning styles in biology. *European Journal of Science Education*, 4, 325-37.

_____ and White, R.T. (1982b) Promoting self-control of learning. *Instructional Science*, 11, 227-47.

Barnes, D. (1969) Language in the Secondary classroom. In D. Barnes, J. Britton, H. Rosen and the LATE, *Language, the Learner, and the School*, Harmondsworth, Penguin.

Bloom, B. S. (1984) The search for methods of group instruction as effective as one-to-one tutoring. *Educational Leadership*, 41(8), 4-17.

Brainerd, C. J. (1977) Cognitive development and concept learning: An interpretive review. *Psychological Bulletin*, 84, 919-39.

Brown, A. L. (1980) Metacognitive development and reading. In R. J. Spiro, B. C. Bruce, and W. F. Brewer (Eds.), *Theoretical issues in reading comprehension. Perspectives from cognitive psychology, linguistics, artificial intelligence, and education*. Hillsdale, NJ,

Erlbaum.
Capon, N. and Kuhn, D. (1979) Logical reasoning in the supermarket: Adult females' use of a proportional reasoning strategy in an everyday context. *Development Psychology*, 15, 450-52.
Dewey, J. (1916) *Democracy and education. An introduction to the philosophy of education.* London, Macmillan (Free Press Edition, New York, 1966).
Fenstermacher, G. D. (1986) Philosophy of research on teaching: Three aspects. In M. C. Wittrock (Ed.), *Handbook of research on teaching*, third edition. New York, Macmillan.
Flavell, J. H. (1981) Cognitive monitoring. In P. Dickson (Ed.), *Children's oral communication skills.* New York, Academic Press.
Hewson, P. W. (1981) Aristotle: Alive and well in the classroom? *Australian Science Teachers' Journal*, 27(3), 9-13.
Kuhn, D. and Brannock, J. (1977) Development of the isolation of variables scheme in experimental and 'Natural Experiment' contexts. *Developmental Psychology*, 13(1), 9-14.
McCormmach, R. (1982) *Night thoughts of a classical physicist.* New York, Avon Books.
Marton, F. and Säljö, R. (1976) On qualitative differences in learning: 1 Outcome and process. *British Journal of Educational Psychology*, 46, 4-11.
Moreira, M. A. (1980) A non-traditional approach to the evaluation of laboratory instruction in general physics courses. *European Journal of Science Education*, 2, 441-48.
Osborne, R. J. and Wittrock, M. C. (1983) Learning science: A generative process. *Science Education*, 67(4), 489-508.
Passmore, J. A. (1975) The revolt against science. In P. L. Gardner (Ed.), *The structure of science education.* Victoria, Longman.
Perry, W. G. (1970) *Forms of intellectual and ethical development in the college years: A scheme.* New York, Holt, Rinehart and Winston.
Piaget, J. (1972) Intellectual evolution from adolescence. *Human Development*, 15, 1-12.
Robinson, F. P. (1946) *Effective study.* New York, Harper and Brothers.
Sinnot, J. D. (1975) Everyday thinking and Piagetian operativity in adults. *Human Development*, 18, 430-43.
Sirotnik, K. A. (1983) What you see is what you get - Consistency, persistency, and mediocrity in classrooms. *Harvard Educational Review*, 53(1), 16-31.
Tisher, R. P. (1984) *What do high and low achievers really think of their learning environments*? Paper presented at the annual meeting of the Australian Science Education Research Association, Melbourne.
Walberg, H. J. (1984) Improving the productivity of America's schools. *Educational Leadership*, 41(8), 19-27.

White, R. T. and Tisher, R.P. (1986) Research on natural sciences. In M. C. Wittrock (Ed.), *Handbook of Research on Teaching.* Third edition. New York, Macmillan.

Zimmerman, B. J. (1983) Social learning theory. A contextualist account of cognitive functioning. In C. J. Brainerd (Ed.), *Progress in cognitive development research. Vol. 3. Recent advances in cognitive-developmental theory.* New York, Springer-Verlag

# 4.3

# Teachers' Conceptions of their Subject and Laboratory Work in Science

John K. Olson

## INTRODUCTION

The good that comes from laboratory activity has always been a feature of the justification of science. Laboratory work is stimulating and rewarding for pupils, so the argument goes. It helps them enjoy science while doing it. Notions about what to study in the laboratory have changed, of course. Where once activities based on science in the home were considered appropriate, now tests for air or water pollutants are done and some consider that discussing related science issues ought to form part of the laboratory work.

The laboratory was seen as a place where students might gain first-hand experience of scientific phenomena and have something to reflect upon. Actual manipulation of equipment typically has been the defining characteristic of the laboratory experience, but ideas about laboratory experience have evolved both in science and in education. 'Think tanks', for example, resemble actual laboratories in the intense focus on problems taking place in specialized settings, reflecting Einstein's idea of thought 'experiments' which emphasize the importance of mental 'work' in science. Science before and after Einstein has grappled with the relationship between ideas and experience in understanding nature.

In science education we can find textbooks which call for more than manipulation of equipment in laboratory exercises. Where does the laboratory end and something else begin?

Teachers tend to look upon discovery-based laboratory work and post-laboratory discussion in rather similar and perplexed ways. The relatively ordered process of talking about the subject matter is set aside to talk about actual experience. Here teachers have problems in coping with diverse observations, with alternative interpretations and with unexpected ideas. Having to treat student ideas seriously and hold their own knowledge more or less tentatively are part of laboratory episodes in which experience and opinions are discussed. In this chapter a consideration will be presented of how teachers cope with science classes in which student experience and opinion are discussed. This will lead to a consideration of the difficulties teachers have in coping with laboratory work and its aftermath, and how their conception of science affects how they cope. The teacher's point of view will be considered in relation to its significance for the curriculum model. The method used has much in common with that approach to understanding human activity represented, for example, by Clifford Geertz (1973). It is an ethnographic approach based on the interpretation of cases.

As we shall see later in the chapter, teachers have important investments in the way they think about laboratory work. Efforts to reform science through changed notions of laboratory work in science will have to reckon with how teachers construe such work, and why they construe it that way. For bound up in teachers' views about their work are resolutions of the dilemmas that everyday teaching poses, and these hard-won and delicately poised resolutions are not soon abandoned in favour of new and untried visions of the science classroom and the laboratory work that takes place there.

An exploration of the meanings teachers attach to laboratory work in science will be illustrated by the cases of eight teachers who worked with the Schools Council Integrated Science Project (SCISP) (Olson, 1982), and a case taken from a major study of science education in Canada (Olson and Russell, 1983).

What sense do these teachers make of laboratory work within the context of their overall responsibilities, and especially with those to do with preparing students for examinations? The influence of examinations on what is taught in science and how it is taught is, of course, well known. Usually that influence is described in *instrumental* terms - that is, in terms of what concepts are learned and what methods of teaching are effective. However, teachers face diffuse and undelimited tasks in their work and it is in their interest to establish clear-cut routines and to conform to certain acceptable and expected notions of what content is to be taught in what order and with what balance between laboratory work and discussion activity. Teachers have developed a certain know-how, a stock-in-trade which allows them to 'get through' the course and prepare students for examinations. Indeed gaining access to students who 'do' examinations is an important job perquisite, for it is with such students that teachers can use their subject and their methods to good effect; allowing them to create a satisfying impression in their students. Acts through which this satisfying impression is created are called *expressive* acts. This notion can be used to understand how the teachers in our cases construed their experience with SCISP. We will be particularly interested to understand why they had trouble with the SCISP notion of laboratory work.

Central to understanding the sense teachers made out of SCISP doctrine is the idea that subjects are for something practical, getting on in the world and getting on in school, and that teachers have considerable 'face' invested in making that system work.

In the last decade studies have revealed an interesting resistance by teachers to the use of innovative ideas in science education and in other subjects. Studies of the role of the teacher in curriculum development document the extent of this unwillingness to reorient, but say less about why this happens, and what should be done about it. They give glimpses of the underlying concerns of teachers which appear to have influenced their use of the new ideas.

Herron (1971), for example, who interviewed users of the Biological Sciences Curriculum Study materials, found that, in general, teachers did not have the same degree of understanding of the nature of science as the designers of the project did, in spite of the fact that many of these teachers had been to institutes at which the philosophy of the project was discussed. Such institutes had apparently failed to do the job. Herron concluded that: 'Teacher perception of new course material ... is a problem that lies at the root of resistance to curricular change'.

Perhaps most revealing was Carlson's (1965) finding that teachers, using programmed instruction involving independent rates of progress, acted to keep students moving more or less at the same rate. Teachers gathered the class together periodically to teach them, even though this was not necessary, since they did not see the use of the programmed instruction as 'teaching'. Carlson concluded that, for some reason, teachers were not prepared to abandon the ways in which they normally organized their teaching.

What do these studies tell us? First, they illustrate the notion that proposed reorientations (major changes) often end up being variations (small scale changes within existing practice), or no change at all. Proposed discovery approaches become didactic and individual progress is retarded. Curriculum innovators appeared to have underestimated the conserving tendencies of teachers and schools.

Are teachers who do not innovate to be considered unimaginative and conservative? Is their unwillingness to change to be thought of as evidence of lack of professionalism? These studies and the ones discussed in this chapter suggest that we need to look again at our assumptions about the work that teachers do, and how new ways of working can come to be. These studies hint at complex, underlying systems of coping that teachers hold on to for reasons we do not understand.

What the teachers do with an innovation can be interpreted as an accommodation between the various demands of the innovative doctrine in the three main domains of the teacher's work and the preferred solutions teachers have already adopted to cope with the demands of their work. Innovative doctrines require, by definition, reformulations of relationships between social systems, technology and goals. They are pressed upon teachers as visions of what might be, and they implicitly, if not explicitly, criticize the existing practices they seek to change. Innovative doctrines propose resolutions amongst the domains of teachers' work different from those that exist. The demands of innovation recreate consciousness of, and highlight already existing, but submerged, perceptions of the satisfactory and unsatisfactory consequences of existing resolutions of dilemmas. Brown and McIntyre (1982), for example, found that teachers were concerned about the effects of guided-discovery methods on their relationship to their classes: particularly class management. They document some of the cost and rewards associated with the Scottish Integrated Science Project.

Smith and Sendelbach (1982) also found that teachers using new materials faced perplexing dilemmas. They observed that teachers departed from guidebook prescriptions for enquiry teaching in science. Through talking to teachers about their planning they were able to appreciate the way teachers dealt with dilemmas associated with discovery vs control in the elementary science classroom.

New ways of conducting laboratory work recreate dilemmas for teachers by forcing them back to choices once made and now forgotten. Many factors can be imagined that influence the degree of awareness of dilemma, but the general point stands. Innovations have the potential, because of what they are and how they are perceived, to evoke at least some reflection - to provide a starting point for a reconsideration of practice through discussion with outsiders, and a basis for systematic reconstruction of the rules that guide practice. Study of innovation thus provides an arena for the understanding of school life itself.

I will return to this point later in a discussion of the curriculum model but now move on to consider the particular demands the SCISP laboratory scheme made on teachers and how they responded.

## LABORATORY WORK AND THINKING PROCESSES

### The SCISP doctrine

Why did teachers have difficulty with the laboratory work in SCISP? We can see from the doctrine of the project that problems stemmed from the integrated design, from the abstract quality of the content defined as problems and patterns, and from the stress on discussion of social issues as part of laboratory work. The comments of teachers help us understand what they were confronted with.

The teachers found that SCISP redefined the way they thought about their subject. The SCISP stress was on thinking processes developed through frequent laboratory experiences, many of which departed radically from what was familiar - laboratory work was to stress thinking processes.

Ann Williams at Mercia School (pseudonyms) was also concerned about the integrated scheme: 'Downplaying one's subject offends one's sense of importance. Quite often you are doing a bit you find quite difficult, [and you say] what on earth did I train four years for?' Richard Simpson, Head of Science at Mercia, put it directly: 'In certain situations where you are asked a question outside your specialist subject, it may make you look a bit of a clown if you cannot give a ready answer.'

Concern about the stress in laboratory work on thinking processes and students' discussion, and de-emphasis of subject matter was common. Bryan Jenkins said: 'As the course developed I found it very difficult to get used to the non-factual aspects of the course and being a conceptual course rather than a factual one, it might have taken me two years to overcome that.' Ann Williams found that the stress on vague processes and the open-ended questions of the laboratory activities left her uncertain as to the progress she was making with the class. The lack of evidence of progress was a recurrent theme in teachers' comments. As she said: 'I found it dreadfully difficult in the beginning; the different points of view, the different angles. [It was] not the way I would have done it ... I'm not sure I'm putting it over properly. The course does leave a feeling of insecurity.' Andrew Scott said that he found the guides 'baffling' and that he 'couldn't take it in'. He said he had a mental block about the organization of the course and found the various aims and cross-referencing confusing.

When asked to discuss how SCISP was organized and what pattern-finding and problem-solving - two key terms within the project - were, none of the teachers (with one exception) could go beyond a few everyday notions. For example, when asked to discuss pattern-finding the following comments were offered: 'You are looking for common ground'; 'A generalization'; 'It's the way we learn'; 'Doing several experiments and finding the results'. The teachers agreed that pattern-finding was more like a way we learn, rather than an aspect of the methods of the scientist. Problem-solving was simply that: having a problem and finding a solution. The teachers did not make distinctions between numerical problems, scientific problems and social problems; problems to be dealt with in the laboratory and ones to be discussed in class.

Teachers spoke about not wanting to look like a fool and about wanting to be sure that they were putting the material over 'properly'. One senses from their comments a concern not to be seen as inadequately preparing their students for the examination; to be seen as reliable and efficient in using classroom and laboratory work to help students succeed in their examinations.

These comments should not be treated lightly. In a system in which external examinations are important and good results crucial for employment and further education, teachers take very seriously the task of preparing students for passing these hurdles; they 'coach' them, and they take pride in being a good coach, and want to be seen so. Indeed these teachers share a common culture based on their concern to prepare their students for the difficult rite of O Levels. Such impressions help them carry out the instrumental tasks they see as central to their work at O Levels.

As we shall see these teachers felt that working with SCISP upset the balance between the instrumental and expressive functions of their work, because initially it was not at all clear how students could be coached to pass the SCISP O Level, and the lack of efficient tools for 'doing' O Level work left teachers concerned about the impression they were having on their students.

Preparing students to take external examinations is the backdrop against which we have to consider experience with the laboratory work in SCISP which involved teachers in discussions with their students - real discussion where the answer was to be discovered through conversation rather than imposed through teacher or text authority. These discussions were intended to foster certain attitude aims that SCISP doctrine held as important.

## Discussion in laboratory work

SCISP teachers were asked to engage their students in discussion based on laboratory work - that is, to do activities interposed in the text in the same way laboratory work often is and intended to perform the same learning function, but without actual manipulation of materials. Students read supple- mentary material or reprinted newspaper articles and the like and discussed it, or they had discussions based on actual laboratory activity. These activities were meant to develop the very habits that the teachers had rated as of low importance, especially for able pupils; that is those most likely to do well at O Level.

Two forms of discussion as a way of teaching can be discerned in the SCISP laboratory activities. One asked the pupils to seek generalizations, what the handbook called 'pattern finding'; what is the case is inferred through an assessment of the data. In the second form of discussion, what ought to be the case is considered; controversial issues emerging from the laboratory work are discussed rather than experimental data themselves. In both forms of discussion the teacher and pupils are guided by standards of inference and argument; no one has the final word by right of position; the influence of the teacher is indirect (Wilson, 1972). The teachers said that their students were concerned about the way laboratory work was being conducted; that is, by the many discussions that were placed in the text where one would normally find 'hands-on' activity in the laboratory. Students did not accumulate books filled with laboratory reports which they could use to revise for examinations. Our discussions with students confirmed these impressions of the teachers. Interviews with Mercia students reflected the views of the other students. They noticed that SCISP was different:

> I used a regular textbook [in addition to SCISP]. What you needed to know was in your head, not in notebooks.
> [The exam] was like an English text.
> [We filled] less than half the number of exercise books in SCISP.
> One didn't have to revise at all.

**Table 4.3.1**: Student perceptions of activities in science class

| How often do you have the following activities? | Very Often | Quite Often | Occas-ionally | Seldom |
|---|---|---|---|---|
| Listening to the teacher summarize work/explain things | SCISP<br>non-SCISP | | | |
| Doing laboratory work | | SCISP | non-SCISP | |
| Discussing social issues | | SCISP | non-SCISP | |
| Having to write essays in which I put my own point of view | | | SCISP | non-SCISP |

To supplement the interview data, SCISP and non-SCISP fifth form students at Mercia and Westdale Schools were questioned about their perceptions of what went on in their science classes. Their responses indicated that significantly different perceptions existed. SCISP students saw more opportunity to express themselves and work on their own (Table 4.3.1), and that, of the influences they felt their course had had on them, the only significant one was having become more concerned about social issues.

These data and the teacher and student comments suggest that SCISP pupils had had more opportunity than usual to express themselves in class. It was this opportunity that created a dilemma for teachers, because they were perceived by some students not to be teaching science in the expected way; that is with the teacher doing the talking and giving extensive notes, or writing up the results of their laboratory activities which were frequent.

Instead, students were being asked to discuss their findings and interpret their data in order to discover patterns. Some students found these discussions difficult. At Midfield, this was especially so. James Edwards described how difficult it was for his students to deal with the *first* investigation.

> One of the first experiments in the third year was to find the patterns; the whole of SCISP was to find the pattern ... Add acids to carbonates; they fizz. Acid plus a carbonate, but then it stops [with marble] ... Now in mixed ability groups, or for anybody who is lower ability, they were finding it difficult to change their idea.

The 'carbonates fizz' example was used by a number of teachers to describe how the SCISP material was organized. It was the laboratory approach based on a combination of a find-out-for-yourself philosophy and numerous questions in the books that the teachers called the 'language' of SCISP. James Edwards said he preferred direct teaching to pattern finding, because it was more efficient; he did not see that discussion and instruction have different functions:

> This finding of patterns is all very well, but it doesn't follow into A Level. You're not searching for patterns when it comes to A Level. You use the patterns you already know.

The dilemma Robert Young experienced was that the SCISP laboratory activities left him with the feeling that the problems pupils had with SCISP reflected badly on him:

> The level at which they had to discuss and the amount of information required first of all to understand and then to approach from a different point of view was beyond their ability ... I think possibly one of the reasons why I felt I wasn't being successful teaching SCISP, was because I didn't have a sufficiently broad background of teaching experience ... I was in the situation where you are asking two questions; am I successful as a teacher and am I successful in teaching the material?

Peter Judge and Andrew Scott considered that their teaching had become more didactic to compensate for a lack of structure. Their comments suggest that a dilemma exists for teachers here; unless the teachers placed their perspective forward quite clearly, the pupils might not get the point of the lesson; yet to do so tends to destroy the 'discovery' experience.

At risk were not only O Level results, but the impression among the students that their teachers knew what they are doing. From our conversations with these teachers we found that they maintained strict control over the point and direction of the discussions they conducted. Having to conduct these discussions as part of the laboratory work of SCISP simply undermined their influence over the lesson.

The teachers found it difficult to understand how they should behave and how to think about their students' behaviour in these lessons. They tended to think of discovery teaching approaches as if they were variants of more familiar teacher directed forms. The following comment indicates the nature of the dilemmas teachers faced in the SCISP laboratory activities:

> It's quite foreign to a lot of science teachers [being a neutral chairman]. They deal with a lot of facts and here we are asking for discussions which could be very open-ended ... It's very difficult to manage [a discussion] with some of them absent, or some have the facts and some don't ... Then you've got pupils at different levels of maturity to discuss something. Whereas some can and they might be mature enough to put forward certain views, but not in a mature manner, laughing about it, giving some stupid sort of view.

What do such comments tell us about how teachers consider their role in discovery-based laboratory work? It is evident that the teachers tend to contrast such teaching with that where they are in charge and able to act in familiar ways. In other words, the familiar role becomes a basis for describing and evaluating the discovery situations.

The teachers tended to appraise discovery situations by reference to more familiar roles, and they tended to see these situations as non-functional variants of teacher-dominated ones. Further, they assumed that all of their teaching served common goals; so, for example, laboratory work and especially those labs which involved much discussion were criticized for not being as efficient as more directed forms of teaching. The teachers assumed that the more 'open' methods were aimed at information transfer; that is, they assumed a conformity of goals for all methods.

How can we make sense of the teachers' tendency to collapse 'open' methods into 'closed' ones? Their influence, as they see it, is assured not only through instrumental acts, but through the way they impress their students: expressive acts. As Peter Judge put it, teachers strive for a 'Good show, well presented'. Influence is based not only on being able to provide the stimulus and expert guidance to help students obtain the credentials they expect to gain, but by authenticating what is transacted, and guaranteeing it. Influence, as the teachers see it, is based not only on understanding their subject but on their ability to convince the students that what is happening is well produced and directed.

Teachers have to meet the expectations of parents, principals, peers and students who expect them to be influential in quite different ways by helping students pursue instrumental goals. By acting as energizer, editor and guide, by identifying a syllabus of content, by seeking career system goals, the teacher is able to adopt a relatively clear-cut approach to meeting these expectations. On the other hand, discussing the results of laboratory work, especially work related to social issues, and with a view to developing thinking skills is a much less clear-cut way for a science teacher to work. The goals and rules of the game are not at all clear. Well might the teacher ask: 'What is the use of my subject matter expertise and my ability to interpret it?' The teachers in this study used their ideas of a proper balance of instrumental and expressive acts to reconstruct the project in their own terms.

Stevens (National Science Foundation, 1978) has noted:

'A disciplinary curriculum and authoritarian teaching are ... easiest for everybody'. The report to which she contributed concluded that teachers did not adopt enquiry methods because they were unwilling to risk laboratory situations in which they might not know the answer, and it suggested that teachers lacked experience in dealing with the questions of thoughtful students on doubtful topics. We can see the same problems existed in the cases we have been considering and why teachers found it difficult to change their ideas about laboratory work in science.

## SEARCHING FOR A SECURE BASIS OF PRACTICE

### Science content as a backdrop to practice

SCISP doctrine urged teachers to change their practice in radical and complex ways. The doctrine urged grand changes in the goals of science education but more importantly in the teaching methods to achieve those goals. We found that teachers, unable to construct workable methods in support of the SCISP goals and unable to construe just what the goals meant, reconstructed the project so that it might function as a system they could operate: the pursuit of O Level passes supported by clear-cut material to be learned. Teachers reconstructed SCISP as they did in order to restore their influence in the classroom.

To understand why teachers abandoned many of the SCISP labs and the stress on pattern-finding and problem-solving, we have to consider the potential consequences of adopting the less specific or more diffuse alternatives.

Wilson (1962) notes that teachers are in the business of socializing children; he defines this process as:

> Motivating, inspiring and encouraging them, transmitting values to them, awakening in them a respect for facts and a sense of critical appreciation; all of this is unspecific. It implies 'what a [person] is' as much as 'what a [person] does'. The role obligation is diffuse, difficult to delimit and the activities of the role highly diverse (p. 25).

He lists a number of features of the role which define its diffuseness: it is not clear what tasks are to be accomplished;

these tasks are not defined in terms of exact manipulations; there is a lack of precise content change; a lack of specialized and defined expertise; a lack of formal limitation to the competencies required; an undelimited commitment is required. Wilson's analysis of the teachers' work leads him to conclude that it is full of conflicts and uncertainties; he notes six of these: a lack of demarcation whereby the person knows he or she has 'done the job'; diverse expectations for the role; potential marginality of humanistic goals in institutions; vulnerability to outside criticism; conflict between career concerns and value commitments.

If we suppose that teachers will choose to avoid or minimize the conflicts and insecurities of their work we might expect teachers to shy away from laboratory schemes which increase the diffuseness of their work and increase the problem of maintaining a balance between instrumental and expressive acts. Rather, we would expect teachers to define laboratory activity in such a way that their expertise can be made available. This analysis would lead us to expect that, given the choice between many of the diffuse activities in the SCISP laboratory scheme and what they normally did, teachers would choose either to avoid the lab activity or modify it.

Teachers in this study modified or avoided activities because they required them to cope with increased work diffuseness. We can understand why teachers did not abandon prior commitments but rather shaped the SCISP laboratory scheme in order to maintain existing commitments. Let us see how this happened. Compared to the preparation for external examinations what SCISP doctrine urged be accomplished was diffuse. Teachers were expected to use laboratory work to foster scepticism with respect to argument, and concern about the uses of science in society. However, how does one promote these aims? How does one know that they have been promoted? Who says these are the aims to be fostered?

In fact, the teachers did not value such aims, and they tended to ignore them. They arranged the subject matter and laboratory work to display their expertise in specialized disciplines, to give students a basis for choosing post-13+ work and to help recruit candidates for A Level work. Subjects were narrowly conceived as sources of information useful in obtaining tokens recognized outside the school. The task to be accomplished was set in terms of passing an examination; the manipulations required were to transmit content which was relevant to the syllabus and old examination questions and the task of the students was clear - learn the material and be able to do well in mock examinations and finally sit the external examination itself. The competencies of the teachers were limited to teaching what was required to pass the examination; the commitment to the pupil was delimited by the specifications of the syllabus.

The language used by the teachers reflected the specific way they saw their tasks. Richard Simpson, for example, in rejecting SCISP as a suitable precursor for A Level said, 'There are serious omissions which have to be crammed into the A Level courses, or crammed into the O Level courses'. Ann Williams said, 'I do like the facts. I do think you have to get the facts in first'. Bryan Jenkins talked about 'hammering' concepts. Cramming material into the syllabus and hammering that into the heads of the students is a very specific way of defining what is to be done in science education.

## Mr. Swift and the syllabus

These teachers are not alone in seeking to delimit their tasks in the classroom. I have found in subsequent research evidence of a similar focus on specific tasks and their definition by means of a syllabus of content (Olson, 1984). It is worth looking at one case where the role of the syllabus in defining the tasks of the teacher is clear-cut. Mr. Swift (a teacher of science to 12 and 15 year olds) considered the syllabus an essential part of his work, and we will look in more detail at the part played by the syllabus in Mr. Swift's work and the tension he felt between covering the syllabus and doing laboratory work based on a discovery approach; a tension similar to one experienced by SCISP teachers.

The parallel between Mr. Swift and the SCISP teachers is close but not exact. Mr. Swift, for example, is a non- specialist middle school teacher. In both cases teachers take their task to cover a syllabus of content in order to prepare their students for further work by gaining credentials. By covering the syllabus these teachers do their work in a 'measured' way. SCISP teachers consider discussion-based laboratory work as time consuming and confusing; Mr. Swift considers that laboratory work involves misbehaviour and waste of time. SCISP teachers were most concerned to ensure that their students obtained good qualifications, Mr. Swift wanted his students to enjoy his subject, but not at the expense of valuable class time. In both cases teachers felt strongly bound to the content which they had to cover, and found laboratory work problematic.

How are we to understand this strength of commitment to the subject matter of the syllabus? It matters that we do understand it, because, as we have seen, efforts to re-think the nature of laboratory work must deal with the way teachers construe the function of the subject matter in schooling and how they see their role in teaching it. It became clear from talking to Mr. Swift that without an 'official' syllabus what is to be taught is unclear, and it is impossible to organize the material into carefully timed parts. For him the danger of drift is constantly present when the work is not under the control of some regulation.

The official syllabus supplied Mr. Swift with a regulating mechanism - the mandate to present the core material of that syllabus to students. Given a budget of limited time and extensive programme of material to cover, the use of time is a critical factor for Mr. Swift in deciding how to proceed. Time becomes a factor influencing not only what is presented, but how the material is presented. With the syllabus authoritatively mandating content to be covered, Mr. Swift is left with the task of deciding how that content might best be dealt with, covering the material in ways that are interesting.

Mr. Swift retains firm control over lessons and does not spend too much time on discussion or 'side-trips', and what might have been usefully included if time had not been of paramount importance has to be omitted; especially laboratory work. Mr. Swift is aware of the dilemmas inherent in the regulation of time by the syllabus but it provides an orderly context for planning - for defining the task to be done and showing what to stress in the time available. This regulation is a mixed blessing in Mr. Swift's view: a source of authority about what to include and a source of pressure to exclude interesting but time-consuming laboratory work. Content information is included; certain time-wasting activities are excluded. The balance isn't perfect.

To pursue in greater depth Mr. Swift's attempts to resolve this dilemma, we asked him to sort statements of science teaching activities, which ranged from highly teacher-controlled to student-controlled and which included various laboratory activities (Olson and Reid, 1982). These statements he arranged in a number of groups according to some underlying construct he had chosen to organize his thinking about them. We then discussed each of these activities in turn in relation to the set of constructs he had used to sort them.

One important construct dimension that emerged from our discussion was that of 'keeping on track' vs 'squandering time'. He said that all of the activities could be organized along this dimension. Teacher-centred activities were seen to be on-track activities: 'I, as a teacher, know where I'm going and I don't want to be thrown off track too much. I have a definite goal to achieve and a definite amount of time in which to achieve it'. The importance of knowing the goal and relating time to its achievement can be seen in how Mr. Swift views an activity in which students are at work doing an experiment to verify a law. As Mr. Swift sees it, he has limited control here. Mr. Swift sees teacher-controlled activities as having a definite goal and a definite time to achieve the goal: 'If that clock says I've got five more minutes to get that done so that they can get their note, I'll eliminate [discussion] and revert to [telling them] ... It's safe. I know where I'm going'. Mr. Swift talked about 'savouring' his lesson time vs having to 'cover' the ground.

The stress on classification, on definition, on the vocabulary gives support to teachers unsure of their knowledge of the subject. The situation many middle school teachers find themselves in is ameliorated through an official syllabus. Is it an accident that Mr. Swift found the syllabus a 'godsend'? Yet that syllabus tends to make it harder for Mr. Swift to incorporate laboratory work in classroom time. Mr. Swift has created a programme which resolves some of the tension between teaching an enjoyable subject yet covering mandated material. While information is transmitted, some of the practical aspects of science are captured in the work he gives outside of class time. His combination of in-school syllabus activity and out-of-class at home laboratory work allows him to pursue instrumental and expressive functions in his teaching. The subject can be fun, yet there is serious work to accomplish as defined by the syllabus, and no time in school must be wasted.

The case of Mr. Swift gives us further insight into how the way the teacher construes the subject affects the way laboratory work is viewed. Critical to understanding how teachers think about laboratory work is an adequate knowledge of how teachers cope with the expressive and instrumental demands of teaching. We have only hinted at the importance of these elements of teaching; it is now time to consider them in detail.

## LABORATORY WORK AND THE FUNCTIONS OF THE SUBJECT

### Expressive and instrumental dimensions

'The subject is a tool in the struggle for power in the classroom - a struggle reflecting that in the larger society' (Young, 1974). This view tends not to pay attention to the complex demands placed upon teachers as teachers - the diffuse and undelimited tasks that Wilson (1962) speaks of. If we take seriously the idea that teachers have to consciously cope with these tasks, we must make sense of their practice, and we cannot do this without consulting them and learning from them what the practice is and to what ends it is directed. It is here that one has to consider the complex interaction between the expressive and instrumental acts of the teacher.

Teachers do not want to do laboratory work in which their efficacy is questionable. The very questioning of their efficacy undermines their ability to be the teacher. It is not surprising if teachers take innovative doctrines and redefine them in order to maintain certain impressions in their students. These impressions permit them to function in the classroom by fostering in their students a willingness to trust the teacher. This is especially critical when significant social tokens like external examination results are at stake. Teachers are able to match methods, student ability and resources in complex ways; simultaneously managing to do what they can to get students over the examination hurdle and creating the impression in their students that they are reliable managers of this process.

SCISP doctrine, and other revisionist doctrines, fail to appreciate what is at stake for teachers when they ask them to abandon well tried approaches in the laboratory, and the critics of the failure of teachers to adopt revised approaches fail to see things from the teachers' point of view.

The instrumental problems are severe, but more severe are the expressive correlates. Analogies with drama are useful here. How can an effective play be put on without props? How can the teacher star in a play whose script he/she has not mastered and whose props are unfamiliar? How can the 'illusion' be maintained? Teaching is an illusive process.

## The teacher and the curriculum model

Laboratory work is one of the tools that teachers have to create an illusion about the subjects they teach. They will do laboratory work in a way that sustains in the minds of their students images about what the subject and what they the teacher are like. This illusion is constructed out of limited things: some few pieces of equipment; limited time; limited activities to do; limited student interest and ability. Out of that a sense of the subject is created, and a stock of knowledge useful for 'getting' the subject is supplied as well; getting it in the sense of seeing what it is all about and obtaining examination credit.

It is not surprising that teachers rely on useful, stable ways of achieving their goals. They are within the means of the school and the ability of the students - the teacher can 'pull them off' without excessive strain. The laboratory work they usually do fits the curriculum, the budget, the students. It does work for the teachers and when something works why throw it away? What if there are better ways to do laboratory work? What if SCISP doctrine is right - laboratory work in science needs to be reconstructed to promote higher order thinking and an awareness of social relevance? What chance does such a doctrine stand in the face of a system which already works?

Typically, the approach of projects like SCISP is to urge revision in line with project doctrine. High fidelity implementation is sought, but there is another way of considering the matter. Challenges like SCISP to existing conceptions of laboratory work are relatively expensive and inefficient. Key elements of the challenge are lost in the welter of sample schemes and the complex politics of reform. Obviously what happens in the laboratory *now* is not the final word, but yet neither are schemes like SCISP. How can existing practices be reconsidered without having to adopt yet another project as a stimulus change?

One way of approaching this problem is to take the teacher's point of view more seriously. What is it that teachers are trying to say about laboratory work? Brown and McIntyre (1982) argue that curriculum makers must pay attention to how teachers talk about their work. They use, as an example of what can happen by default, the differences between the way teachers viewed guidance (in guided-discovery) and a curriculum development team. They probed teacher conceptions of guidance and found teacher conceptions at variance with the 'official' view of the curriculum makers. Similarly Smith and Sendelbach (1982) found that one of the reasons elementary science teachers departed from procedures suggested in the teacher's guide was the inherent difficulties in communicating to teachers why certain procedures were desirable. They found that the most reliable influence of the programme material appeared to have been on the kinds of activities carried out and on lower level knowledge addressed. What can we learn from these studies about communication in curriculum? If innovators wish to communicate their ideas to teachers, they must address the instrumental and expressive concerns of teachers. Brown and McIntyre (1982), in their account of teacher translation of innovative ideas, suggest that teachers come to understand the meaning of innovative proposals through coping with their practical implications. Over time teachers might also consider more general ideas associated with the new practice.

The model (Figure 1.2.1) shows teacher inputs. This should be understood as taking account of the belief systems of teachers as an important element because it is respectful of the commonsense theories of teachers, and sensitive to their problems. Thus there should be feedback from instructional practices to the planning process so that these processes can be sensitive to what is important in the teacher's world.

## CONCLUSION

As we see in these cases, the problems teachers have in using innovative ideas can be traced to a failure of the innovators to take seriously one of the rules of giving advice: they did not find out about the life of the people they presumed to advise. Instead of advising, they exhorted, they advocated, they enjoined: each of these is a quite different act from advising and each supposes a different set of communicative rules. They failed to do the research that advice giving requires.

Innovation will be a more practical process once productive dialogue exists between old practices and new ideas. This is the lesson for change we take from these cases.

## REFERENCES


Brown, S. and McIntyre, D. (1982) Costs and rewards of innovation: taking account of the teacher's viewpoint, in Olson, J. (Ed.), *Innovation in the science curriculum*, London, Croom Helm.

Carlson, R. (1965) *Adoption of educational innovations*, Eugene, University of Oregon.

Geertz, C. (1973) *The interpretation of cultures*, New York, Basic Books.

Herron, M. (1971) On teacher perception and curriculum innovation, in Weiss, J. (Ed.), *Curriculum theory network*, Toronto, OISE.

National Science Foundation (1978) *Case studies in science education*, Urbana-Champaign, University of Illinois.

Olson, J. (1982) Dilemmas of inquiry learning, in Olson, J. (Ed.) *Innovation in the science curriculum*, London, Croom Helm.

_____ (1984) Mr. Swift and the clock: teacher influence in the classroom, *Teacher Education*, October, 13-26.

_____ and Reid, W. A. (1982) Studying innovations in science teaching: the use of repertory grid techniques in developing a research strategy, *European Journal of Science Education*, 4, 193-201.

_____ and Russell, T. (1983) *Science Education in Canadian Schools: Vol. III,Case Studies of Science Teaching*, Ottawa, Science Council of Canada.

Smith, E. and Sendelbach, N. (1982) The program, the plans and the activities of the classroom: the demands of activity-based science. In Olson, J. (Ed.), *Innovation in the science curriculum*, London, Croom Helm.

Wilson, B. (1962) The teacher's role: a sociological analysis, *British Journal of Sociology*, 13, 15-32.

Wilson, J. (1972) *Practical methods of moral education*, London, Heinemann.

Young, M. F. (1974) Notes for a sociology of science education, *Studies in Science Education*, 1, 51-60.

# Part 5

# COMMONPLACES: CONTEXTS AND POLICY ISSUES

# 5.1

# School Laboratory Work: Some Issues of Policy

**John Ainley**

## INTRODUCTION

Compared to other facilities in schools, science laboratories are often seen to represent a substantial investment of the capital resources available to schools. Consequently the provision of special facilities for science raises policy issues concerned with their effects; whether the provision of specialist facilities influences the way science is taught and whether different types of laboratories have different influences on teaching. Such questions are often raised in the context of costs and are linked with the implication that the desired goals might be achieved in other ways (see Bates, 1978). In the present economic climate it is important to be able to identify the goals of laboratory work in science, and to suggest the sort of facilities which best help students progress towards those goals. The argument becomes even more critical in the face of evidence that laboratory work and hands-on science activities are used less frequently than most science educators would consider desirable (Blosser, 1980, 1983).

## FACILITIES AS A LIMITING FACTOR

The standard of facilities can operate as a limiting condition on science teaching and learning in that the absence of adequate facilities can preclude the inclusion of certain activities. This is especially true of laboratory work. Beisenherz and Olstad (1980) reported that there were differences in the extent to which high school Biology teachers in two areas used laboratory activities, that a lack of materials, equipment, and facilities were seen as major impediments to laboratory activities; and that in the area where fewer activities were conducted lack of facilities and equipment was seen more frequently as an impediment. Dark and Squires (1975), Robertson (1962), and Kerr (1964) have also reported evidence that insufficient laboratory facilities, over-use of laboratories, a shortage of qualified teachers, a lack of laboratory assistants, and large classes impair the effectiveness of laboratory work. Holley (1974) reported on factors which teachers saw as constraints on their work as part of a wider study of senior physics in England. Among the heads of physics departments who responded, a lack of resources, such as laboratories and apparatus, was the second most frequently cited constraint on laboratory work. External factors such as examination requirements were mentioned more frequently as constraints than were resources but student characteristics, such as student knowledge and interest, and school organizational policies concerned with scheduling and general emphasis on practical activities were rated as less important.

Conversely, there is some evidence that the removal of impediments can promote laboratory work. Tamir (1976) noted that the provision of easy access to laboratory materials, living organisms, and chemicals through special supply centres was influential in promoting laboratory work in high school Biology in Israel. Evidence of the effects of removing impediments to laboratory work can sometimes be masked by the presence of other impediments. In this case most of the other influences on the teaching of Biology (e.g. the curriculum, and the forms of student assessment which were used) were also directed towards the encouragement of laboratory work.

## THE WIDER INFLUENCE OF FACILITIES

Facilities can influence teaching and learning processes through the symbolic message which is conveyed as well as by imposing physical limitations on science teaching and learning activities. In practice it is hard to disentangle the wider influence of the type of facilities from the pragmatic limitations which they might impose. However, a number of studies suggest some wider influence through the examination of relationships other than direct limitations.

Englehardt (1968) investigated a number of aspects of potential relationships between characteristics of science rooms and science teaching methods. Several different types of measure of the standard of facilities were used: the availability and location of rooms, aspects of room quality (e.g. the presence of certain fittings, the size of sinks), and room design (separate classrooms and laboratories or dual purpose classroom laboratories). Some 500 teachers in 59 schools in the New England states of the United States were interviewed about teaching practices, and their responses were related to the author's observations about the characteristics of the rooms in which they taught. It was found that some characteristics of the architectural space were associated with teaching practices. Classroom laboratories, as opposed to separate laboratories, were associated with greater use of enquiry methods in science laboratory work especially when teachers had sufficient preparation time, easy access to the rooms, and taught mainly in the same room. Other associations which were reported included those between the provision of individual laboratory space and the frequency of individual projects, the provision of suitable sinks and the frequency of laboratory work in biology and earth sciences, and the proximity of the library and the use of library assignments. The study by Englehardt suggested an association between certain facilities and enquiry teaching methods but was not able to separate the influence of facilities as a limiting factor from their influence as suggestive space.

Two important features of the study by Englehardt were the consideration of specific features of rooms as well as of general room design and the examination of dependent variables (e.g. use of enquiry methods, individual projects, laboratory work, and library assignments) which could be potentially influenced by the nature of the facilities available in contrast with more distant measures of generalized achievement in science. In addition the study by Englehardt obtained information from teachers about how frequently they used certain teaching methods and independently observed the characteristics of the science rooms in which they taught. This contrasts with studies which asked teachers' opinions of the adequacy of the facilities available to them.

Norman (1969) investigated overall design differences of various types of biology rooms and their suitability for various teaching practices. Four laboratory designs were considered:

split lecture and laboratory, perimeter tables, central fixed tables, and central movable tables. In the study, teachers were asked to rate the suitability of their laboratory design for each of three methods of instruction: small group instruction, large group instruction, and independent study. In the analysis there was a statistical control for the effects of recency of construction. The study results contrasted with those of Englehardt in that the split design was rated best for all methods and the perimeter design was rated as second best for all methods. These differences could be attributable to the ways in which the studies were conducted and the fact that one was restricted to biology while the other included a range of science subjects. The study by Englehardt asked teachers how frequently they used certain methods whereas that by Norman asked teachers to rate the suitability of their rooms for different methods. Such a difference may be important given that participants are often unaware of environmental influences on their behaviour. Perhaps even more importantly the two studies were examining different types of teaching methods.

Davis (1972) examined the impact of a United States federal funding programme (National Defense and Education Act Title III) on science education in one state. Funds available could be used to acquire laboratory and other materials to assist teaching in various subjects. In practice, science subjects used a large proportion of the available funds in completely or partially equipping a substantial number of school laboratories. In a review of studies of the impact of NDEA funds in other states Davis concluded that the provision of equipment and materials had improved patterns of teaching in science. The study conducted by Davis supported that conclusion. Primary evidence was the opinion of science teachers and school system supervisors. Science teachers regarded the new materials as facilitative in three main ways. First, the materials were considered to make the teacher's work more convenient and effective. Second, additional materials were claimed to increase the amount and type of student experimental work, and provide an enriched classroom atmosphere. Third, it was considered that the additional materials enabled the science curriculum to be broadened. These changes were claimed to have been reflected in increased student knowledge and interest in science among students.

The studies outlined so far have used information from teachers, either reports of what happened in science classes or opinions about facilities, as a source of data. A study conducted as part of an evaluation of the Australian Science Facilities Programme used student reports of science lessons as the main source of data in examining the association between the standard of science teaching facilities and the quality of science education experienced by students in Year 9 (Ainley, 1978a,b). Students reported their views by completing a questionnaire concerned with six aspects of their science lessons: involvement, organization, stimulation through variety, encouragement to explore, practical work, and textual teaching. The questionnaire consisted of 42 items describing science lessons to which students could respond on a Likert scale, so that groups of items provided scores on six subscales corresponding to these aspects of the learning environment. These aspects of the science learning environment had been considered to be theoretically linked to the standard of facilities available and evident in the intentions of the Australian Science Facilities Programme. A little fewer than 3000 students from 105 science classes, chosen from schools spanning a wide range of science facilities from poor to excellent, completed the questionnaire. Class means were used as the unit of analysis.

In general it was found that bettter facilities were associated with what students perceived as an enriched learning environment and more active forms of science learning. An enriched learning environment was one in which there was greater involvement in purposeful activity, better organization and more stimulation in the variety of methods and materials which were used. Student perceptions of an environment which was well organized were most strongly associated with good quality rooms and also supported by the abundance of apparatus. Being frequently in science rooms, using rooms rated as of good quality, and having sufficient apparatus were all associated with students reporting greater involvement in learning activities. Stimulation through variety was most strongly associated with abundant apparatus.

Student reports of active forms of learning defined as encouragement to explore and more experimental work were also associated with better science facilities. Encouragement to explore was reported as higher among students who experienced better quality science rooms with sufficient apparatus where there were laboratory assistants to help teachers prepare and organize. Greater emphasis on practical work was reported when apparatus was more plentiful. None of the indices of facilities was associated with the emphasis on a use of textbooks as a medium in the learning of science. The use of textbooks was independent of the quality of facilities available and not related to the emphasis on the two other forms of learning: encouragement to explore and practical work.

A number of interaction effects were also detected in four of the aspects of science teaching studied: organization, stimulation through variety, encouragement to explore, and practical work. These suggested more detailed interpretations and extensions of some of the main results described above. Organization was enhanced by the presence of laboratory assistants when classes were held frequently in science rooms but not so strongly if that was not the case. Organization was also enhanced by the presence of laboratory assistants when the supply of apparatus was limited but not when it was abundant. When interactions were examined it appeared that a greater abundance of apparatus resulted in greater stimulation through variety especially when coupled with ample provision of laboratory assistants, and with good quality rooms. Similarly the effect of the quality of rooms on stimulation through variety was greater when classes were less frequently in them. This last observation was interpreted as reflecting the greater use by some teachers of resources outside the laboratory. A similar pattern was found for encouragement to explore. When rooms were of good quality there was greater encouragement to explore if the proportion of lessons held in science rooms was lower. Again it seemed that the sort of activity involved in this aspect of science teaching may have been facilitated by the use of resources outside of science rooms.

The study by Ainley also involved interviews with science teachers and observations of science classes in action in schools with different levels of provision of facilities. Two of the cases reported in that part of the study help to illustrate the ways in which the quality of facilities might influence the teaching of science. In the first case the science coordinator had been at a secondary school for some time. That person was able to compare the effect of new science rooms with previous conditions. The coordinator noted that prior to the scheme science was conducted in old rooms 'to which one might occasionally bring a bucket of water and a portable burner'. The science coordinator observed that 'all classes could have all of their science in a science room' and that 'activity-based programmes which we run now would not have been possible under the old situation'. The same teacher elaborated that since the new rooms had been occupied the science staff had begun writing their own programmes to suit local need. These were sometimes used in preference to an activity-based curriculum package widely available to schools through the Australian Science Education Project (ASEP) which is described in Chapter 6.1. In response to a question it was stated that the new rooms had been a stimulus to this activity and that there had been no point in doing this beforehand. It was emphasized that the school was not restricted to student experiments, that teachers were expected not to simply have students work through a unit as written but to punctuate their teaching with varied methods including more formal demonstration lessons and discussions. The argument was not that a great amount of material was essential for worthwhile science teaching but that being able to draw on material for particular parts of a programme enabled the presentation of a richer range of science experiences.

A second case was a newly established school which had moved from temporary science accommodation to a new suite of modern, well-designed science rooms. An experienced teacher in that school reflected on some subtle differences which the change had made in terms of efficiency, a reduction in the amount of time wasted, and the more responsive attitude of the students. In contrast a teacher in a school with poor quality science rooms and a shortage of apparatus observed that a lack of simple basic equipment such as test tubes and beakers impeded the use which could be made of the Australian Science Education Project materials (see Chapter 6.1). A colleague observed that the cramped conditions of the rooms and the lack of equipment meant that any practical exercise involved a great deal of preparation.

Overall the comments by teachers in the interviews did not suggest a dramatic transformation of learning activities in response to the quality of facilities available but more subtle changes which were reflected in the views of students. These small but important changes included the removal of barriers to activities which were desired, changing the attitudes and motivations of teachers, and changing the responsiveness of students. The comments by teachers suggested changes in the modes of teaching and learning science towards a greater use of laboratory work and exploration as vehicles for learning.

## LABORATORIES AND CURRICULUM DEVELOPMENT

During the 1960s and 1970s a number of new science curricula were introduced in Australia, Britain, and the United States (Ramsey, 1972a,b). Even though there were important differences between the curricula in different sub-disciplines, at different levels, and at different times over the period, all emphasized to varying extents student experiments of an investigatory type, the use of a variety of teaching aids such as films and television, and less didactic methods of teaching. Over the same period there were, in those countries, attempts to improve the physical facilities available for science teaching in schools. Where authorities had introduced Biological Sciences Curriculum Study (BSCS) Biology, there had been concurrent attempts to improve laboratory facilities. Similarly, Tamir (1978) suggested that the introduction of BSCS Biology in Israel had an impact in terms of improving laboratory facilities in schools. Tamir (1976, p. 308) noted that the extended period of development of the BSCS materials for Israel 'gave time to prepare facilities that proved essential to implementation: laboratories, equipment, and supplies, centres for the distribution of living materials and audiovisuals'.

In general, it appears best to view both the curriculum developments and the improvement of facilities as having a common genesis in a concern with providing a practical base for the teaching of science. Baas (1973) interpreted the developments in science room design as reflecting a concern for 'individual students' involvement in the methodologies and results of the scientific process'. In a general sense the changes in curricula could be interpreted as reflecting a similar concern.

### Science room design

The views about design expressed by Baas (1973) were supported by another study of the type of science facilities in schools throughout the United States by Novak (1972). The study identified an evolving pattern from separate lecture and laboratory space with massive fixed furniture, through integrated classroom laboratories, to flexible open space provision. Novak observed that this pattern of evolution appeared to be associated with developments in curricula, educational technology, and teaching roles.

The trend from formal laboratory to flexible teaching space seems to have continued. A recent study for the National Science Teachers' Association in the United States suggests that, among several important trends in what were nominated as excellent science teaching programmes, was a move towards an increasing use of the science laboratory in the wider environment for students (Penick and Yager, 1985). The laboratory remained important in all the programmes identified as exemplary, but it was not a place to go routinely to perform predetermined exercises. An emphasis on science room designs which were multidisciplinary and provided for several modes of learning remained a key feature of the conditions for good science teaching recommended by the National Science Teachers' Association in the United States in 1984 (Showalter, 1984).

Throughout North America, science accommodation received close attention during the 1960s and 1970s. Publications from state and provincial authorities and other sources made recommendations about the type of science facilities which would suit the new curricula. Writers such as Edsall and Templeton (1960), Martin (1960), Richardson (1971), and Schlessinger (1963) analysed the new courses and concluded that the laboratory was of central importance to most and that flexible designs would be most useful. Documents from state authorities emphasized the importance of sufficient laboratory space in the form of combined classroom laboratories which could be adapted for teaching a range of science content using a variety of teaching styles (e.g. California State Department of Education, 1964; Minnesota State Department of Education, 1971). The Ontario Department of Education (1968) also argued for flexibility in design to permit different teaching styles and changing requirements of organization and curricula. It added that laboratories should allow for effective teaching in different sciences. The Texas Education Agency (1980) elaborated the types of design of well-equipped classroom-laboratories which would allow all classes in Years 7 to 12 to be laboratory-oriented and consistent with the science curricula which were seen as appropriate for students of that age.

In Australia a similar pattern of development in the design of science rooms was evident over this period. A major federally funded programme, known as the Australian Science Facilities programme, intended to provide science laboratories and equipment in secondary schools commenced in 1963 and continued through to 1975 (see Ainley, 1978a). Over the 12 years of the programme some $123 million was allocated by the federal government to both government and non-government secondary schools. The committee which advised on the provision and design of science teaching facilities under this programme encouraged a move away from separate rooms or areas for laboratory work and lecturing to dual-purpose rooms with large fixed work benches. In its publications it argued that:

> Courses now aim for better integration of laboratory work with the theory of the subject. Such courses may be given in a dual purpose laboratory in which a class meets for all its science lessons, allowing complete flexibility in the type of activity to be followed at any time. (Commonwealth Advisory Committee, 1973, p. 5)

The newer designs incorporated movable working and writing areas with services located at pillars in the body of the room, or at perimeter benches. A series of observers of science room design in the Australian states noted a common emphasis on the development of more flexible designs often explicitly related to the perceived needs of the then modern science curricula (Betjeman, 1974; Fish, 1974; Hall, 1974; Prosser and Woolley, 1974; Robins, 1974; Turner, 1974). The rationale for the development of more flexible science teaching spaces was summed up as follows:

> The demand is towards rooms where practical activities associated with any of the sciences can be conducted safely by pupils; where group work and discussions can be readily catered for; and where, in general, there are fewer limitations on the particular teaching approach or approaches a teacher may choose to use at any particular time. (Field, 1974, p. 35)

In Britain, the Industrial Fund for the Advancement of Scientific Education in Schools stimulated reconsideration of what constituted appropriate science facilities. Because of its financial contributions its standards and guidelines became widely influential. It stressed the central role of the laboratory in science teaching, and it favoured rooms which could be used for either class teaching or student experimental work (Savage, 1958). Support for increased specialization of design in the later years of school was reflected in both the Industrial Fund recommendations and the Oxford School Development Project (Department of Education and Science, 1964). The Oxford Project had also argued that separate provision of laboratory and lecture space had merit, but in the face of opposition from teachers restricted that recommendation to provisions for the sixth form. In general, there was a greater tendency to retain specialized laboratory designs in Britain than in either Australia or the United States, possibly reflecting a greater emphasis on the separate teaching of the branches of science throughout the secondary schools of Britain compared to the other two countries.

In its recommendations of conditions for good science teaching, the National Science Teachers' Association of the United States argues that 'systematic effective learning by large numbers of students at a given time requires the provision of learning spaces specifically designed for that purpose' (Showalter, 1984). It recommends that science rooms should include spaces for laboratory work, for group instruction, and for individual use. Significantly, the NSTA statement of conditions for good science teaching in 1984 extended beyond secondary schooling to include a consideration of areas for science learning activities within elementary school classrooms.

## Standards for facilities

In the literature about planning science facilities there was wide support for the proposition that there should be sufficient science rooms in a school to accommodate readily all science lessons. This remains a key recommendation of the National Science Teachers' Association in its statement of conditions for good science teaching (Showalter, 1984). To allow this in practice several organizations have suggested planning of requirements for science rooms to allow for the rooms to be vacant for some of the time in each week. Typically the suggestion was for science rooms to be allowed to be vacant for about one third to one fifth of the total teaching time; corresponding to a usage rate of between 70 and 80 per cent. This was to allow for preparation, cleaning up, and the other constraints on timetable organization which could prevent science lessons being scheduled in science rooms. Different organizations adopted slightly different recommendations regarding usage rates, possibly reflecting differences in financial resources or the availability of laboratory assistants.

In Britain these planning principles were enunciated by the Industrial Fund (Savage, 1958), became the policy of the Science Teachers' Associations in Britain (The Four Associations, 1964), and were accepted by the Oxford Project. It opted for an occupancy rate of 80 per cent if sufficient laboratory staff were provided. In Australia although the various state authorities adopted different criteria for allocating science rooms to government schools, the Advisory Committee to the federal government adopted the principle that all science lessons should be held in science rooms and that planning should be based on a maximum usage rate of 75 per cent.

## ADOPTION AND IMPLEMENTATION OF CURRICULA

In the study of new curricula a distinction is frequently made between adoption and implementation. In brief, while new curricula may be adopted by schools in a formal sense it does not necessarily follow that those materials will be used or implemented in the way in which they were intended. It appears that the physical facilities available exert a stronger influence on the way a curriculum is implemented than on its adoption.

### Adoption

In Britain curriculum materials developed through the Nuffield Science Teaching Project, which were intended to generate an increased emphasis on learning through laboratory investigation, have been the focus of a number of studies. One study by the Schools Council (1969) reported the results of enquiries directed to Local Education Authorities about the use of a range of Nuffield materials. The explanations of the respondents for delays by schools in introducing Nuffield science courses were grouped in three categories. These were listed in order of importance as lack of finance, demands on resources (laboratories, laboratory assistants, and teachers' time), and reservations about the proposed teaching methods. Although these views were obtained from administrators somewhat removed from teaching they were possibly reflecting opinions expressed to them by teachers.

Kelly and Nicodemus (1973) reported on the views of a group of teachers attending briefing sessions on Nuffield A-level Biology. Teachers who stated that they were considering adopting the course were compared with those only considering it for trial. The two groups differed significantly in the facilities available to them in terms of preparation time, equipment, and laboratory space, those planning to adopt the course having the better facilities. Of course there were other differences between the groups, particularly related to experiences during teacher training, but the results suggested that school conditions could influence the decision to adopt a course of study, or to modify the way it was implemented. A difference between the adoption and implementation of curricula was evident in an evaluation of Nuffield O-level Biology materials by Kelly and Monger (1974a,b). No association was reported between the decision to adopt the course and the standard of facilities available. However, the authors did draw attention to apparent inconsistencies between the objectives and content of the course, teachers' understanding of the objectives and content of the course, and the actual teaching of the course. Among possible reasons postulated by the authors for these inconsistencies was that lack of suitable facilities may have influenced the conduct of the course.

## Implementation

Jenkins (1967) reported the views of about one hundred chemistry teachers, who had attended an in-service education course, about the introduction of Nuffield O-level Chemistry. The most frequently mentioned obstacle was a concern about the suitability of the course as preparation for further studies. The next most frequently mentioned obstacle was the inadequate preparation time which, presumably, reflected the level of laboratory assistance available. Lack of suitable facilities was the fourth most frequently mentioned obstacle to implementing the new course and was included by one third of the teachers. Jenkins (1967, 1971a) argued that inadequate facilities were mentioned less frequently than might have been expected. In part, this could be attributed to the fact that teachers attending an in-service course could be expected to be more enthusiastic, and more able to overcome obstacles (Jenkins, 1971b) than their colleagues who did not attend. A similar study of a group of Biology teachers indicated that

Biology teachers were more concerned than Chemistry teachers that a lack of appropriate facilities impeded the implementation of Nuffield Biology (Jenkins, 1967).

Similar results have been reported from studies of junior secondary science curricula. In studies of investigative laboratory work in a Canadian province (Wideen and Batt, 1974) and of an integrated science curriculum in Scotland (Brown *et al.*, 1976) it was reported that teachers placed considerable importance on resource constraints when deciding whether to use the curriculum. The integrated course placed greater demands on resources than did the teaching of separate science subjects in a traditional way, and in schools where apparatus was limited and laboratories were scattered the course was viewed with disfavour. Subsequently Brown (1977) wrote that while such concerns may be less important than philosophical issues to curriculum developers, they may be very important to the teachers who use the materials. The study by Brown *et al.* indicated that the implementation of new curricula of the type studied was encumbered when new demands were made on already inadequate resources. These results were supported by the observations made in a case study investigation by Hamilton (1975). The report of that study described the way in which the Scottish Integrated Science Scheme was implemented in two schools. One of the schools had poor science teaching facilities with poor and scattered science laboratories and a shortage of apparatus. In combination with other problems relating to lack of communication between science staff, patterns of school organization, and teachers' absences these factors resulted in a pattern of teaching at variance with the intentions of the scheme. Worksheets had become the syllabus instead of being support materials, and 'as teacher demonstrations were often substituted for pupil practical work, class teaching had become the dominant activity' (Hamilton, 1975, p. 187). Brown *et al.* and Hamilton used different methods in their studies of the introduction of the same new course but the conclusions were similar: the ways in which schools and teachers implemented the new curriculum was influenced by the resources available.

Overall, the results of the studies reviewed suggest that although the standard of laboratories and equipment may not always be the most important influences on the decision to adopt a curriculum which emphasizes laboratory work, it can shape how successfully the programme is implemented. Sometimes under poor conditions modification and compromise may be such that what emerges no longer faithfully reflects the intentions of the programme. Possibly the extent to which the standard of facilities impinges on the way a curriculum is implemented depends on the nature of the discipline involved, the extent of other support which is available, and the age level to which the programme is directed. One of the difficulties in generalizing from studies of curriculum change is that not many studies have considered resources in a systematic way. Furthermore, enthusiastic teachers who are able to overcome resource obstacles are the most likely to be involved in adopting a new programme in its early stages and be involved in studies of adoption. Despite the uncertainties involved it does appear that lack of resources can limit the way in which curricula based on extensive laboratory work are implemented in practice.

## CONCLUSION

The preceding discussion in no way refutes the proposition that 'good teaching can occur in any conditions'. There are sufficient examples of active involved learning occurring in poor science rooms to support that. However, it does appear the physical facilities are one component which influences the teaching and learning of science and the views which students hold of their science education at school.

Recently, and specifically related to the teaching of science, the National Science Teachers' Association expressed the view that:

> Research can show that without adequate laboratory facilities and materials most students cannot learn biology in any meaningful way. Conversely, one cannot assume that meaningful learning will take place with such facilities and materials. (Showalter, 1984, p. 1)

Overall, it seems that having well designed and equipped science laboratories can facilitate the attainment of some of the goals of science teaching, but the facilities of themselves are not sufficient.

## REFERENCES


Ainley, J. (1978a) 'The Australian Science Facilities programme: A Study of its Influence on Science Education in Australian Schools' Hawthorn, Vic., Australian Council for Educational Research.

_____ (1978b) 'An Evaluation of the Australian Science Facilities programme and Its Effect on Science Education in Australian Schools', unpublished Ph.D. thesis, University of Melbourne.

Baas, A. M. (1973) 'Science Facilities', Educational Facilities Review Series No. 11, ERIC Clearing House on Educational Management, ED-071-144.

Bates, G. R. (1978) The role of the laboratory in secondary school science programs, in M. B. Rowe (Ed.). *What Research Says to The Science Teacher*, Washington D.C., National Science Teachers' Association, pp. 55-82.

Beisenherz, P. C. and Olstad, R. G. (1980) The use of laboratory instruction in high school biology. *The American Biology Teacher*, 42(3), 166-75.

Betjeman, K. (1974) Science facilities in Western Australian secondary schools. *Australian Science Teachers' Journal*, 20(3), 23-32.

Blosser, P. E. (1980) *A Critical Review of the Role of the Laboratory in Science Teaching*, Washington D.C., National Institute of Education, ED 206-445.

_____ (1983) What research says; the role of the laboratory in science teaching. *School Science and Mathematics*, 83(2), 165-9.

Brown, S. A. (1977) A review of the meanings of, and arguments for, integrated science. *Studies in Science Education*, 4, 31-62.

_____ McIntyre, D., Drever, E. and Davies, K. J. (1976) *Innovations in Scottish Secondary Schools*, Stirling Educational Monographs No. 2, Department of Education, University of Stirling.

California State Department of Education (1964) *Science Curriculum Development in the Secondary Schools*, Sacramento, California State Department of Education.

Commonwealth Advisory Committee on Standards for Science Facilities in Independent Schools, Commonwealth of Australia (1973) *The Design of Science Rooms*, Canberra, AGPS.

Dark, H. G. N. and Squires, A. (1975) A survey of science teaching in 9-13 middle schools. *School Science Review*, 56(196), 464-78.

Davis, J. (1972) 'An assessment of changes in science instruction and science facilities initiated by NDEA Title III funds used for high school science in Tennessee between 1965 and 1970', unpublished Ed.D. thesis, University of Tennessee.

Department of Education and Science (1964) *Designing for Science*, Oxford School Development Project, London, HMSO.

Edsall, L. F. and Templeton, H. (1960) *Planning Science Facilities for Secondary Schools*, Albany, State University of New York, ED036985.

Englehardt, D. F. (1966) Space requirements for science instruction in Grades 9-12, Cambridge, Mass., Harvard University, Graduate School of Education. ED022353.

_____ (1968) 'Aspects of spatial influence on science teaching methods', unpublished Ed thesis, Harvard University, Graduate School of Education, ED024214.

_____ (1970) *Planning the Teaching Environment: Secondary Science Facilities*, New York, Englehardt and Legget Inc, ED045307.

Field, T. W. (1974) Science facilities in independent schools: a decade of Australian government involvement. *Australian Science Teachers' Journal*, 20(3), 33-9.

Fish, G. (1974) Secondary school science facilities in new schools in Tasmania. *Australian Science Teachers' Journal*, 20(3), 41-5.

Hall, I. M. (1974) Victoria: science labs in state post primary schools. *Australian Science Teachers' Journal*, 20(3), 105-6.

Hamilton, D. (1975) Handling innovation in the classroom: two Scottish examples, in W. A. Reid and D. F. Walker (Eds.), *Case Studies of Curriculum Change*, London, Routledge and Kegan Paul, pp. 179-207.

Holley, B. J. (1974) *A-Level Syllabus Studies: History and Physics*, Schools Council Research Studies, London, Macmillan.

Jenkins, E. W. (1967) The attitude of teachers to the introduction of Nuffield Chemistry. *School Science Review*, 49(167), 231-42.

_____ (1971a) The implementation of Nuffield O-Level Chemistry courses in secondary schools. *Educational Research*, 13(3), 198-203.

_____ (1971b) The implementation of curriculum development projects. *School Science Review*, 53(182), 203-8.

Kelly, P. J. and Monger, G. (1974a) The evaluation of the Nuffield O-Level Biology course materials and their use - Part I. *School Science Review*, 55(192), 470-82.

_____ and Monger, G. (1974b) The evaluation of the Nuffield O-Level Biology course materials and their use - Part II. *School Science Review*, 55(193), 705-15.

_____ and Nicodemus, R. B. (1973) Early stages in the diffusion of the Nuffield A-Level Biology materials. *Journal of Biological Education*, 7(6), 15-22.

Kerr, J. F. (1964) *Practical Work in School Science*, Leicester, Leicester University Press.

Martin, W. E. (1960) Facilities, equipment, and materials for the science programme, in N. B. Henry (Ed.), *Rethinking Science Education*. The Fifty-ninth Year Book of the National Society for the Study of

Education - Part 1, Chicago, University of Chicago Press.

Minnesota State Department of Education (1971) *Designs for Science Facilities*, St. Paul, Minnesota State Department of Education, ED083721.

Norman, J. T. (1969) 'An investigation of the suitability of various types of Biology laboratory designs for certain instructional practices', unpublished Ph.D. thesis, Michigan State University.

National Science Teachers' Association (1970) *Conditions for Good Science Teaching*. Recommendations of the Commission on Professional Standards and Practices of the NSTA, Washington D.C., National Science Teachers' Association.

Novak, J. D. (1972) *Facilities for Secondary School Science Teaching: Evolving Patterns in Facilities and Programs*, Washington D.C., National Science Teachers' Association.

Ontario Department of Education (1968) *Science Laboratories for Secondary Schools*, Toronto, Ontario Department of Education, ED085905.

Penick, J. E. and Yager, R. E. (1985) Trends in science education: some observations of exemplary programs in the United States. *Australian Science Teachers' Journal*, 31(3), 28-34.

Prosser, B. and Woolley, T. (1974) Open plan science in South Australia. *Australian Science Teachers' Journal*, 20(3), 57-64.

Ramsey, G. A. (1972a) Curriculum development in secondary school science, Part 1. *Quarterly Review of Australian Education*, 5(1).

_____ (1972b) Curriculum development in secondary school science, Part 2. *Quarterly Review of Australian Education*, 5(2).

Richardson, J. S. (1971) Rooms for developmental science courses, in National Science Teachers' Association, *School Facilities for Science Instruction*, NSTA, Washington.

Robertson, W. W. (1962) 'A critical survey of laboratory work in science in New South Wales secondary schools', unpublished M.Ed. thesis, University of Sydney.

Robins, G. (1974) Science facilities in Queensland state secondary schools. *Australian Science Teachers' Journal*, 20(3), 5-8.

Savage, G. (1958) Planning of school science blocks. *Australian Science Teachers' Journal*, 4(2), 3-12.

Schlessinger, F. R. (1963) *Science Facilities for Our Schools*, Washington, D.C., NSTA.

Schools Council (1969) The Nuffield science teaching project. *Education in Science*, 32, 20-3.

Showalter, W. M. (1984) *Conditions for Good Science Teaching*. Washington, D.C., National Science Teachers' Association.

Tamir, P. (1976) The Israeli high school Biology project: a case of adaptation. *Curriculum Theory Network*, 5(4), 305-15.

_____ (1978) The impact of BSCS on laboratory facilities for biology students in Israeli high schools. *International Evaluation Journal*, Princeton, N.J., Educational Testing Service.

Texas Education Agency (1980) *Planning a Safe and Effective Learning Environment for Science*. Austin, Texas, Division of Curriculum Development, Texas Education Agency.

The Four Associations (1964) The provision and maintenance of laboratories in grammar schools. *School Science Review*, 41, 145.

Turner, J. (1974) The planning of science accommodation in high schools. *Australian Science Teachers' Journal*, 20(3), 9-22.

Wideen, M. F. and Batt, R. L. (1974) An ex-post facto evaluation of the implementation and use of a province wide junior high school science programme, Paper presented to the annual conference of American Educational Research Association, ED095198.

# 5.2

# Evaluation of Student Laboratory Work and its Role in Developing Policy

**Pinchas Tamir**

## INTRODUCTION

The purpose of this chapter is twofold:

1. to describe a variety of means and ways of evaluation of student learning in the school laboratory;
2. to highlight the relationship between evaluation and policy making with special reference to learning in the laboratory.

As we have seen in Chapters 3.1, 3.2, 3.3 and 3.4 the laboratory can play an important role at all levels of science education. Figure 1.2.1 indicates that one of the questions implied by the model is, to what extent are the goals and objectives assigned to learning in the laboratory actually being achieved? It will be shown that the laboratory does not only provide unique learning experiences (see Chapters 6.1, 6.2, 6.3), but also requires special assessment procedures. These special procedures will be described and illustrated with pertinent examples. Teaching in the laboratory requires special skills, is time consuming and quite expensive. Unless we can show clearly that learning in the laboratory carries important benefits, it is doubtful whether educational systems and schools will be willing to allocate time and means to it. The relationship between evaluation and policy goes in two directions: on the one hand, policy determines the kind of evaluation that takes place in schools; on the other hand, the results of evaluation are used as a basis for decisions regarding a variety of school-related policies, such as distribution of students into study groups, promotion to a higher grade level, selection of instructional materials and instructional strategies, setting the conditions which facilitate success in external examinations and directing the efforts invested by students and teachers in their studies. In this chapter we shall attempt to show how the interaction between policy and evaluation works with regard to student learning and evaluation in the laboratory.

## OBJECTIVES OF STUDENT WORK IN THE LABORATORY

The most widely used framework for evaluation of student attainment is the taxonomy of educational objectives in the cognitive domain (Bloom, 1956) and in the affective domain (Krathwohl, Bloom and Masia, 1964). While this framework can be used and has been applied to science students, it has soon been realized that for many purposes the framework is too general and consequently it fails adequately to reflect certain important objectives and behaviours related to the learning of science. There are at least two kinds of limitations to the use of the 'general' taxonomies. The first is that the various abilities (i.e. comprehension or synthesis) depend a great deal on the specific subject matter, as well as on the context. One may perform well at the level of application while dealing with one topic in one context and at the same time experience difficulty in applying concepts related to another topic in another context.

Failure to consider the context is of special importance to science education which is routinely associated with different contexts, such as verbal, laboratory and outdoors learning. There is enough evidence to suggest that the 'practical mode' (see below) is a distinct mode of performance, which deserves special consideration in evaluation of science achievement. Undoubtedly the outdoors context also has some unique characteristics which deserve the attention of evaluators. As we review the evaluation literature we shall attempt to highlight the effects of subject matter and context as much as existing data permits us.

The second kind of limitation of the general taxonomies is that they refer to mental abilities which may be implied by the nature of the task, but they do not deal directly with important objectives such as enquiry skills, both intellectual and psychomotor. The need to design specialized taxonomies for particular disciplines has been recognized by many evaluators. Some have modified the general taxonomies to adapt them for their specialized needs (e.g. Wilson, 1971, for mathematics and Klopfer, 1971, for science). Others have designed entirely different taxonomies, (e.g. Assessment of Performance Unit, 1981; Tamir, Nussinovitz and Friedler, 1982) for performance in practical laboratory tests).

## THE NATURE OF THE PRACTICAL MODE

Olson (1973) observes that human learning is always mediated or specified through some form of human activity. Both knowledge ('knowing that') and skills ('knowing how') are acquired by using one or more of the following modes: 'direct experience', 'modelling or observational' and 'symbolically coded' (i.e. information transmitted through the media of speech, print, pictures and films). Any educational experience involves both knowledge and skill. Because different experiences generate different kinds of skills, it is important to recognize that the means of instruction exert a decisive effect on the acquisition of skills. Therefore, the crucial, yet largely overlooked, issue for instruction becomes one of deciding which skills should be cultivated. In our schools we use mostly symbolic systems. Consequently, while 'schools aspire to facilitate the acquisition of both knowledge and skills, they are reasonably successful in serving the goals pertaining to the acquisition of knowledge, but they serve poorly the educational goals pertaining to the development of skills. Yet it is these skills that are primarily responsible for the generalizable effects of experience' (p. 41).

Table 5.2.1 attempts to summarize Olson's arguments by presenting the relative contributions to knowledge and skills of four major components of education: goals, modes of learning, instructional media and the assessment of outcomes.

How does the laboratory relate to the different educational components? As far as educational goals are concerned the laboratory is certainly expected to provide for the development of motor and intellectual skills as well as problem-solving abilities and affective outcomes. The major learning mode is direct experience, based on instructional media characterized by performance acts with direct feedback. While typical paper and pencil tests are, as suggested by Olson, 'high' in knowledge

**Table 5.2.1:** The relationship of knowledge and skills to goals, modes, media and outcomes

| Educational Component | Knowledge | Skills |
|---|---|---|
| Educational goals: | | |
| Cultural heritage | High | Low |
| Motor skills | Low | High |
| Intellectual skills | Medium | High |
| Problem-solving ability | Medium | High |
| Affective outcomes | Medium | High |
| Modes of learning (forms of experience): | | |
| Direct experience | Medium | High |
| Modelling and observations | Medium | Medium |
| Symbolically coded | High | Low |
| Instructional media: | | |
| Performance acts with direct feedback | Medium | High |
| Mass media (speech, print, TV, films, pictures) | High | Low |
| Present-day schooling | High | Low |
| Assessment of outcomes: | | |
| Paper and pencil achievement tests | High | Low |
| Real-life situations | Medium | High |
| Technique-oriented practical tests | Low | High |
| Enquiry-oriented practical tests | High | High |

and 'low' in skills, practical tests of any kind are judged by us to be 'high' in skills. However, there is a fundamental difference between the two types of practical tests - while technique oriented tests (see Chapter 3.1) are 'low' in knowledge, enquiry oriented tests (see Chapter 3.2) are 'high' in both knowledge and skills. It has been argued elsewhere that the school laboratory occupies a unique position since it offers a balanced mix of modes of learning and provides ample opportunities for the development of skills. This uniqueness has been designated the practical mode (Tamir, 1972a).

The distinctiveness of the practical mode is restricted neither to the involvement of the learner in manipulation, nor to the activation of psychomotor skills. We argue that even intellectual skills such as planning an investigation, formulating hypotheses and interpreting data are different when carried out as part of an investigation going on in the laboratory. Our argument is supported by empirical data. For example, the Israeli matriculation examination in biology includes planning of investigations in two contexts: one in the laboratory, where students actually carry out the experiment, and the other as part of a paper and pencil test. The correlations between the scores of these two seemingly similar tasks have ranged over the years between 0.3 and 0.4. Similarly, the Assessment of Performance Unit (1984) reports significant differences between the nature of the planning in a paper and pencil test on the one hand, and that taking place in the context of performing a practical investigation, on the other.

It has repeatedly been shown that performance in the practical mode is only weakly correlated with performance on paper and pencil tests (Ben Zvi *et al.* 1977; Comber and Keeves, 1973; Robinson, 1969; Tamir, 1972a, 1975). Students who do well in the laboratory do not always do well in content (paper and pencil) examinations (Hearle, 1974), since 'practical work involves abilities both manual and intellectual which are in some measures distinct from those used in non-practical work' (Kelly and Lister, 1969, p. 122). It may be concluded that the practical mode is a unique mode of instruction as well as a unique mode of assessment.

These findings have important implications for policy making. Firstly, it is obvious that paper and pencil tests cannot substitute for practical tests. Secondly, the use of results in theory examinations as a basis of moderation of teachers' internal assessment of practical work, as advocated by certain examination boards in the UK (see Kempa, 1986, pp. 86-92) is totally unjustified. Even when the correlation between the scores in the theory paper and in the internal assessment is as high as 0.6 (see Kempa, 1986, p. 89) the moderation is not justified because only 36% of the variance is overlapping. The use of a theory paper to moderate practical scores is no more justified than the use of an IQ test to moderate the results of an achievement test.

Practical examinations should be used not only for the assessment of individual students' achievement, but also for the purpose of curriculum evaluation. Hundreds of studies comparing method A with method B, or one instructional medium with another, have resulted in non-significant differences. An explanation often offered for this finding is that the measuring instruments have not been sensitive enough to identify the differences. This, of course, implies that the researcher believes that such differences do exist. Olson offers an alternative explanation. He appears to believe that we obtain these non-significant differences in so many studies because there are no differences. This is so because 'these studies typically assess only the knowledge conveyed - the level at which these systems converge - and overlook the skills developed - the point at which they diverge' (p. 36). In other words, if the educational goal is knowledge, 'learning as now measured is largely independent of the details or means' (p. 31). On the other hand, when it comes to skills, one should find significant effects of 'details and means'.

This prediction has been realized. Highly significant differences were obtained between high school biology students who followed an enquiry-oriented laboratory-based curriculum and a control group that spent a similar number of hours in the laboratory but had followed a non-enquiry curriculum in which all laboratory exercises were at Herron's (1971) level zero (see Chapter 6.2). It was obvious that the two curricula had diverged with regard to the development of skills. The students in the enquiry-oriented curriculum, when required to work out a novel problem in the laboratory, were much more proficient in such skills as observing, designing a controlled experiment, and recording and interpreting results (Tamir, 1974). This was achieved with no loss in knowledge. In fact, the students who followed the enquiry-oriented curriculum achieved better on content paper and pencil tests as well (Tamir and Jungwirth, 1975).

## EVALUATION STRATEGIES

Several studies have shown that the relationship between the quality of the products of practical work and the quality of procedures used in it is quite weak. For example, Buckley (1970) compared scores based on results of exercises in A-level Chemistry with those based on direct observations of students' performance and found an average correlation of 0.25. Similarly, a low correlation ($r$ = 0.33) was found between scores obtained by internal teacher-based assessment which focused on processes and those obtained by a formal external product-based practical examination (Kempa, 1986, p. 89).

Several reasons may account for these discrepancies:

1. 'It is possible for a relatively minor error in one of the subtasks to affect adversely the final outcome, even if all

other subtasks have been carried out satisfactorily' (Kempa, 1986, p. 73);

2. In the case of internal teacher-based assessment, the scores are based on several observations spread over the year and students' performance may indeed change over time; on the other hand, a practical test is a one-time event, typically occurring at the end of the course. Policy makers should be aware of the weak correlations between the two approaches. In certain situations it may be possible and desirable to combine the two approaches.

## Continuous teacher-based assessment

Presumably, the most authentic way of evaluating practical performance may be accomplished by the teacher who unobtrusively observes individual students during normal lab activities and rates them on specific criteria. It is authentic because it is done under natural conditions or at least closer to them than external examinations. The assessments can be recorded for each student over an extended period of time or they can be made on one selected laboratory investigation. Normally, only a few students may be evaluated during a given laboratory lesson. This approach has been adopted by a number of examination boards in the UK (e.g. Joint Matriculation Board, 1979; University of London, 1977), who require the teachers to assess each student on several occasions throughout the year in order to cover a variety of tasks and skills. Most commonly the following outcomes have been assessed: planning and design; manipulative skills; observation, measurement and data recording; interpretation of experimental results; responsibility, initiative and work habits (Lunetta, Hofstein and Giddings, 1981).

There is one means of continuous evaluation, however, which may be quite useful if properly employed: the laboratory report. Students should be required to keep laboratory notebooks in which they (a) write their procedures and findings during the lesson, and (b) summarize and analyse their results and try to make sense out of them at home. Checking these laboratory reports may be an excellent means of evaluation. Laboratory reports are part and parcel of practical work, their routine assessment is certainly expected by the students and hence may be regarded as highly unobtrusive, much more so than that performed by teachers who set aside special laboratory tasks for assessment of their students' work by direct observation.

It should be noted, however, that the teacher's assessment, which focuses on performance, evaluates processes, assessment, of students' reports, is an (indirect) evaluation of products. Hence one is not a substitute for the other. Even though laboratory reports and other informal means are quite useful, most teachers have used and will continue to use tests as an important means of student evaluation. Evaluation of laboratory work is no exception.

## Paper and pencil tests

The tendency in the USA and elsewhere has been to test for many practical objectives by paper and pencil tests (e.g. Korth, 1968; Ruda, 1979). In the few cases where practical tests were used, they tended to focus on performance categories, mainly manipulation techniques and description of observations. In the UK there has been a long tradition of practical work followed by practical examinations. However, even there the same tendency is found to some extent.

A great number of paper and pencil tests which purport to measure processes of science and enquiry skills have been designed and used over the years (Ben Zvi *et al.* 1976; Biological Science Curriculum Study, 1962; Brumester, 1953; Fraser, 1980; Golmon, 1975; Kruglak, 1958; Swanson and Ost, 1969; Tenenbaum, 1971). However, as already explained, the nature of the skills assessed by paper and pencil tests is qualitatively different from that assessed by practical tests.

## Vicarious experience tests

In these tests the students do not manipulate equipment, materials or organisms, but do make observations on phenomena (e.g. films) and may even employ enquiry skills and follow a whole investigation. For example, the Biological Science Curriculum Study single topic loop films have been used as tests of enquiry skills in Israel (Tamir, 1976). Similarly, Oliver and Roberts (1974) presented students with filmed experiments in chemistry who had to respond to questions following each episode in the experimental sequence. Based on their trials they concluded that the introduction of visual elements into the assessment makes it possible to extend the range of scientific behaviours being assessed. The microcomputer is a new, potentially powerful means which may be used to simulate scientific investigations, thereby serving as an evaluation tool. For example, some of the programs developed in genetics (e.g. Kinnear, 1982) may be ideal for assessment purposes. (See also Chapter 5.3.)

Tests of this kind are not widely used and their potential and actual benefits and limitations are yet to be discovered by research. However, based on the characteristics of these tests, they may be regarded as occupying a middle position between paper and pencil tests on one hand, and practical tests on the other hand. They also have unique advantages. For example, with regard to observations, in practical tests, students who prepare a slide and observe it under the microscope may see different things because of real differences among inidividual slides. If, on the other hand, a video tape of the slide is presented, all examinees will be seeing the same picture and the assessment of the quality of their observations may be easier and more reliable. Another kind of advantage relates to the potential of extending the repertoire of enquiry-oriented practical tests. Such tests are limited to experiments which can be completed in a relatively short time (one to two hours). If we use computers, for example, either alone, or in combination with some actual practical activities, we can incorporate investigations which normally take days or even weeks, in areas such as microbiology or genetics, by using time-lapse procedures.

## Practical laboratory tests

As postulated above the practical mode is a unique mode of performance. Hence, there is no real and complete substitute for practical tests. This means that practical tests can and should be used to assess all the outcomes expected of learning in the laboratory although, as we shall see, different kinds of practical tests assess different kinds of skills.

Five arguments are offered as justification for the use of practical examination: (1) Grobman (1970), summarizing the state-of-the-art in the USA, wrote: 'there has been little testing which requires actual performance in real situations, or in a simulated situation approaching reality ... to determine not whether the students can verbalize a correct response but whether [they] can perform an operation, e.g. a laboratory experiment or an analysis of a complex problem. ... This is an area where testing is difficult and expensive, yet since in the long run primary aims of projects generally involve doing something rather than writing about something, this is an area which should not be neglected.' (2) The relative importance assigned to different assessment procedures determines how the student learns and what efforts he invests in different learning tasks. (3) Assessment procedures may have profound effects on instruction. For example, the use of open-ended laboratory examinations as part of the matriculation examination in high school biology in Israel. It has changed dramatically the laboratory instruction has become significantly more prominent and much more enquiry-oriented. (4) Kelly and Lister (1969) suggested that 'the need for some measurement of pupils' performance in practical work is based on the desirability of reflecting the important role of practical work in courses they take'. (5) Lastly and most importantly: 'practical work involves abilities, both manual and intellectual, which are in some measure distinct from those used in non-practical work' (Kelly and Lister, 1969).

Based on the arguments and evidence presented above we define a practical examination as a task which requires some manipulation of apparatus or some action on materials and which involves direct experiences of the examinee with the materials or events at hand. Practical examinations may be administered individually or to groups.

*Individually-administered examinations*

Individually administered examinations involve students who perform the required tasks and an examiner who observes and/or guides the performance and assigns marks. Three different procedures will be described.

In the first, the examiner observes the performance and assigns marks following a prepared checklist. Usually the examiner neither asks questions nor gives directions, but rather allows the examinee to follow written directions and hence assessment is confined to the observed behaviours. The checklist-based assessment is especially suitable for direct assessment of technical skills and quality of performance. It is recommended for use in internal teacher-based assessment (Kempa, 1986, pp. 78-81). The Assessment of Performance Unit (1984) has made extensive use of checklists, for the assessment not only of techniques and manipulative skills, but also of the experimental conditions employed by the examinees in particular investigations, such as adequately setting the treatments, accurately following directions, or keeping time as required.

**Figure 5.2.1:** An observation checklist (Tyler, 1942)

The teacher's goal is to see whether the student is able to operate a microscope so that a specimen present in a culture is located.

| | | |
|---|---|---|
| a. Does the student wipe the slide with lens paper? | YES | NO |
| b. Does the student place a drop or two of culture on slide? | YES | NO |
| c. Does the student wipe cover glass properly? | YES | NO |
| d. Does the student adjust cover glass adequately? | YES | NO |
| e. Does the student wipe off surplus fluid? | YES | NO |
| f. Does the student place slide on stage adequately? | YES | NO |
| g. Does the student look through eye piece and hold closed one eye? | YES | NO |
| h. Does the student turn to objective of lowest power? | YES | NO |
| i. Does the student adjust light and concave mirror? | YES | NO |
| j. Does the student adjust diaphragm? | YES | NO |
| k. Does the student use properly coarse adjustment? | YES | NO |
| l. Does the student break cover glass? | YES | NO |
| m. Does the student locate specimen? | YES | NO |

The second kind of individually-administered practical test is an oral interview, which is carried out partly during the execution of the experiment by the pupil and partly right after the completion of the investigation. The interactive nature of the oral examination enables the examiner to probe deeply into students' understanding and, hence, this kind of test is useful as a diagnostic instrument capable of identifying misconceptions. The Assessment of Performance Unit have used oral interviews in combination with other procedures in assessing their category 'performing investigations'. They explained their choice of assessment strategies as follows:

> The problems lie in finding a way to take advantage of the richness of information the answers provide and to interpret the performance as correctly as possible. In some cases it is possible to assess the performance by observation using a checklist or set of categories; in a few

> cases, the pupils write and give a fair account of what they have done; in other cases, it may be necessary to use a combination of these two kinds of information plus additional data from interviewing pupils when they have finished. (Assessment of Performance Unit, 1978, p. 21)

An example of a task using an oral interview is given in Figure 5.2.2.

The third kind is an oral examination based on concrete phenomena or materials. Examples are identification of parts in a prepared dissection, or the kind of oral included in the matriculation examination of high school biology students in Israel, who carry out an ecological project for several months and then bring selections of organisms that they have studied to serve as objects for their oral examination (Tamir, 1972b).

In principle, oral examinations can be used to test almost any outcome. Their disadvantages are high cost and low reliability. However, when tests have to be tailor-made for individual students, such as the case of the oral examination linked with individual projects, they constitute the only feasible alternative. While the one (examinee) to one (examiner) practical examinations described above have their place as an important research tool and as a means of evaluation under special circumstances in schools, they are too expensive to become a routine assessment procedure in most schools, in which group-administered examinations are usually to be preferred. Unlike the one-to-one examination where assessment is based on direct observation and oral probing, group examinations are based on written responses. Three types of group-administered practical examinations are described below, all of which have been used in the matriculation examinations in Israel as well as in high schools and universities in the UK. They have been used in other countries such as Japan, Hungary and the USA to some extent, but many countries have avoided them because of the difficulties involved in their design, administration and scoring.

**Figure 5.2.2:** A task during a practical examination (Assessment of Performance Unit, 1978)

**Materials:** Three dropping bottles labelled P (distilled water), Q (acetone in water), R (citric acid in water). Dry cobalt chloride paper in desiccator; blue litmus paper; six clean test tubes; safety spectacles.

**To the student:** You are given three clear liquids labelled P, Q and R, one of which is just water on its own. Follow the instructions below to find out which is just water.

a. When cobalt chloride paper is put in a liquid which contains water it changes colour from blue to pale pink.
b. When blue litmus paper is put in a liquid which is acid it changes colour from blue to red.

1. Test each liquid, one after the other, using a clean test tube and fresh piece of the indicator paper each time, first by the cobalt chloride paper and then by the blue litmus paper. Record your results in the table below.
2. Smell each liquid and tick in the table below which of the liquids has a smell.

| Test | Liquid P | Liquid Q | Liquid R |
|---|---|---|---|
| Cobalt chloride paper | | | |
| Litmus paper | | | |
| Smell | | | |

3. Which liquid is just water? - Give your reasons.

The first type of group-administered examination to be described involves the use of a dichotomous key to identify the name of an unrecognized object such as rock or an organism. This type of examination has been described as follows: 'The ability to identify an unknown plant with the aid of a key is considered as an important skill in biology. While developing this skill some corollary objectives may be achieved, such as developing observational skills, getting acquainted with the principles of taxonomy, getting to meet and observe plants, experiencing the diversity of type and the unity of patterns; one of the major themes of biology' (Tamir, 1972c, p. 62).

The following procedures may be used. A group of

students, usually no more than thirty, is seated one student per desk. A key, magnifying glass, needles, razor blade, pencil and paper, and a plant bearing flowers, fruit and roots are provided. On some occasions students are asked to give a full description of the plant referring to the unique characteristics of roots, stem, arrangement of leaves, shape of leaves and their edges, arrangement of flowers, structure of flower, type of fruit. In order to keep track of the sequence used by the examinee and his or her ability to identify correctly the different features of the organism, the student is required to record the numbers of items followed as well as the descriptions in the text which fit the organism under examination. A detailed scoring key is used according to which the student loses a pre-agreed mark for each mistake. 50% of the marks are assigned to correct identification of the family, genus and species, while the other 50% are assigned to the correct recording of sequence and features. The examination lasts 30 to 45 minutes.

The second type of examination, which is often used in the UK and in college science courses elsewhere, is designated as a 'circus', or a series of stations. Each table in the laboratory room constitutes a station. Each station provides equipment, materials and instructions for a particular task. The student gets between 5 and 25 minutes depending on the nature of the task. For example, in the framework of the lower level of matriculation examination in biology in Israel examinees are required to perform three tasks, each lasting 25 minutes. One examiner can control six sets of three students each, altogether 18 students. Examinees perform tasks which ask for written reports so that marking is, by and large, based on the written report. In certain tasks, the examiner is asked to make occasional checks. For example, in a task which requires the use of a microscope the student is instructed to call the examiner when he has completed mounting a slide for observation. The examiner checks the preparation, lighting and adjustment and assigns a mark which will constitute part of the overall mark of this particular task. Figure 5.2.2 presents an example of a station task for 13-year-old students.

The third type is the most innovative practical test currently used as external examination in Israeli secondary schools. It was designated an enquiry-oriented practical examination. The test problems were selected and designed with several considerations in mind; notably that they should pose some real and intrinsically valuable problems for the students, and that the problems should be novel to the examinee, but the level of

**Figure 5.2.3:** An enquiry-oriented laboratory test (used in the biology matriculation examination in Israel)

**Materials:** a toad, beakers, a thermometer, crushed ice, bunsen burner, table salt, a watch or a stopper.

**To the Student: Part A**

Observe the toad on your table and design an investigation of the toad, using some or all of the materials on your table. The investigation should involve quantitative measurements, the results of which can be reported by making a graph. It should be completed in about 90 minutes.

1. What is the problem you are going to investigate?
2. What hypothesis do you intend to test?
3. Design an experiment to test your hypothesis.
   a. What is the dependent variable? How will you measure it?
   b. What is the independent variable? How will you change it?
   c. Describe your procedure in detail.

Hand Part A to the examiner and obtain Part B.

**To the Student: Part B**

Even though your design may be appropriate, please follow the instructions below which will enable you to test the effects of different temperatures on the toad's activity.

4. The activity of the toad will be measured by counting the throat movements. What is the assumption behind this measuring procedure?
5. Place the toad in the 400 ml beaker in about 4 cm of water. Cover the beaker and insert a thermometer through the cover. Practice counting the throat movements during 30 seconds. If the movements are too fast, put the beaker with the toad into a large beaker containing crushed ice to lower the temperature. Record your results.
6. Measure the activity of the toad at 20°C, 10°C, 5°C. Repeat each count three times. Record your results.
7. What do you think will happen to the toad if you lower the temperature to -5°C? Explain.
8. Perform the observation of the toad at -5°C. Describe your procedure step by step. (Don't be afraid. The toad will not die.)
9. Record the results at -5°C. Take out the small beaker with the toad and put it on the table.
10. Summarize the results in a table.
11. Draw a graph presenting the results.
12. What happened to the toad when returned to room temperature.
13. How would you explain the toad's behaviour at different temperatures? Is this behaviour of value in the toad's adaptation to the external environment? Explain.
14. Would you expect a similar behaviour with a mouse? Explain.

difficulty and the required skills should be compatible with the objectives of and experience provided by the curriculum. An example is presented in Figure 5.2.3.

In order to standardize the assessment of students' responses in this open ended enquiry-oriented practical examination, a Practical Tests Assessment Inventory (PTAI) was developed (Tamir, Nussinovitz and Friedler, 1982) and since its introduction has been regularly used in assessing the practical matriculation examinations in biology in Israel. The use of PTAI has not only increased the reliability of assessment but, most importantly, has provided detailed information about the performance profile of students for the whole country as well as for each of the participating schools. This information is of great value to curriculum designers, the inspectorate, administrators, teachers, students and college professors who are interested in their students' entry characteristics.

This is not the place for too many details regarding PTAI. Nevertheless, it is important to mention that the categories and levels were determined empirically on the basis of an analysis of hundreds of students' test papers. The categories indicate clearly the nature of the outcomes that are being assessed by the enquiry-oriented practical tests used in the matriculation examinations. The results provide accurate and detailed information which can be used to build achievement profiles. Such information is important not only for grading but also for diagnostic purposes. We see, for example, that although most students can make observations and describe them adequately, about one-fifth still do not distinguish between observation and interpretation. We can also observe that planning an investigation in the context of the practical setting poses serious problems to many students. These problems can now be pinpointed. For example, about 30% of the students can neither formulate a problem nor formulate a hypothesis.

However, while in the year 1976 10% formulate a problem and in 1980 11% formulate an assumption instead of an hypothesis, in 1984 the figures come down to 2% and 1% respectively. While about 60% can identify correctly the dependent variable as part of planning an investigation, close to 20% do so in terms of the measuring technique (e.g. number of bubbles per minute) instead of the actual variable (e.g. the rate of photosynthesis). About 10% of the students confuse the dependent and independent variables. Close to half of the students have difficulties in fitting an experiment to the problem or hypothesis to be tested. As to communication of results,

50% to 60% can adequately draw a graph of their results; yet half of them fail to provide an adequate title to the graph. The examples just mentioned illustrate some of the benefits which may be derived from the use of PTAI.

How has PTAI actually been used? For the matriculation examination, the examination committee prepares a detailed scoring key for each test problem. As a first step all the items in the test problem are categorized into the appropriate categories of PTAI. Following that, each level within each category is allocated a certain number of points. The evaluator has to judge the level of each student answer. Having done so, the score is determined by the key. The high inter-rater agreement obtained led to an important policy decision: each test is now graded by one evaluator instead of two, a procedure which saves time and money.

### Other evaluation strategies related to the laboratory

We have concentrated on the evaluation of students' outcomes in terms of cognitive and psychomotor skills. Yet evaluation in the laboratory is much broader. Other evaluation activities which have important implications for monitoring laboratory work in a particular course are discussed by Hegarty-Hazel (1986) and Chapters 6.2, 6.3. These include opportunities to learn, classroom observation and attitudes and interests.

## THE IMPACT OF ASSESSMENT ON SCHOOL PRACTICES

Assessment and evaluation are extremely powerful means of affecting what is taught in schools and how it is taught. The following are examples of such effects brought about by changes in the matriculation examination in biology in Israel. A new form of matriculation examination was introduced in 1969. This new form is comprised of a paper and pencil component which accounts for 60% of the grade and a practical component which accounts for 40% of the grade.

***Examinations in schools***. All the biology teachers are using practical examinations as part of their internal assessment. Five enquiry-oriented laboratory tests are designed each year. These tests are published in booklets which are available to teachers and students and are widely used in schools. The use of different practical tests has not been confined to years 11 and 12 but has permeated to lower grade levels as well.

***Instruction and learning***. The impact of the matriculation examination has not been restricted to teachers' examinations. The published test problems are often used as laboratory exercises incorporated into the daily learning. However, the most important effect of the practical matriculation examination has been on time allocation. Students and teachers indeed devote at least 40% of the class time to different kinds of practical experiences, including field trips. The oral examination facilitates students' seriousness in working on their individual ecological projects which are carried out mostly out of school and on top of the regular timetable of the school.

***The involvement of teachers***. The involvement of teachers takes place in a number of ways: some teachers participate in the actual design and development of laboratory tests. Many teachers serve as examiners for the different parts of the practical test: teachers from one school are assigned to test students in other schools. Not only do these teachers benefit from the guidance sessions which are held each year in preparation for the examinations, but, significantly, they gain tremendously by interacting with students who study the same curriculum from other teachers in different settings. The grading of the different tests is done by teachers under close supervision - another golden opportunity for the self-development of teachers. Lastly, as described above, teachers always make special efforts to participate in in-service training days devoted to the previous year's matriculation examinations. The discussion on these occasions often stimulates second thoughts among teachers as to what they have been doing and what can be done in their teaching. These occasions contribute considerably to the socialization of novice teachers. They also provide an opportunity for misunderstandings to be clarified and allow for fruitful exchanges between teachers and the project's staff.

***Equipment and learning media***. In two studies (Tamir, 1978a; Dreyfus, 1979), it was shown that as a result of the laboratory examinations, not only are students performing laboratory investigations regularly, but, almost without exception, schools have acquired well-equipped laboratories which improve yearly in terms of materials and facilities. Also, most schools employ laboratory technicians who facilitate learning in the laboratory. The importance of facilities in enhancing learning and achievement is dealt with in Chapter 5.1.

***Attracting students.*** Fifteen years ago biology was a relatively unattractive topic in high school, usually avoided by the more talented students, who elected to major in physics or chemistry. Today it is considered as a high level, highly respected, science course.

***Impact on college.*** University professors have begun to realize that students majoring in high school biology are better prepared for studies in the university. First year courses are now taking into account the preparation of students in high school biology as evidenced by their matriculation examination (Tamir, 1978b).

***Teacher-training exercises.*** A unique feature is the teacher-training exercises based on the answers of examinees in the matriculation examinations. The answers of students to the enquiry-oriented laboratory investigations were collected and used for exercises, each dealing with one particular enquiry skill, such as: problem formulation, hypothesis formulation, designing experiments, designing controls, performing observations, measuring, identifying the dependent variable, identifying the independent variable, designing graphs, interpreting and explaining. The fact that these exercises are based on actual answers of students, correct and incorrect, routine and creative, usual and unexpected, makes them exercises an important means of helping teachers identify the deeper level of their students' learning, achievement and understanding. Some teachers have been using these exercises with their students and have found them highly profitable and rewarding.

## IMPLICATIONS FOR POLICY DECISIONS AND IMPLEMENTATION

In the previous pages we presented an overview of evaluation of student work in the laboratory. To a large extent we have drawn upon the experience of the Israeli High School Biology Project (IHBP) which has deliberately used examinations as a means of carrying out educational policy. We are ready now to highlight the implications of the IHBP experience regarding the use of student assessment in general and laboratory examinations in particular for implementing policy decisions and guiding an educational system in a particular direction.

Testing may, and often does, have negative effects on learning. This is not the place to analyse these negative effects, nor to describe the potential positive effects. Our basic premise is that, at least for the foreseeable future, testing in schools will continue and hence our interest is to try to make the best use of it. It is also the case that tests can play an important role in promoting excellence on the one hand, and facilitating equal opportunities, namely, equity, on the other hand. This 'double edge' effect is particularly true with regard to the use of external examinations such as the matriculation examinations in Israel which are set centrally by agents of the Ministry of Education. When the new high school biology was introduced to Israeli schools in the mid-1960s, biology was taught mainly by book, chalk and talk, stressing memorization of descriptive information, and generally considered a third-class subject elected as a specialized matriculation option by relatively few. In the mid-1980s biology has become a highly prestigious subject, devoting considerable time to student work in the laboratory and outdoors, featuring science as enquiry and stressing major ideas and principles of modern life science; it is elected as a specialized matriculation option by one-third of the student cohort and 55% of the students admitted to the highly selected medical schools have matriculated in high school biology.

In our judgement, this positive development has been a result of continuous effort on the part of the policy makers who established a comprehensive support system in which the practical matriculation examinations have played a decisive role.

We have already described the three parts of the practical matriculation examination, namely identifying an unknown organism with the aid of a key, the oral examination based on an individual ecological project, and the enquiry-oriented laboratory test. These three parts of the practical account for 40% of the total matriculation score (as compared with 10% to 20% in other systems in other countries which employ practical examinations). The decision to allocate such weight to the practical reflects the importance of the objectives and learning experiences associated with it and has had tremendous impact. It may be concluded that considerable attention should be given to the tests used by teachers as well as by external examiners in order to benefit from their potential positive effects. More specifically:

1. A variety of test formats should be used.
2. The tests should reflect the whole array of objectives and learning experiences.
3. In science a substantial weight should be allocated to enquiry-oriented practical tests.
4. Teachers should be involved in decisions about the nature of external examinations, in the design and administration, in the process of assessment, and in the analysis of results.
5. Rather than wasting time and energy in trying to abolish matriculation examinations because of their poor quality, invest in designing high quality valid examinations thereby upgrading achievement, promoting excellence and facilitating equality of opportunities.

The contributions of the continuous use of practical matriculation examinations to curriculum development implementation are:

1. Clarifying the objectives and intentions of the developers;
2. Providing models for learning activities;
3. Ascertaining the place of practical work in the implemented curriculum;
4. Providing high quality learning activities;
5. Regulating the relative investment of effort in different components of the curriculum;
6. Facilitating the development of special instructional materials to remedy identified weaknesses and deficiencies; for example, a model entitled *Basic Concepts of Scientific Inquiry* (Friedler and Tamir, 1984, 1987), or a module entitled *Plant Identification* (Avidor and Tamir, 1985). For an elaboration see Tamir (1981).

The practical matriculation examinations have had substantial impact on the actual processes of instruction.

1. Influencing teachers' questioning behaviours;
2. Motivating teachers and students to work in the laboratory;
3. Demonstrating that laboratory work can be challenging and rewarding;
4. Improving the conditions of learning by influencing schools to acquire better facilities or to hire laboratory

technicians;
5. Enhancing diversity of instructional modes and creative teaching strategies.

The impact of the practical examinations on teacher education has been mentioned in the previous sections and elaborated by Tamir (1983). The practical examinations may influence teacher education as follows:

1. Providing useful experience to pre-service and in-service programmes;
2. Providing new material for teacher education exercises aimed at helping teachers in identifying student misconceptions;
3. Teaching prospective and practising teachers innovative and interesting assessment strategies;
4. Familiarizing teachers with the rationale and measures used in external examinations pertaining to the practical components of the programme they are supposed to teach;
5. Alerting teachers to the skills they need to develop for teaching in the laboratory; and
6. Training teachers in designing high quality practical tests, in their assessment and in analysis of their results.

## REFERENCES


Assessment of Performance Unit (1978) *Science Progress Report 1977-78*, Elizabeth House, London.

_____ (1981) *Science in schools: Age 11*, Report No. 1, Her Majesty's Stationery Office, London.

_____ (1984) *The assessment framework of science at ages 13 and 15*, APU Science Report for Teachers No. 2, Department of Education and Science, London.

Avidor, O. and Tamir, P. (1985) *Plant Identification*, The Israel Science Teaching Center, Hebrew University, Jerusalem (in Hebrew).

Ben Zvi, R., Hofstein, A., Kempa, R. F. and Samuel, D. (1976) The effectiveness of filmed experiments in high school chemical education. *Journal of Chemical Education*, 53, 518-20.

Ben Zvi, R., Hofstein, A., Samuel, D. and Kempa, R. F. (1977) Modes of instruction in high school chemistry. *Journal of Research in Science Teaching*, 14, 433-9.

Biological Science Curriculum Study (1962) *Processes of Science Test*, Boulder, Colorado.

Bloom, B. S. (Ed.) (1956) *Taxonomy of Educational Objectives, Handbook I: Cognitive Domain*, David McKay, New York.

Brumester, M. A. (1953) The construction and validation of a test to measure some of the indicative aspects of scientific thinking. *Science Education*, 37, 131-40.

Buckley, J. G. (1970) 'Investigation into assessment of practical abilities in sixth form chemistry courses', M.Sc. thesis, University of East Anglia.

Comber, L. and Keeves, J. (1973) *Science Education in Nineteen Countries*, Almqvist, Stockholm.

Dreyfus, A. (1979) *Evaluation of the Impact of the BSCS on the Laboratory Facilities in Schools*, The Israel Science Teaching Center, Hebrew University, Jerusalem (in Hebrew).

Fraser, B. J. (1980) The validation of a test of inquiry skills. *Journal of Research in Science Teaching*, 17, 7-16.

Friedler, Y. and Tamir, P. (1984) *Basic Concepts of Scientific Research*, Israel Science Teaching Center, Hebrew University, Jerusalem (in Hebrew).

_____ (1987) Teaching basic concepts of scientific inquiry to high school students. *Journal of Biological Education*, 20, 263-70.

Golmon, M. (1975) Assessing laboratory instruction in biology. *Iowa Science Teachers Journal*, 19, 4-10.

Grobman, H. (1970) *Development Curriculum Projects: Decision Points and Processes*, F. E. Peacock, Itasca, Illinois.

Hearle, R. (1974) The development of an instrument to evaluate chemistry laboratory skills. Paper presented at the Annual Meeting of the National Association of Research in Science Teaching, Chicago.

Hegarty-Hazel, E. (1986) Monitoring, in D. Boud, J. Dunn and E. Hegarty-Hazel (Eds.) *Teaching in Laboratories*, SRHE-NFER Nelson, Guildford.

Herron, M. D. (1971) The nature of scientific inquiry. *School Review*, 79, 171-212.

Joint Matriculation Board (1979) The internal assessment of practical skills in chemistry: Suggestions for practical work and advice on sources of information.

Kelly, P. J. and Lister, R. (1969) Assessing practical ability in Nuffield A-level biology, in J. F. Eggleston and J. F. Kerr (Eds.) *Studies of Assessment*, English Universities Press, London, 129-42.

Kempa, R. (1986) *Assessment in Science*, Cambridge University Press, Cambridge.

Kinnear, J. (1982) Computer simulations and concept development in students in genetics. *Research in Science Education*, 12, 89-96.

Klopfer, L. (1971) Evaluation of learning in science, in B. S. Bloom, J. T. Hastings and G. F. Madaus (Eds.) *Handbook of Formative and*

*Summative Evaluation of Student Learning*, McGraw-Hill, New York, pp. 559-642.

Korth, W. W. (1968) *Life Science Process Test: Form B*, Educational Research Council of America, Cleveland, Ohio.

Krathwohl, D. R., Bloom, B. S. and Masia, B. B. (1964) *Taxonomy of Educational Objectives, Handbook II: Affective Domain*, McKay, New York.

Kruglak, H. (1958) Evaluating laboratory instruction by use of objective type tests. *American Journal of Physics*, 26, 31-2.

Lunetta, V., Hofstein, A. and Giddings, G. (1981) Evaluating science laboratory skills. *The Science Teacher*, 48, 22-5.

Oliver, P. M. and Roberts, I. F. (1974) Filmed sequences and multiple choice tests. *Education in Chemistry*, 132-3.

Olson, D. R. (1973) What is worth knowing and what can be taught. *School Review*, 82, 27-43.

Robinson, J. (1969) Evaluating laboratory work in high school biology. *The American Biology Teacher*, 31, 236-40.

Ruda, P. (1979) *A Chemistry Laboratory Practical Examination*, Cleveland Hill High School, Checktowaga, New York (unpublished).

Swanson, A. B. and Ost, D. H. (1969) *Operations of Science Test*, Science Education Center, University of Iowa, Iowa City.

Tamir, P. (1972a) The practical mode: A distinct mode of performance in biology. *Journal of Biological Education*, 6, 175-82.

_____ (1972b) The role of the oral examination in Biology. *School Science Review*, 54, 162-5.

_____ (1972c) Plant identification: A worthwhile aspect of achievement in biology. *The Australian Science Teachers' Journal*, 18, 62-6.

_____ (1974) An inquiry oriented laboratory examination. *Journal of Educational Measurement*, 11, 25-33.

_____ (1975) Nurturing the practical mode. *The School Review*, 83, 499-506.

_____ (1976) The Israeli high school biology: A case of curriculum adaptation. *Curriculum Theory Network*, 5, 305-15.

_____ (1978a) The impact of the BSCS on laboratory facilities for biology students in Israeli high schools. *International Evaluation Newsletter*, 19, 20-3.

_____ (1978b) An inquiry oriented biology course which builds on high school preparation. *Assessment in Higher Education*, 4, 3-21.

_____ (1981) The potential of evaluation in curriculum implementation, in A. Lewy and D. Nevo (Eds.) *Evaluation Roles in Education*, Gordon and Breach, New York, pp. 341-56.

_____ (1983) Equality, equity and excellence: The role of external examinations, Paper presented at the First International Conference on Education in the 90s, Tel Aviv.

_____ and Jungwirth, E. (1975) Students' growth and trends developed as a result of studying BSCS biology for several years. *Journal of Research in Science Teaching*, 12, 263-80.

_____ Nussinovitz, R. and Friedler, Y. (1982) The design and use of practical tests assessment inventory. *Journal of Biological Education*, 16, 42-50.

Tenenbaum, R. S. (1971) The development of the test of science processes. *Journal of Research in Science Teaching*, 8, 123-36.

University of London (1977) Advanced level chemistry: Notes of guidance and report form for the optional internal assessment of special studies, London.

Wilson, J. W. (1971) Evaluation of learning in secondary school mathematics, in B. S. Bloom, J. T. Hastings, G. F. Madaus (Eds.) ,*Handbook of Formative and Summative Evaluation of Student Learning*, McGraw-Hill, New York, pp. 643-96.

# 5.3

# Developing and Improving the Role of Computers in Student Laboratories

**Michael T. Prosser and Pinchas Tamir**

## INTRODUCTION

Over recent years there has been increasing interest in and concern with the resources required for the introduction of computers into student laboratories. This interest and concern has been shown by the increase in the number of papers describing particular developments and applications of computers in the professional science education literature. When developing a policy for their introduction two major issues, among others, need to be addressed. The first is a consideration of the ways computers can be used in student laboratories and what sorts of aims they can meet. That is, computers should be introduced to meet specific aims which either are not being met at present or which can more efficiently and effectively meet present aims. The second is a consideration of the means of monitoring their introduction with the aim of further developing and/or refining such policy.

## USE OF COMPUTERS IN STUDENT LABORATORIES

There are three broad ways in which computers can be used in student laboratories. These are (1) teaching and learning about the use of computers in science laboratories, (2) teaching and learning with the help of computers in student laboratories, (3)

**Table 5.3.1:** References to the use of computers in student laboratories

| Use of Computers | Scientific Discipline<br>Physics | <br>Chemistry | <br>Biology |
|---|---|---|---|
| Teaching About Computers | | | |
| Interfacing | Millar *et al.*, (1984) | Warren (1984) | Watson (1984) |
| | Sparkes (1982) | Salin (1982) | Coder (1985) |
| Analysis | Chonocky (1982) | Warren (1984) | Coder (1985) |
| | Millar *et al.* (1984) | Rosenberg (1985) | Lu (1985) |
| Teaching With Computers | | | |
| Instructional/Drill Practice | Sparkes (1982) | Gerhold (1985) | Butcher *et al.*, (1983) |
| Revelatory/Simulations | Sparkes (1981) | Butler *et al.* (1979) | Wood (1984) |
| | Chonocky (1982) | Gerhold (1985) | Day *et al.*, (1985) |
| | Walker *et al.* (1985) | | |
| Conjectural/Model Building | Cox (1983) | | Anderson (1984) |
| Emancipatory/Word Processing, Data Base, Spreadsheet | Feinberg *et al.*, (1985) | Rosenberg (1985) | Lu (1985) |
| Computer Management | | May *et al.*, (1985) | |

computer management of teaching and learning in student laboratories.

Table 5.3.1 shows an analysis of these broad areas and their sub-areas with a selection of articles available in the professional science education journals. We have tried to select reasonably representative articles which report the use of microcomputers in the laboratory. As will be further explained in the next section, we have focused in this chapter on individuals or on departments (up to about 40 staff) developing innovations and monitoring and evaluating these innovations rather than whole systems. With this in mind, the articles listed in Table 5.3.1 are probably more appropriate for individuals and departments than for whole systems.

## Teaching and learning about computers

One of the areas that seems to be receiving substantial attention at both secondary and tertiary levels is learning how computers can be used in the science laboratory. There appear to be two major uses: (1) interfacing computers to laboratory equipment to both control experiments and collect data from experiments, (2) the graphical or statistical analysis of data from experiments. While the first use may be more suited to tertiary laboratories, the second is appropriate to both the secondary and tertiary levels.

A very good introduction to these uses may be found in a series of articles in *Bioscience* under the title 'The Biologist's Toolbox'. These articles have been written with the novice user in mind and would provide a useful introduction for the use of computers in the physical and chemical sciences as well as the biological sciences. Before looking at some examples of the use of computers for these purposes, a warning by Dessy (1982) may be appropriate. He warns against purchasing hardware for these purposes with an eye to the initial purchase price only. He suggests that software costs 'represent at least 25% of any installation, and peripherals another 18-20%' (Dessy, 1982, p. 322).

In interfacing the computer to laboratory equipment its use as an analogue to digital converter has received substantial attention. In physics education Millar and Underwood (1984), for example, discuss the use of a BBC microcomputer in the investigation of capacitor discharge at secondary level. In biology at the secondary level Watson (1984) describes how a microcomputer can be interfaced to temperature and light sensors to monitor temperature and light in an urban ecology course. While the primary focus needs to be on the students' ability to use equipment involving interfaces, in later tertiary years an increasing focus will need to be on interfacing techniques themselves.

In using the microcomputer to aid the analysis of experimental data, the data can be that received from an analogue to digital converter interfaced to the microcomputer and stored for analysis later or analysed in real time. The data can also be fed into the computer sometime after the experiment is completed. There are a range of statistical and graphical packages available which can be used to analyse the data saved. One example of such analysis is described by Warren (1984), who had interfaced a microcomputer to a calorimeter and wrote a graph-fitting procedure enabling the computer to draw the best graph to fit the data. Coder (1985) describes the use of a statistical routine using data stored in a spreadsheet (see later) to analyse data to see whether rats fed on different diets gained weight differentially.

*It seems clear that both secondary and tertiary students need to develop an awareness and more skills in the use of microcomputers in science laboratories as part of the laboratory programme. Such skills need to be one of our aims for laboratories.* There are a large number of examples of such developments in the practitioner science education literature and warnings about possible hazards. With this in mind a gradual controlled introduction of some interfacing and data analysis experiences into science laboratories is warranted. Such developments need to be carefully thought about at the initial planning stage and carefully monitored during initial implementation. The last section of the chapter will focus on such monitoring procedures.

Technical skills associated with collecting and analysing data were discussed in each of the three chapters in Part 3 of this book. In her chapter (Chapter 3.1), Hegarty-Hazel discusses the development of skills associated with data acquisition and data and error analysis, and indicates that these are important laboratory skills to be learnt by students. As these are now being routinely handled by computers the use of computers for these purposes needs to be an important aim in our laboratories. Klopfer, in his chapter (Chapter 3.2) on developing enquiry skills, discusses the need for processing experimental and other data as an important component of the development of scientific enquiry skills and, again, a knowledge of how computers can contribute to these tasks in the laboratory would seem to be an integral part of developing scientific enquiry skills. Finally, in their chapter on the development of attitudes to science and scientific attitudes (Chapter 3.4) Gardner and Gauld remark on the importance of the physical conditions of the student laboratory for the development of positive attitudes towards science teaching and on variety as an instructional condition which may encourage the development of positive attitudes to science. Both of these would be enhanced by the appropriate introduction of computers to obtain and process information from at least some of the experiments in the laboratory.

## Teaching and learning with computers: retrospective

We now turn to what may be considered the more traditional uses of computers in education, teaching and learning with the aid of computers. There are a number of monographs which deal with various aspects of this matter (e.g. Rushby, 1979; Maddison, 1983; Lathrop and Goodson, 1983). These monographs deal with the selection, use and development of computer assisted learning materials. Generally these books have tended to focus on drill and practice, tutorial and simulation software, an exception being Rushby's book. As there seems to be a substantial range of material available covering the development of software, we will not deal with these issues. Rather we will explain what we mean by the different ways of teaching with computers in student laboratories and briefly outline some examples.

In doing this we will distinguish between retrospective and prospective uses of computers. The retrospective uses will be those that have been available for some time (i.e. instructional and revelatory), and prospective uses will be those only now being developed but of increasing importance (e.g. conjectural and emancipatory).

Obviously research on the use of computers in science teaching has focused on the retrospective uses. Meta-analysis has been used to summarize the results of this research. Meta-analysis by Kulick, Kulick and Cohen (1980) on the effectiveness of computers for instructional and revelatory uses at the college and university levels has shown only small gains in achievement and attitudes but substantial savings in instructional time. These results were, in general, confirmed by another analysis by Willett, Yamashita and Anderson (1983). While these analyses focused on the use of computers in science teaching generally, not on laboratories in particular, nonetheless it does not seem that the use of drill and practice and simulation courseware in the laboratory would lead to substantial improvements in achievement. However, the savings in instructional time may be of substantial benefit in laboratory teaching and this will be discussed later.

*Instructional (drill and practice and tutorial modes)*

Drill and practice and tutorial software is probably the most commonly available software, although it does not seem to have had much impact on student laboratories. In drill and practice, students are led through a structured set of problems, with the student not progressing to subsequent problems until he/she can solve earlier problems. In the tutorial mode, students are led by questions and answers through a piece of subject matter. In both cases, students have little or no control over the material that is presented to them, their path through the material or the rate at which they progress through that material. Many aspects of these materials are similar to programmed learning techniques.

Drill and practice and tutorial modes have a place in the student laboratory. In Chapter 3.1 on the development of technical skills, Hegarty-Hazel discusses the importance of students developing an overview of technical skills routines. Drill and practice and tutorial modes are well suited to this purpose. In the area of developing scientific enquiry skills, computers in this mode may well be useful in efficiently developing, in students, skill in processing experimental data and error analyses which can then be used in the enquiry process. Gardner and Gauld, in their discussion of instructional conditions important to the development of positive attitudes to science (Chapter 3.4), suggest that time pressure and a lack of integration of theory and laboratory may be factors in developing poor attitudes to science. Drill and practice and tutorial coursework can quickly and efficiently provide the student with an overview of the theory appropriate to the experiment. As well, the savings in instructional time associated with this use of computers may reduce the time pressures on students in the laboratory.

Sparkes (1982) expresses some concern with drill and practice and tutorial modes in physics laboratories, arguing that they can easily turn into a page-turning exercise for the student. There seem to be few other descriptions of the use of these modes in science laboratories, although there are many examples from science teaching generally (see Gerhold, 1985; Butcher and Murphy, 1983).

*Revelatory (simulations)*

Rushby (1979) defines revelatory computer-assisted learning

as 'A form of CAL in which the user is guided through a process of discovery so that the subject matter and the underlying theory are progressively revealed to him [sic] as he proceeds through the CAL package' (Rushby, 1979, pp. 118-19). Examples of this are simulation exercises which are in some respects analogues to real-life situations.

This is the form of computer-assisted learning which in the past has had most application in student laboratories. It is particularly useful for laboratory exercises that require sophisticated and expensive equipment or for exercises that require long periods of time to complete. Sparkes (1981) describes some simulation exercises in mechanics which are based upon ideas by Feynman *et al.* (1965). Chonocky (1982) describes simulation exercises in student laboratories in which experimental results are combined with theory. One example is of students being asked to 'investigate the similarities between the fluctuating results of a sample of N repetitive, fixed interval measurements of radioactivity events counted by a G-M tube-scaler, and the results of N repetitive "games" involving a randomly generated set of simulated coin flippings' (Chonocky, 1982).

In chemistry, Butler and Griffin (1979) describe a laboratory course in which they have integrated a number of simulations. Two examples are (1) acid-base filtration in which the students were asked to obtain the normalities of three strong acid unknowns and (2) rate of reaction in which the students were asked to obtain the value of the specific rate constant for some combinations of pH and temperature. Wood (1984) describes some simulation examples in biology which 'are intended to supplement and complement the actual practical exercises rather than replace them'. He continues 'the computer simulation is important because the classroom situation, often involving a single three-hour practical, is insufficient to generate enough meaningful data' (Wood, 1984, p. 309). He describes three simulations of practical microbial genetics experiments and a simulation of the mapping of DNA molecules using restrictive endonucleases.

In terms of the intended learning outcomes discussed in Part 3 of this book, simulations may well be the most useful application of computers in student laboratories. For example, Hegarty-Hazel discusses the importance of reading and mental rehearsal and practice before physically attempting technical skills such as using analytical balances, pipettes, etc. These are the sorts of tasks to which appropriate simulations can make a substantial contribution. In his chapter on the development of enquiry skills (Chapter 3.2) Klopfer discusses the importance of developing skills in using common laboratory equipment, of observing and describing phenomena, etc. Computer simulations can provide practice in these skills. Further, in discussing student learning and the development of scientific knowledge (Chapter 3.3) Atkinson discusses the problem of students linking theory and practice, and the fact that laboratory exercises do not seem to provide this link. It was the enhancement of this relationship between theory and practice which was a major objective of many of the simulations outlined above. The use of these simulations with laboratory hands on exercises may well provide this link. Further, in discussing student attitudes towards science teaching, Gardner and Gauld (Chapter 3.4) argue that cognitive challenge and the integration of theory and laboratory are important to the development of positive attitudes. Again, as these are the aims of many computer simulations, their use may be expected to play a useful part in developing positive attitudes.

## Teaching and learning with computers: prospective

In this section we focus on developments which have very good prospects in the immediate future.

### *Conjectural (hypothesis testing)*

Rushby defined conjectural computer-assisted learning as 'the use of the computer to assist the student in his [*sic*] manipulation and testing of ideas and hypotheses. It is based on the concept that knowledge can be created through the students' experiences; its emphasis is on the students' exploration of information on a particular topic' (Rushby, 1979, p. 114). The major difference between this form and the previous one is that here the student is expected to construct, in part, the model him/herself rather than exploring a model programmed into the computer. In the past this form does not seem to have been as popular as the revelatory or simulation form of computer-assisted learning in the laboratory, although it is particularly well suited to student laboratories.

One way of how such software could be developed is the use of data bases, allowing students to explore hypotheses or ideas of their own and test them against the information in the data base. One example of this is described by Anderson (1984). In this example the data base comprises all the observations on sea birds made by a scientist during the 1982 Australian Antarctic Research Expedition together with meteorological information, time, ship's position and activity. With this data base the students are able to form their own questions and answer them by exploring the data base. In this way students construct their own ideas about bird life observed during the expedition.

Cox (1983) describes a 'dynamic modelling system' developed for the 'Computer in the Curriculum Project'. He describes some software developed for mechanics in which students type in a set of equations for variables they wish to investigate such as: force, distance, mass, etc. The program then allows the students to build up complicated physical systems. While this form of computer-assisted learning is relatively under-utilized, its use in developing scientific enquiry skills and positive attitudes towards science needs to be carefully explored. Klopfer, in his chapter, discusses enquiry skills associated with asking appropriate questions, and forming and testing appropriate hypotheses, skills which are at the heart of conjectural uses of computer-assisted learning. Gardner and Gauld, in their discussion of development of scientific attitudes argue strongly that students need to have some control over laboratory procedures. Conjectural uses of computers are designed to give students such control.

This form of computer-assisted learning does not seem to have been discussed as such in the practitioner science education journals. However, it is a use which may be of particular value. Its objectives are ones which traditional laboratories have not been very successful in meeting.

*Emancipatory (word processing, data bases, spreadsheets)*

In this form of computer-assisted learning the students use the computer as a tool to relieve them of some of the more repetitive tasks associated with their work. As well, they develop many of the skills associated with computer literacy such as word processing, use of spreadsheets, etc.

A very good introduction to the use of word processing, data-base and spreadsheet software in the sciences is given in the series Biologist's Toolbox in *Bioscience 35, (3, 5, 6)* While word processing software has been in wide use for some time, data-base and spreadsheet software is somewhat more recent and less well known. Lu (1985) describes data-base programs as programs which 'deal with data that fall into units, like addresses in an address book or literature citations in an article. Each unit such as a single citation is called a record. Subunits within a record, the author or date for example, are termed fields' (Lu, 1985, p. 181). For some of the data-base programs, e.g. d-Base III, it is possible to obtain statistical packages that interface with and operate on data in the data-base, e.g. AB-stat. On the other hand 'A spreadsheet program sets up a matrix of cells in rows and columns. The cells function somewhat like a book-keeper's ledger: you can enter text, numbers, or formulas and incorporate values in other cells as variables. Changing the value in any cell automatically changes all dependent cells' (Lu, 1985, p. 181).

Feinberg and Knittel (1985) describe uses for spreadsheet programs in physics laboratories which include applications in data handling, error analysis and experimental design. The data handling and error analysis examples involve the determination of the acceleration of a falling object due to gravity and the design of a common-emitter amplifier. Rosenberg describes an example of the use of spreadsheets in chemistry in the 'calculation of the activity coefficient of lead in amalgams from measurements of the potentials of cells without liquid junction' (Rosenberg, 1985, p. 1521). In both of these areas, use is made of the program's ability to automatically change the values in dependent cells.

In each of the above cases we have several columns of figures, and on each figure in the same column the same transformation needs to be performed (e.g. each figure in the time column of the free-fall example needs to be squared). The use of the spreadsheet programs can save substantial time for students during repetitive calculations.

This use of microcomputers in student laboratories is one which is likely to be of increasing value. No new software needs to be written (one of the major problems with other uses of computers), the students are learning skills which will be very useful in their future work, and they are saving substantial amounts of time which can be devoted to other laboratory activities.

## Computer management

The use of computers in teaching laboratories to help management functions can range from an individualized laboratory in which the computer guides the student through various aspects of the laboratory experience to storing and collecting assessment results. A brief review of the professional science education literature indicates less interest in the use of computers for this purpose than in teaching about and teaching with computers. Nevertheless with the development of user-friendly data-base and spreadsheet software it seems that substantial amounts of teachers' time used for keeping records and other such tasks can be saved.

Sparkes (1982) briefly discusses some management aspects. He suggests that the handling of student records and control of stock and chemicals will be more efficient in the future, although he does express some concern with the 'reliability' of floppy discs and suggests that hard disc technology is far more reliable. May (1985) describes an example of the use of computers in assessment of students in chemistry laboratories. As part of their reports to their tutor, students have to attach a computer print-out, verifying the calculations they have made. The verification is done by the computer based upon information fed to it by students. The students also enter their numerical answer. If the answer calculated by the student does not match that calculated by the computer, the student is informed and is asked to redo the calculation. It is claimed that this system, in areas where the tutor may need to check through large amounts of tedious numerical calculations, is very time efficient.

In this part of the chapter we have tried to outline the various ways that computers can be used in science laboratory instruction illustrated by several examples taken from the practitioner science-teaching literature. More general discussions of the uses described in this chapter can be found in Rushby (1979), Maddison (1983), and Lathrop and Goodson (1983).

## Longer term developments

There are two areas for the application of computers in student laboratories which in the future are likely to make a significant impact. They are the use of intelligent tutoring systems and videodiscs. Both of these will be commented on in this section.

The first of these was described in an interesting scenario by Leo Klopfer.

> Here is a scenario of what an intelligent tutor installed on a personal computer could do - Imagine that the computer software has the capability to recapture the special features of the classical apprentice system of teaching and that it can do this in a cost-effective way. Teaching in an apprentice system is highly individualized, and its best characteristic is that each apprentice always receives from the master exactly the kind of information and guidance he needs to develop his knowledge and skills to the fullest. The master is a true expert in her trade or subject matter, knowing not only its general principles and major trends but also a plethora of specific information. Her expertise includes a panoply of tested strategies and procedures suitable for instruction in different aspects of the subject for different purposes and on different occasions. For instance, a science master guiding the development of apprentices in ecology would know various procedures for teaching and the facts and concepts of ecology, strategies to teach a particular ecological principle or theory, and strategies for teaching apprentices how to induce ecological principles and how to test them empirically. The science master would know how to use direct instruction, discovery approaches, simulations, drill, Socratic dialogues, problem solving strategies, and even the Suzuki method. In addition a successful master in an apprentice teaching system has detailed knowledge about each apprentice's stage of development and rate of progress, and she employs this knowledge to tailor the instruction for each individual apprentice. For instance, the successful science master would know, as Louis Agassiz knew, the best time to set an apprentice on the completely unguided task of observing and describing a single preserved fish specimen for a two-week period (Shaler, 1909/1963). Finally, an ideal master not only possesses all the knowledge described but also is wise, so that she properly motivates and encourages each apprentice to persevere in developing his competence.
> (Klopfer, 1985, pp. 7, 8)

Klopfer went on to describe research being conducted at the University of Pittsburgh's Learning Research and Development Center aimed at developing such systems. Areas in which development work is progressing are in electricity, hydrostatics and geometrical optics. Klopfer suggests that such software will be available in about five years' time and is likely to be able to be run on today's Apple Macintosh with hard discs.

One of the problems at the moment in teaching with the aid of computers is the generally poor quality of the courseware that is available. Much of the available software can be described as 'page turners'. If software such as that described above becomes readily available, then software quality will be improved dramatically.

The interactive videodisc combines the benefits of video and microprocessors. The videodisc stores information in a similar way to a normal floppy disc except that the information is read to and from the disc by a laser. The videodisc can store large amounts of information, for example O'Shea and Self (1983) state that 'one side of such a disc can hold 10 billion bits of information or about 30 minutes of film'. The major advantage of the videodisc is that, like the floppy disc, information is obtained from the disc by random access and any single frame is retrievable in 2-3 seconds. When the access to this information is controlled by a microcomputer, the benefits of interactive videodisc become obvious.

If these benefits were combined with intelligent tutoring systems, a very powerful instructional medium would be likely to result. McAleese (1987) describes a project with these aims. In his model, the computer stores 'expert' knowledge of the subject matter, elicits from the student his/her present knowledge state, checks the student knowledge against the expert knowledge, assigns an appropriate interactive videodisc program to the student and continually monitors the student's knowledge state as the program progresses. The major advantage of this system is that it focuses on and monitors the student's present knowledge state, and as such large amounts of information can be stored on the videodisc it is able to present efficiently appropriate courseware to the student. In the paper he describes a system being developed to teach techniques of staining mammalian tissue as part of a laboratory skills program.

## SOME PRECAUTIONARY COMMENTS

In an occasional report entitled 'Conventional classrooms as environments for computer-based education' Walker asserted that:

> It is by no means a foregone conclusion that the educational potential of microcomputers will be realised within schools. It is not clear how the use of computers for education can be reconciled with traditional practice of long standing schools and classrooms. There seem to be significant tensions if not outright inconsistencies between the environment required by the new technology if it is to be used successfully and the environment optimal for traditional education practice. If an accommodation cannot be reached between the computer and the classroom, it is unlikely that computers will find wide use in these traditional settings. (Walker, 1984, p. 1)

While we share Walker's assertion we would like to go one step further and ask: assuming that such accommodation can and will be achieved, what are some dangers and hazards to be considered in order to avoid potential ill effects that may accrue as a result of using microcomputers in the science laboratory? Following are some of the limitations and hazards which should be considered as we use microcomputers in the science laboratory:

1. The dialogues in simulations and similar programs are limited. For example, in the emergency patient simulation the student can merely select the investigation or treatment, and cannot discuss his decisions, nor can he ask why treatments have their various effects. (Hartley and Lovell, 1977).
2. While simulation programs can provide illustrations that are difficult to give by any other means, the students cannot in any real sense, place their own construction on the problem and set out their own methods of solution. 'Practically all the program languages which are used do not permit this.' (Hartley and Lovell, 1977, p. 42).
3. Different students benefit differently from different kinds of feedback. For example, in one educational experiment undergraduate students were asked to plan experiments concerning reaction kinetics and the determination of equilibrium constants. Students were randomly allocated to three computer-based treatments. In the first the student was merely told if his responses were correct or not. In the second, the program evaluated the student's response, located the error, and provided information by which the student could see how the correct answer had been arrived

at. However, there was no check that the student had attended to or comprehended the feedback message. The third treatment was similar to the previous one except that the student had to demonstrate his understanding by typing a satisfactory answer. It was found that on the average the first treatment was least satisfactory for learning but took less time. However, the more able students in the first group learned almost as much as students with similar ability in the other two groups (Hartley and Lovell, 1977).

4. With regard to laboratory applications Arons cautions that if the computer short circuits insights, if it simply makes available any results for analysis or 'confirmation', it is educationally sterile or even deleterious, particularly in introductory courses (Arons, 1984, p. 1051).
5. While the computer can relieve students from tedious numerical calculations and while its graphic display offers considerable advantages over programmable hand calculators, 'where extensive numerical computation is done without directing students' attention to analysis and interpretation of methods and results, the effort is largely wasted' (Arons, 1984, p. 1051), and may promote rote learning.
6. With regard to simulations, Arons (1984) cautions that 'some of the effort being devoted to simulations is producing undesirable materials'. As a rule, 'if the phenomena are readily accessible, then it is better to expose the beginning students to the actual phenomena than to simulation thereof' (p. 1052). This implies that often in order that students can benefit from computer simulation they had better have some experience with the real phenomena before being engaged in simulation thereof.
7. There is a danger that continuous learning with computers may cultivate dependence on the reinforcement provided by the computer so that students will not be able to engage in genuine independent learning, asking their own questions, checking their own reasoning for internal consistency, etc.
8. Displacement of classrooms by computers will be a disaster since the classroom gives the student contact with distinguished intellects, motivates through human relations, encourages conversation, argument and discussion, and creates a unique social system which can be neither substituted nor simulated by computers.

We have listed a number of limitations and potential hazards. We have previously referred to some problems in developing applications aimed at teaching about the use of computers. Many others could most probably be added. The conclusion to be drawn is not to give up the use of microcomputers in classrooms, but rather to use them judiciously in a careful and adaptive manner. In the next section, we will describe some procedures which can be used during the planning and implementation phases to obtain this result.

## MONITORING AND IMPROVING THE USE OF COMPUTERS IN STUDENT LABORATORIES

Early attempts at providing information for decision making and ongoing policy development focused on either an objectives model or an experimental-comparative model of curriculum evaluation. In the objectives model (see Tyler, 1950) decisions about the quality or worth of a curriculum were made on the basis of student achievement of specified objectives. In the experimental-comparative model (see Taba, 1962) one curriculum is compared with another, exercising experimental control and systematic variation. While evaluations based upon these models may have produced many defensible reports, as Prosser (1984) noted, decision makers and policy developers took little notice of such reports for a variety of reasons. During the 1960s and 1970s new models for curriculum development and evaluation were developed and used, taking explicit account of the particular needs of decision makers in particular contexts. Cronbach's (1963) criticisms of the earlier approaches were instrumental in encouraging this reconceptualization of curriculum evaluation. His criticisms were that curriculum monitoring and evaluation needed to focus on: (1) decisions that decision makers and policy developers had to make; (2) decisions that needed to be made while the curriculum was being implemented; (3) the characteristics of particular courses rather than comparisons with other courses.

As a result of criticisms and suggestions such as these, a range of models was developed (e.g. Stufflebeam's Context-Input-Process model (1973); Parlett and Hamilton's Illuminative model (1976); Stake's Responsive model (1975) and Guba and Lincoln's Naturalistic model (1981)). These models focused on the needs of decision makers and policy developers, attempting to supply appropriate and timely information.

At the same time as these developments were taking place Schwab (1969; 1971; 1973) was arguing for what he termed a deliberative approach to course design and improvement. He argued that curriculum decisions cannot be made on the basis of research results alone, but that such results need to be carefully examined and interpreted in the particular context in which the innovation takes place. Educational decisions in this sense are practical and not theoretical. In this deliberation four matters designated as the commonplaces need careful consideration. These are: learners, teachers, content and context. See also Chapter 1.2 and Figure 1.2.1. When attempting to introduce and improve upon the use of computers in student laboratories each of these commonplaces requires careful collection of appropriate information and consideration. That is, decisions about the future development of the role of computers do not follow a number of well defined simple steps, but they involve a complex, deliberative process with the defining and redefining of the questions and issues involved, discussion and debate between those involved in the decision-making process and careful consideration of the consequences of proposed changes. With this in mind evaluation procedures aimed at identifying problems and issues from the perspective of the four commonplaces need to be considered. This approach to curriculum decision making and policy development is consistent with the developments in curriculum monitoring and evaluation described earlier. In the remainder of this chapter we wish to outline some aspects of a monitoring and ongoing policy development process drawing on various aspects of these models.

When discussing procedures appropriate to collecting information and identifying issues about these commonplaces the level of analysis of the discussion is a difficulty. That is, it is possible to focus discussion on the single teacher in a single classroom or on whole departments at the secondary or tertiary level, introducing computers into their laboratories to a whole school system. In order to provide an appropriate focus we wish to concentrate on a departmental level. That is, for example, a biology department in a tertiary institution or the science department in a secondary institution. Thus there may be five to ten courses being taught and between, say, fifteen and thirty teachers involved. Focusing at this level means that the discussion should also have practical implications for evaluation both at the single classroom level and at the system level.

If the evaluation is designed to identify and provide information about issues associated with the four commonplaces, then the issues and problems are unlikely to be predetermined or decided in advance. So the first, and probably the most important task is to identify issues and problems from each of the perspectives. In the remainder of this chapter we will discuss the problem of the identification of issues and problems from the perspective of the learner, teacher, content and context.

The issues identified and information collected may be considered to be 'subjective' or 'objective' and the procedures for identifying the issues and collecting information can be systematic or non-systematic. It seems to us that in evaluation of this sort both subjective and objective information should be collected and analysed but the procedures for collecting and analysing that information should be as systematic and as public as possible.

During the review aimed at providing information for deliberation by decision makers there are three broad areas of information collecting: (1) developing a description of the innovation in practice, (2) identifying issues and problems and (3) developing interpretations of and collecting further information about problems.

Thus descriptions of the innovation from the perspective of the teacher and teaching, the student and learning, the content and the context will need to be developed. The sources may include a combination of: (1) interviews of teachers and students; (2) interviews of departmental administrators; (3) classroom observation of teaching; (4) content analysis of software programs, experimental descriptions, etc.

Two instruments which may assist the user in carrying out content analysis of science software will be mentioned here. The first, designated as the Microcomputer Software Evaluation Instrument, was published in the USA by the National Science Teachers' Association (1983) for the purpose of examining the merits of any software package intended to be used in science instruction. The instrument provides a sensible process and basic criteria for judging science software packages. It focuses on four characteristics, namely: (a) policy issues, e.g. the cost is exorbitant for what it delivers; (b) subject matter standards, e.g. the science content is free from errors; (c) instructional quality, e.g. the student using the program is passive and does little more than punch keys occasionally; (d) technical quality, e.g. the program's graphic displays are crisp and clear.

The second instrument, designated as Computer Software Inquiry Inventory, was designed by Zohar and Tamir (1986) for the purpose of assessing the enquiry nature of laboratory simulations. It is based on the Laboratory Assessment Inventory (Tamir and Lunetta, 1978) but modified to the needs of software content analysis. It also uses Herron's (1971) levels of enquiry. It includes five major areas each comprised of several questions to which a 'Yes' or 'No' response is assigned. The areas are: general features, nature of investigation, advantages of the computer program over actual lab investigation, enquiry skills involved (planning, performance, reporting, application) and concepts related to enquiry. Examples of questions are: (1) does actual performance require too much time? (2) is the student required to formulate an hypothesis? (3) is the student required to explain the nature of the control? (4) does the program present concepts related to design of experiments?

Descriptions of the innovation are very important for the monitoring process because the innovation as implemented is unlikely to be the same as that described in the planning documents (Barham and Prosser, 1985). The interpretation of the problems and issues is necessarily made against the background of a description of the innovation.

It is also likely that the innovation will look different when viewed from the perspective of the four different commonplaces. For example, the teachers' description and the students' description may well be quite different.

Issues and problems associated with the implementation of computers in science laboratories can be identified using a number of procedures. Among them the following are suggested: (1) interviews of teachers and students; (2) questionnaire surveys of teachers and students; (3) classroom observation of teaching.

At this stage, interviews should give the opportunity for the teacher or student to identify issues and problems of concern to them with little or no structure from the person conducting the interviews. It may well be worthwhile asking for descriptions of typical teaching/learning sessions from the viewpoint of the teacher and learner. The sorts of questions asked in both the interviews and the questionnaires should be as open as possible. Well tried questions at this stage are: (1) describe what happens in a typical session; (2) what is the best thing about ...? (3) what is the worst thing about ...? (4) what is in the most need of improvement ....?

Classroom observation using a systematic schedule and recording of observers' impressions can also be very beneficial. Differential use of the computer by various groups in the class an example of the sort of issue which can be identified here.

A content analysis of the 'best things', 'worst things' and 'things most in need of improvement' is also likely to produce several issues. A comparison of the description of the sessions from the viewpoint of the teacher, learner and observer can also highlight differences in perspectives which may be important.

Interviews or surveys of teachers and students asking each group to list or describe the sort of things students learn about computers or learn from the use of computers is one way of focusing on content issues. As well as content analysis of the results of students' work using the computers focusing on both correct and incorrect conceptions can also highlight problems with content. Another way of identifying issues associated with the content is to analyse the content in terms of the aims and objectives.

In attempting to identify issues associated with the context it may be necessary to interview administrators, heads of departments, etc., asking questions such as: (1) at what rate will the resources allow computers to be introduced into the department? (2) what sorts of computers have students had experience of within other departments? (3) what sorts of physical requirements may be needed to introduce computers? (4) how will maintenance of the computers be dealt with? (5) how does the use of computers in this student laboratory relate to other parts of the students' programmes? Clearly there is a whole range of context questions dealing with resources, room requirements, relationships to other courses, etc., which need to be identified.

Once the issues associated with each of the commonplaces have been identified the group of people responsible for the implementation and improvement of the innovation need to discuss the issues and problems in detail and reach some consensus on those which are of highest priority and on which further interpretations or information may need to be collected.

This may mean re-examining each of the commonplaces with the aim of collecting and analysing more objective information or further interpretations. For example, problems relating to students' prior experience of the use of computers may have been identified from statements from some of the teachers and students. But at this stage the magnitude of the problem or the range of prior experiences may not be known. A detailed questionnaire with fixed responses may be developed in order to obtain this detailed information. On other issues there may be enough information about it within the deliberation group or included in the information collected in an earlier phase.

Once this sort of information has been collected and analysed, the group may meet again to reconsider the issues and try to reach some conclusions about them. For many issues the conclusion may be obvious, or it may not be possible to wait for detailed information to emerge from the monitoring. The group will have to make the best professional judgement based upon the information in hand.

This process may cycle through several years as the innovation grows or becomes more complex. If such a graduated increase in size or complexity is adopted, then if at any stage it is thought the innovation is too large, complex or problematic then it can easily revert to the earlier successful stage with little or no loss of resources or effort.

Having described in general terms review processes, we will comment briefly on review procedures appropriate to the three areas of use of computers in laboratories. Clearly teaching about the use of computers and with the help of computers involves the four commonplaces. Using computers to help manage learning in the laboratory involves the teaching and context commonplaces and may involve the student commonplaces if the management system includes feedback to students. It would probably not include the content commonplace. The context commonplace for management may be broader than for teaching about the use of computers and teaching with computers. If the management system for the laboratory part of the course is integrated with a management system for the course as a whole or for the department, then the range of participants who will have views on the success or otherwise of the innovation will be enlarged. A more detailed description of such a process can be found in Barham and Prosser (1985).

## CONCLUSION

In this chapter we have explored the various ways computers may be used in student laboratories now and in the future. We have also argued for a gradual introduction of computers with careful and systematic monitoring and sustained deliberation. Computers do have and will continue to have an important role in student laboratories.

A major implication to be drawn is that computers should be looked upon as another tool which may improve both instruction and learning. In the context of the laboratory they may be used sometimes as a substitute for actual hands-on experiences, but more often to supplement, complement and enrich experiential learning along with actual laboratory experiences.

## REFERENCES


Anderson, J. (1984) Computing in schools. *Australian Education Review*, 21, Australian Council for Educational Research, Hawthorn.

Arons, A. (1984) Computer based instructional dialogs in science courses. *Science* 224, 1051-6.

Barham, I. and Prosser, J. (1985) Review and redesign: beyond course evaluation. *Higher Education* 14, 297-306.

Butcher, P. and Murphy, P. (1983) Tutorial CAL and biology education. *Journal of Biological Education* 17(1), 43-50.

Butler, W. and Griffin, H. (1979) Simulations in the general chemistry laboratory with microcomputers. *Journal of Chemical Education* 56(8), 543-5.

Chonocky, N. (1982) Microcomputer data management in an introductory physics laboratory. *American Journal of Physics* 50(2), 170-6.

Coder, D. (1985) The biologists toolbox: Anova with Super Calc. *Bioscience* 35(5), 306-7.

Cox, M. (1983) *Evaluation and Dissemination of Science Software*. Paper presented at the Japan/UK Science Seminar, November, 1983.

Cronbach, L. (1963) Course improvement through evaluation. *Teachers' College Record* 64, 672-83.

Day, M., Henderson, P. and Wood. A. (1985) GENMAP - a microbial genetic computer simulation. *Journal of Biological Education* 19(1), 67-70.

Dessy, R. (1982) Chemistry and the microcomputer revolution. *Journal of Chemical Education* 59(4), 320-7.

Feinberg, R. and Knittel, M. (1985) Microcomputer spreadsheet programs in the physics laboratory. *American Journal of Physics* 53(7), 631-4.

Feynman, R., Leighton, R. and Sands, M. (1965) *The Feynman Lectures in Physics, Vol. 1*, Addison-Wesley, New York.

Gerhold, G. (1985) Computers and the high school chemistry teacher.

*Journal of Chemical Education* 62(3), 236-7.

Guba, E. and Lincoln, Y. (1981) *Effective Evaluation.* Jossey-Bass, San Francisco.

Hartley, J. and Lovell, K. (1977) The psychological principles underlying the design of computer based instructional systems, in Walker, D. and Hess, R. (Eds.) (1984) *Instructional Software*, Wadsworth, Bellmont, Ca.

Herron, M. (1971) The nature of scientific enquiry. *Science Review* 79, 171-212.

Klopfer, L. (1985) *Intelligent Tutoring-systems in Science Education: The Coming Generation of Computer-based Instructional Programs.* Paper presented at the US-Japan Seminar on Science Education, Washington, D.C.

Kulik, J., Kulik, C. and Cohen, P. (1980) Effectiveness of computer-based teaching: a meta-analysis of findings. *Review of Educational Research* 50, 525-44.

Lathrop, A. and Goodson, B. (1983) *Courseware in the Classroom* Addison-Wesley, Reading, Massachusetts.

Lu, C. (1985) The biologist's toolbox: Microcomputers in biology: Hardware and software profile. *Bioscience* 35(3), 178-82.

Maddison, J. (1983) *Education in the Microelectronics Era.* Open University, Milton Keynes.

McAleese, R. (1987) Computer-based authoring and intelligent interactive video, in Osborne, C. *International Yearbook of Educational and Instructional Technology 1986/1987.* Kogan Page, London.

May, P, Murray, K. and Williams, D. (1985) CAPE: A computer program to assist with practical assessment. *Journal of Chemical Education* 62(4), 310-12.

Millar, R. and Underwood, L. (1984) Using the analogue input part of the BBC microcomputer: Some general principles and a specific example. *School Science Review* 66, 235, 270-9.

National Science Teachers' Association (1983) *Microcomputers Software Evaluation Instrument,* Washington D.C.

O'Shea, T. and Self, J. (1983) *Learning and Teaching with Computers.* Harvester Press, Brighton.

Parlett, M. and Hamilton, D. (1976). Evaluation as illumination, in Tawney, D. (Ed.) *Curriculum Evaluation Today: Trends and Implications,* Macmillan Education, Basingstoke.

Prosser, M. (1984) Towards more effective evaluation studies of educational media. *British Journal of Educational Technology* 15(1), 33-42.

Rosenberg, R. (1985) The spreadsheet. *Journal of Chemical Education* 62(2), 140-1.

Rushby, N. (1979) *An Introduction to Educational Computing,* Croom

Helm, London.

Salin, E. (1982) Computer interfacing for chemists. *Journal of Chemical Education* 59(1), 53-6.

Schwab, J. (1969) The Practical 1: A language for curriculum. *School Review* 78, 1-23.

_____ (1971) The Practical 2: Arts of the eclectic. *School Review* 79, 493-542.

_____ (1973) The Practical 3: Translation into the curriculum. *School Review* 81, 501-22.

Sparkes, R. (1981) Microcomputers in physics. *Physics Education* 16, 145-51.

_____ (1982) Microcomputers in science teaching. *School Science Review* 63, 224, 442-57.

Stake, R. (Ed.) (1975) *Evaluating the Arts in Education: A Responsive Approach*. Merrill, Columbus.

Stufflebeam, D. (1973) An introduction to the PDK book: Educational evaluation and decision making, in Worthen, B. and Sanders, J. *Educational Evaluation: Theory and Practice*. Wadsworth, Belmont.

Taba, H. (1962). *Curriculum Development: theory and practice*. Harcourt, Brace and World, New York.

Tamir, P. and Lunetta, V. (1978) An analysis of laboratory inquiries in the BSCS yellow version. *The American Biology Teacher* 40, 353-7.

Tyler, R. (1950) *Basic Principles of Curriculum and Instruction*. University of Chicago, Chicago.

Walker, D. (1984) *Conventional Classrooms as Environments for Computer-based Education*. Occasional Report, 6. Study of Stanford in the Schools.

Walker, D. and Bailey, D. (1985) Concept development by CAL or stimulation by simulation, in Bawden, J. and Lichtenstein, S. (1985) *Student Control in Learning: Computers in Tertiary Education*. Centre for the Study of Higher Education, Melbourne.

Warren, J. (1984) Calorimetric analysis, course fitting and the BBC microcomputer. *School Science Review* 66, 235, 270-9.

Watson, J. (1984) Using a computer to monitor temperature and light. *Journal of Biological Education* 18(1), 57-64.

Willett, J., Yamashita, J. and Anderson, R. (1983) A meta-analysis of instructional systems applied in science teaching. *Journal of Research in Science Teaching* 20(5), 405-17.

Wood, P. (1984) Use of computer simulations in microbial and molecular genetics. *Journal of Biological Education* 18(4), 309-12.

Zohar, A. and Tamir, P. (1986) Assessing enquiry characteristics of science computer software. *Science Education* (In Press).

# 5.4

# Practical Work and the Laboratory in Science for All

**Peter J. Fensham**

## INTRODUCTION

'Science for All' is a contemporary slogan - and challenge - for science educators around the world in the 1980s. After a period of substantial depression in school science education in the United States, new funds were announced in 1983 for another attempt to improve and extend the effectiveness of science education in that nation's schools. Despite general recession and cutbacks in educational funding in Britain and New Zealand, each of these countries has major projects under way in the field of school science education. In Britain, an extensive five-year project, the Secondary Science Curriculum Review, began in 1982. In New Zealand, Learning in Science spent the years 1979-81 working on the lower secondary years and has now been extended for a further period to work at the primary school level. In Malaysia a new primary curriculum is being developed and there is a major debate about the place that science should play in it. In March 1982, the Commonwealth Office of Education hosted a workshop in Cyprus of science curriculum experts with the theme 'Science and the World of Work'. In this workshop the focus was on what science *all* learners at school should have the opportunity to learn before they leave school. That is, the focus was shifted from extended secondary education (and from those who stay on at school and vie for places in tertiary education) to the period of schooling that all or most children in any given society now have. In some countries represented at the workshop this period coincides with elementary or primary education only, since secondary education is a privilege for a few or costly and hence excluding most. For countries where a full secondary education is mandatory or occurs for most of each age group, the shift of focus is to science in the future citizenship role of *all* students and so again, it is away from the role science may have in schooling as preparation of some students for future science-related careers.

The Asia and Pacific Region of UNESCO in 1984 put before its member states a number of optional programmes for its emphasis in the next five years. After basic literacy, Science for All was chosen as a top priority. In this case, the whole school population was again the target group in the formal systems of schooling, but the Asian programmes also seek to identify a number of other target groups in the wider youth and adult communities for non-formal science education (UNESCO, 1983).

As we think about Science for All as the contemporary target for curriculum development in science education and, in this chapter, of what it means in particular for laboratory work, it is important that we remember that there have been earlier attempts to do the same thing. Indeed, if there is to be any chance at all of success in meeting the contemporary challenge, it will be because we have learnt from the mistakes of the last 25 years rather than because we think we can be successful where our immediate predecessors were not.

The framework for this chapter is that the science education that will serve to provide Science for All will need to be different from the sort of science education with which we are familiar. Furthermore, I will argue that science for an elite group (most current science education in schools) is inimical or antithetical to a Science for All. These two important social demands on science education within the overall schooling system cannot, on this thesis, be met by one type of science education.

This conflict view of the curricula (e.g. Middleton *et al.*, 1986, and Fensham, 1985) that will serve different social demands is by no means readily accepted. Some authorities do not recognize that these demands are competing and the many attempts since the mid-1970s to try to make science curricula more 'socially relevant' are indicative of a willingness to recognize competing demands, but then to try to find a compromise within the one curriculum that will contribute to these competing demands.

Enough evidence that all was not well with school science education had accumulated by the late 1970s in those countries which were first into the curricular reforms to warrant the thesis that has just been stated (Tamir *et al.*, 1979). It was at least enough to enable the new projects of the 1980s to be established and generously funded. However, it would not be right to suggest that all these new projects accept or are based on the conflict thesis above (AAAS, 1985).

If there had not been such a massive effort (with at that time previously undreamed of resources) put into the new science curricula of the 1960s and 1970s, it would be possible to argue against the conflict thesis. That is, the failure of these curricula to contribute a Science for All was due to this or that aspect of their development and implementation being overlooked or done badly. The very seriousness of these efforts not once, but sometimes also through major revisions after an initial few years of use of the first versions, inclines me to the conflict position. We have not been short of effort, we have been trying to do the impossible.

A number of social analysts of curricula have also argued that when there are competing interests in schooling, the curriculum of social reproduction will be dominant (e.g. Young, 1971; Connell *et al.*, 1982). In the case of science, this means the science education of an elite - the traditional preparatory type of science education via academic disciplinary study.

## HOW DOES SCIENCE FOR ALL DIFFER FROM ELITIST SCIENCE EDUCATION?

Words like scientific literacy have been used in conjunction with the Science for All slogan and sometimes (as for example, in the FUSE Project, Showalter, 1974) lists of characteristics of a scientifically literate person have been developed. Thus, taking some examples from the FUSE Project (Showalter, 1974), the scientifically literate person (1) uses processes of science in solving problems, making decisions, and furthering his or her own understanding of the universe; (2) understands and appreciates the joint enterprise of science and technology and the interrelationships of these with each other and with other aspects of society; (3) has developed a richer, more satisfying, and more exciting view of the universe as a result of science education and continues to extend this education throughout life; (4) has developed numerous manipulative skills associated with science and technology.

The sorts of learning objectives that the best intentions of the science curricula of the 1960s and 1970s had tended to be not very different from this sort of list regardless of whether they were for the upper secondary elites (Nuffield O and A levels, CHEM Study, PSSC, etc.) or for lower level secondary students, most of whom would certainly not go on with the study of science.

A more helpful approach to what Science for All might be about is to think about the location of learners in relation to science as a human invention and social institution (Fensham, 1986). The dominant approach in science education thus far has been in line with the dictum that the child at school learning science was not (and should not be) different in kind from the research physicist (Bruner, 1960); i.e. teachers are to be engaged in the task of inducting their students at whatever level of schooling into the knowledge, behaviours and ways of thinking of scientists. Science, in this approach, has been defined by scientists and by research scientists as that. The knowledge is ordered and expressed and given priorities that it has in the sub-world of science and in the work of research scientists. The behaviours are versions of the skills scientists carry out, and the ways of thinking are those that (in retrospect more easily than in actuality) are often ascribed to scientists as 'the scientific method'.

A quite other approach begins to emerge if science is seen as a social sub-system inhabited by a small minority during their working hours who are known as scientists. This sub-system is constantly (and indeed increasingly so) interacting with the rest of society and its knowledge and practices do influence, and hence are important, for all those other citizens whose life spaces are apart from the science sub-group. If such a view is accepted, then the learners in Science for All are located firmly outside of science and the task of their teachers is to extract from the sub-system of science those things - knowledge, understanding, skills, etc. - that will enrich the lives of these citizens and make them more able, personally and socially, to live in ways which optimize human life *and* the well-being of the limited environments in which human existence occurs. Teachers of science are thus not inductors of their learners into science, but couriers and translators who bring from science in digestible forms (ultimately non-scientists' understandings) their best estimate of what science has to offer to these learners.

The former approach defines the science of science education from the inside. The latter defines it from the real worlds of non-scientists, i.e. from outside, looking at science. This reverse view means that research scientists, who are largely responsible for providing the insiders' designations of what science is, are not likely to be very useful as architects of, or references for, what constitutes the content to be learnt in Science for All.

The two perspectives do provide different criteria for judging the worth of various possible science contents. In the case of elitist science the criteria are essentially ones which are part of science's sub-system. For example, concepts which are basic to the sciences, or skills that are commonly used in the sciences. In the case of Science for All the criteria of worth will be rooted in the learners or in the interaction between the learner and the everyday world in which her life is spent.

## THE ROLES FOR LABORATORY WORK IN ELITIST SCIENCE EDUCATION

The framework of induction into science makes sense of the apportionment of science curricula in the 1960s and 1970s to the levels of schooling and the role of practical work within them. The model in Figure 1.2.1 is useful as a way of incorporating the variety of roles of laboratory work.

Secondary schooling was associated with the learning of conceptual science, that is, those building blocks of more or less current scientific generalization and explanation of phenomena. Just enough of the factual properties of the phenomena were to be included to provide a minimal logical basis for the introduction of the concepts at their various levels of qualitative and quantitative abstraction. The role of laboratory work was to engage the students as often as was practicable in experience of these factual properties, which were to be presented in their logical sequence, so as to be consistent with the concepts and their definitions. So deliberately were these secondary science courses an introduction to the conceptual logic of science that none of them was committed to the guided discovery role of the laboratory work (which might have involved omission and 'discovery' of certain topics).

Unlike what was perhaps previously a learning priority in

school science, viz., the actual skills of using some instruments and laboratory processes (see Figure 1.2.1), the priority for laboratory work had become its alleged association with conceptual learning. Ausubel (1968) in the 1960s attacked this assumption, with little effect on the curriculum project team's intentions. Ausubel's position gained support from the behaviour of many teachers in schools in various countries as they found that the guided discovery experiences were not necessary for many students in learning conceptual knowledge, and in some cases actually led students into quite unexpected, unintended and unhelpful (for the expected outcome) pathways of thought. The other traditional role for practical work was to reinforce the theoretical learning by enabling the student to use its ideas and formulae in laboratory-contrived situations. In varying degrees this role persisted with the new curricula, although it was not a notable intention of them.

This primarily conceptual dimension of scientists' science in secondary schooling turned out to be very effective for the preparatory and selective tasks that the societal demand for elite scientific manpower required from secondary schooling. Particularly in the physical sciences, its heavily sequential nature, its mathematical associations, its 'interminable' character (there are always more complex aspects of a concept to learn), and the fact that it cannot be substituted by an aptitude for its practical or laboratory side, meant that only relatively few students could persist long enough in a successful track to succeed.

At the primary levels of schooling the new science curricula (often the first to bear this title) were strikingly different from the Nature Study which they displaced at least in a number of countries. Several different approaches can be distinguished, but at another level they all set out to teach the processes of science that the conceptual emphasis squeezed out of the secondary school.

In the purest forms these process curricula were literally the teaching of process. Thus Science - A Process Approach (the AAAS Project) defined a number of scientific processes like classifying, measuring, inferring and predicting. Suitable content in different countries was chosen to teach these intellectual skills, but the real intention was the acquisition by young students of these tools of science. Science 5-13 in Britain was a similar project with a highly complex teaching/learning sequence.

Several other projects (e.g. ESS in the USA and Nuffield

Primary Science in Britain) stressed the processes of asking questions about phenomena and then trying to set up ways of exploring their aspects so that answers might be forthcoming. Again the topics or content of the phenomena were just vehicles for learning to be discoverers or investigators as scientists were supposed to be. Finally, another set of curricula for this level (SCIS and A Concept Approach) did include conceptual knowledge of science but now with a strong emphasis on a pattern of processes that prepared the way for the concepts to be introduced. This last set of curricula were an attempt, away from the conceptually overloaded courses in secondary school, to do what the secondary curriculum projects would like to have been able to do, i.e. teach science processes in relation to concepts.

The intention of the practical, or hands-on, experience in these elementary science curricula was to provide practice in these processes which would enable learning of them to occur. The learning intentions were derived from scientists' rationalization of their behaviour, and where concepts were involved, the phenomena and topics were chosen as good vehicles for what were said to be important concepts in science, rather than that they were public phenomena of importance (Science 5-13 may be an exception here).

In general the evaluative studies of these process curricula in practice have not been encouraging (Tamir *et al.*, 1979). Teachers find the suggested roles difficult to sustain with the small groups of students and to be in conflict with class teaching and their control procedures. Furthermore if the emphasis is on process learning, why not use social contexts which are more familiar to them and hence easier for them to motivate learners about? Finally, elementary school teachers, despite the stated emphasis on process as the content for science learning, find it hard not to think of science as that long list of facts, rules, concepts and principles that so overwhelmed them in their own secondary education that they chose not to pursue it (particularly physical sciences) any longer than was necessary.

Students in classes where process curricula are being used tend to report that what they are learning about is the details of the particular context being used rather than the processes themselves. Processes seem always to be subservient to content. Despite their unquestioned reality as part of the repertoire of scientists' behaviour, they are relatively inessential for successful learning of science, as any senior secondary student or undergraduate in science could testify. Their success as students has rarely been dependent on their ability to answer process questions on examination papers, or to carry out open ended investigations in a laboratory which involve the use of these processes.

## LOWER SECONDARY SCHOOLING

At the lower secondary level there are a few examples of curriculum projects that tried to move away from disciplinary science. Environmental Science in the USA extended the process emphasis of the primary projects by encouraging lateral and imaginative thinking. Its suggestions for practical work like 'Take a photograph of something that has never been photographed before' or 'Think of something new' were divorced from content and were more easily used in pre-service training or in in-service training of science teachers, to jolt them in their disciplinary ruts, than they were by teachers in their science classes.

The Australian Science Education Project in the early 1970s produced nearly 40 modular units that ranged from traditional disciplinary ones like Forces, Plants, Water, Cells and Electric Circuits to others like Places for People, Signals without Words, Petroleum, Sticking Together, and The Australian Scene that cannot be identified with disciplines or traditional topics. Among the ten most popular units nine were of the disciplinary type and the tenth was Males and Females - a unit dealing with human reproduction (Owen, 1976).

Those that fitted existing disciplinary syllabuses or conformed to disciplinary assumptions about what is worthwhile science were purchased most commonly. Those at the novel or inter-disciplinary end of the range had much less interest shown in them. The constraining influence of traditional disciplinary knowledge was highlighted by the relatively poor sales of units like Metals and Polymers, both of which were straight science ones but involving topics that were not usually taught in the dominant school sciences, physics, chemistry and biology.

## A WAY FORWARD

There is little doubt that most science curricula in schools will continue to use laboratory work for all of the learning objectives that have just been discussed as part of what I have called Elitist Science, to distinguish it sharply from Science for All. To the extent that teachers and their curriculum advisers can extend successful learning of these objectives to anything that could be described as a majority of students, then these efforts will be valid components of and worthy of the name Science for All. Since much has been written about laboratory work that is aimed at these objectives, I intend in the rest of this chapter to take a more polarized position, to push myself to make an alternative case for laboratory work as part of Science for All. Accordingly, all of the above approaches will here be associated with Elitist Science and I will seek to suggest quite different roles for laboratory work that are consistent with the framework outlined above for Science for All.

## The secondary years

The task of suggesting alternative science education at the secondary level is in one sense very difficult, and in another sense easy. It is difficult because of the very established tradition of science in these years of schooling. It is a high prestige field of study. It (together with mathematics) opens doors to more fields of further study and vocations than any other combination of subjects. We who are faced with responding to the challenge of Science for All are all successful products of secondary science's traditional form. In a very real sense we are where we are because it was Elitist Science and somehow we were among that elite. We cannot escape having been socialized into its basic values. Accordingly, one of the major problems of Science for All is that its managers will have to develop science education in ways which are quite other than the ways it attracted them.

The easy side of science education for the citizen, at the secondary level, is that science and technology are now interactive with an ever-widening range of societal aspects of the lives of all students, most of which are now amenable contexts for scientific exploration and comment, if we can only recognize it. The shelter, the diet, the sport, the leisure, the transport, the communications, the arts, the clothing, the work, and the wider environment of Australians and of the citizens of most countries are in visible processes of change as a result of applications of science and technology. In this sense, the outer boundaries of science and society are richer and more variegated territories for learning than its inner more complex levels of speciality. To borrow from Stephen Leacock:

> Ignorance, in its wooden shoes, shuffles around the portico of the temple of learning, stumbling among the litter of terminology. The broad field of human wisdom has been cut into a multitude of little professional rabbit warrens. In each of these a specialist burrows deep, scratching a shower of terminology, head down in an unlovely attitude which places an interlocutor at a grotesque conversational disadvantage.

In another sense the learning of this sort of science should also be easier since it is working much more where the learner is, rather than in an area of thought that is foreign to her/him. Furthermore, this sort of learning from science ought to be simpler in character without the abstractness of the conceptual emphasis. It should also be inherently more attractive to the learners, beginning as it does with their own significant interactions with the world.

To check out these claims the members of the Cyprus Workshop in 1984 set up two criteria against which to check whether they could identify broad areas for learning that could be part of a science education for all students.

1. The content (science knowledge and its associated skills) should have social meaning and be useful for the majority of learners, or
2. The content (science knowledge and its associated skills) should assist learners to share in the wonder and excitement that have made the development of science such a great human and cultural achievement.

By 'social meaning and usefulness', the first criterion meant that this learning would have the potential of directly enabling the learners to enhance and improve their lives beyond the school, in their families and as citizens of the society in which they live. This criterion is strongly pragmatic and the learning of any content chosen under it would be readily recognized as useful by the learners, parents and community authorities.

The second criterion again requires a careful choice of natural phenomena, e.g. those that are easily available to the learners and their teachers and will provide content for learning (when well taught in these basic levels of schooling) that is characterized by wonder and excitement. This second criterion rules out those many 'wonderful' aspects of science which are perforce reduced in schools to prose descriptions in school science textbooks, and that can only be learned by the reception of verbal knowledge passed on from the teachers to their pupils. Wonder to the scientist is not a guarantee of wonder to these learners.

It was not intended that the first criterion of social usefulness could be met if the essential 'use' of the content knowledge to be learned was that it would be a basis for further science learning in the later years of schooling. This sequential aspect of much science learning may, of course, be important in the sort of science those exploring a career in science may study later in their schooling. Rather, this criterion was meant to be concerned with justifications that are more relevant in life external to school and in the comparatively short term. Sequential usefulness is a bonus but not a justification in itself.

These two criteria enabled the workshop to list twelve broad topic fields, within each of which the members, thinking of their very different social situations, could readily see that there were a number of examples of mini-curricula that would meet the criteria. These topic fields are listed in Table 5.4.1.

A number of examples of the mini-curricula within these topic fields were discussed at the workshop. It may seem that this approach is heavily knowledge-centred and that it has discarded process and skills altogether. This is by no means the case, as the details of these examples would illustrate. Practice and encouragement of the acquisition of intellectual processes (like careful observation, classifying, control of variables, prediction, etc.) and of practical skills (like measuring, testing, construction, planting, etc.) were described regularly as basic aspects of learning of these science topics in school. The skills and processes are, however, always associated with, or embedded in, the learning of meaningful knowledge content. The knowledge content, as chosen, is to have ready justification and acceptance from both the scientific and public communities. This in turn, it is argued, will lend more credence and status to the processes and skills that are very much part of the intended learning. The outcome of this workshop thus did not argue for abandoning science processes and skills as a lost cause for the basic science education of all students. Rather it suggested that such learning goals are a lost cause unless they are associated with a knowledge content that has a strong claim in its right to be *scientific knowledge of public worth.*

**Table 5.4.1:** Topic fields for basic science education in schools (Cyprus Workshop)

| | |
|---|---|
| 1. | The senses and measurement as extension of the senses |
| 2. | The universe |
| 3. | The human body |
| 4. | Health, nutrition and sanitation |
| 5. | Food |
| 6. | Ecology |
| 7. | Resources (natural and man-made) |
| 8. | Population |
| 9. | Pollution |
| 10. | Use of energy |
| 11. | Technology |
| 12. | Quality of life |

e.g. Topic 10. Efficient use of energy resources is a pressing personal and social need everywhere. All countries need to lower their dependence on imported energy resources and to reduce wastage. Understanding of efficiency in energy use can lead to an enhanced quality of personal and family life.

At this universal level of science education, the treatment of the content to be learned within these topics will need to match the levels of ability and the interests of the learners. With the socially useful focus for the learning outcomes, the content to be learned will be informed by (or based on) scientific knowledge itself. For example, some more recent approaches to teaching units within the broad topic field of Nutrition have taken this approach. Rather than aim at knowledge of carbohydrates, proteins, vitamins, fats, minerals, etc. (common and appropriate in more advanced science education) as what is to be learned under Nutrition or Foods, the knowledge now to be acquired under Nutrition is that nutritious diets are more likely if each day's meals draw from each of the four groups: meat and fish, cereals and bread, fruit and vegetables and milk and milk products. Provided the process and practical skills of using this classification are also learned, learners with this applied scientific knowledge are in a better position to improve their nutrition than many more advanced students who have acquired some knowledge of the chemistry of foods but have often not reached the point of making it operational as nutritious diet. This view of Nutrition leaves the Food topic in the list in Table 5.4.1 to deal with the important aspects of sources, production, preservation and distribution of various basic and relevant foods.

This exercise in the Cyprus Workshop was encouraging because there do seem to be many content areas that can be plumbed for both practical and non-practical learning of a Science for All. However, much more detailed thought needs to be given to how these become transformed into curricular activities for learning at various levels.

If the proportion of students who are to succeed in learning science (even when it is defined in ways like that just suggested) is to be substantially increased to anything like a significant majority, practical modes of learning are almost certainly necessary and not just for the practical aspects. In other words, the practical in science education will need to be a way of learning what is to be learnt and not just a part of what is to be learnt.

Two possible approaches to laboratory work in the secondary years that have been explored by some teachers are now described. Both clearly lie within the framework that has been set up for Science for All in this chapter.

## Technology as an integral part of Science for All

From inside science there is a rather sharp distinction between science and technology, between a pure science and its applied form which provides science's link with technology. Particularly in the past, technologies often developed first through trial and error and the inventiveness of humankind confronted with practical problems of survival. The science on which these technologies were based was often understood only much later. Quite often it was a technological breakthrough that provided the spur and direction for the efforts of the scientists of the day. More recently, much modern technology has come as a result of basic scientific work that, as least initially, was not related to potential social uses.

From positions in society, outside of science, these distinctions are no longer evident. If students in lower secondary school or older citizens are asked for examples of science, they are more likely to give answers that research scientists would regard as technological in character rather than as scientific.

This blurring of science and technology, as far as citizens are concerned, is a useful basis for planning Science for All and its practical work. If science educators do not accept that it is their task in a Science for All to teach the association between science and technology there is certainly no other subject area in the curriculum of schooling which can or will include it. Indeed other subject teachers, like the citizens above, also see technology as within the bailiwick of science teachers. Some obvious and useful associations can occur between science teachers and technical, art, and craft teachers as the teaching of technology occurs. A foreshadowing of these quite novel sorts of liaison which would move science from the academic pole of the school curriculum towards its practical pole was the briefly exposed statement in the ASE publication in Britain, *Alternatives for Science Education,* which suggested the contexts for the future of science education were the laboratory, the field and *the workshop* (ASE, 1979).

Many science educators and some curriculum groups are now putting their mind to the problem of how to combine technology with science teaching. One probably useful criterion to apply to these attempts, if they are to be seen as possible avenues for Science for All, is the sequence of the scientific knowledge to be learnt. If the knowledge sequences are still essentially those in the science disciplines themselves, then Elitist Science is still dominant. If the knowledge sequence is determined by the technology or issue being studied, then it will inevitably associate knowledge that comes from quite separate traditional topics and even from several science disciplines, since technologies are almost never dependent on a single set of disciplinary concepts.

One approach that seems likely to be fruitful for Science for All is suggested by the model in Figure 5.4.1. Practical exercises that provide experience of these three sorts of problem-solving could form a basis for a good deal of Science for All. Where possible the way in which a technology that solves a real world problem (and all commercially successful technologies do this, otherwise they remain undeveloped patents or novelty inventions), and thus by-passes the more specialized approaches that require scientists' knowledge and skills, should be made explicit.

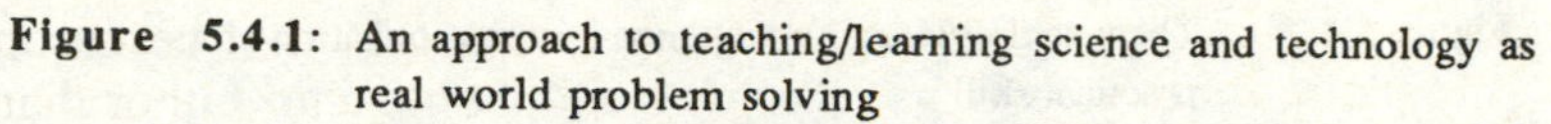
**Figure 5.4.1**: An approach to teaching/learning science and technology as real world problem solving

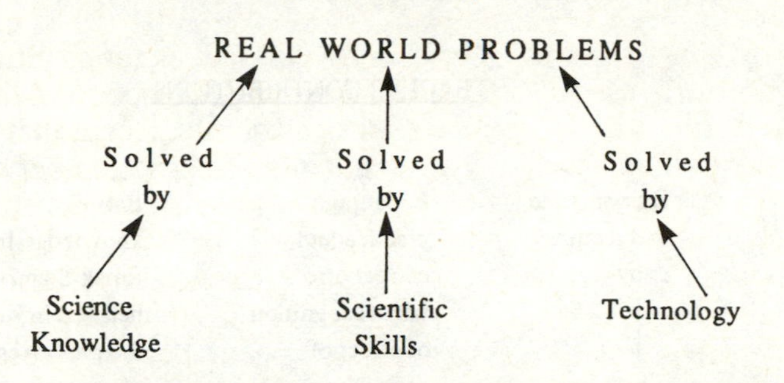


Another presentation of the same idea is shown in Figure 5.4.2 where a teaching/learning sequence for a practical skill is shown. From the success that driving instructors have with the very students with whom we are concerned in Science for All, we know that a very large proportion of an age group can learn a skill to a fair level of competence if the skill is desirable enough. Science for All will need to identify these desirable skills for the different levels of schooling, and practical work should then be devoted to the sequence of demonstration, copy, practice, correction, more practice, that is indicated in Figure 5.4.2. With the confidence of considerable achievement it may turn out that surprising numbers of students will be receptive to learning the scientific bases of skills that they now have acquired. Even if they do not go on to learn the scientific background of these skills, they are likely to view science quite differently when it has been experienced as a source of useful learning for themselves.

## Product worth

Consumer associations have for several decades now made extensive use of scientific processes of investigation to test the claims of various commercial products, and to compare alternative brands of the same product for their qualities and relative costs. Although some attention has been drawn to these sources by science curriculum writers, they have remained marginal to mainstream science education because they do not relate well to disciplinary science.

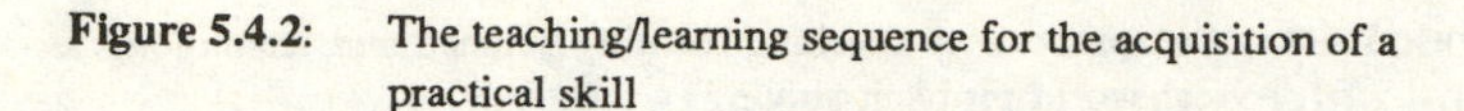

**Figure 5.4.2:** The teaching/learning sequence for the acquisition of a practical skill

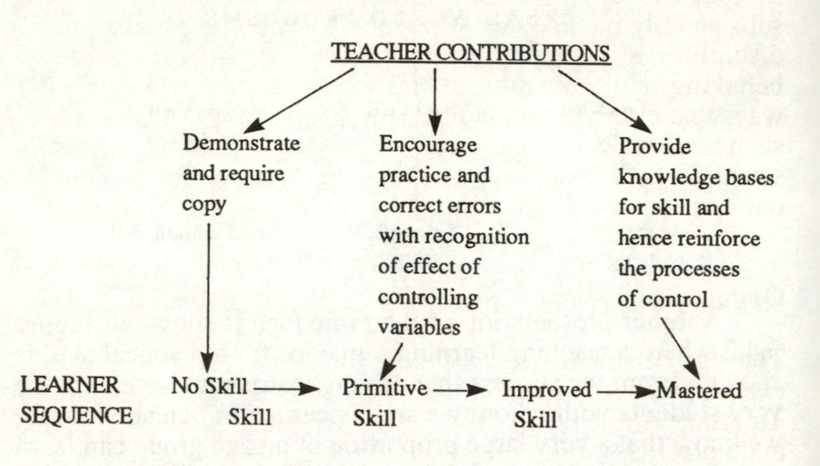


In a Science for All this publicly oriented type of scientific investigation would seem to have a quite central place. School science could include extensive class investigations into the claims and relative merits of products that are relevant to its various age levels. Basically these sorts of tests involve measures of performance under various conditions. For comparative studies the establishment of standard conditions is required. These conditions need to be noted carefully just as for a laboratory experiment in pure science. However, now the variables defining the conditions are not usually the familiar abstracted ones of laboratory science, but ones that are associated with the real world where these products may be used.

## Teachers of Science for All

The teachers of Science for All will need assistance and assurance about the roles the above suggestions require them to play. Secondary school teachers of science have generally studied a mixture at the tertiary level of science and education. In Europe the proportion of science is often high whereas in North America the proportion of education is usually higher. In Australia there are often three years of science and one of education, but even so the science is still largely at the level of text book knowledge, and practical work has rarely extended beyond set exercises to individual open-ended problems or investigations. These sorts of science teachers have been very substantially inducted into the stable knowledge of one or more disciplines of science, but they have not had experience of behaving and practising in relation to this knowledge in the ways that either research or applied scientists do. The plight of such teachers is that they are shielded by an umbrella of *stable* scientific knowledge and what they transmit to their students is one step worse - *stale* scientific knowledge.

## Organizing science practical work in schools

The use of scientific equipment and materials in schools is expensive in capital outlay and in maintenance and replacement. To reduce these costs, most school systems organize students to do practical work in small groups rather than as individuals.

In addition to the cost savings a number of potential benefits are often thought to be associated with this way of organizing students in practical work. There are obvious possibilities for peer instruction since the small groups will encourage articulation of ideas and procedures among the members.

A Science for All could mean even more practical science learning in schools. The small group organization at first sight seems to be the only feasible approach so it is important to ascertain the effectiveness of the present use of groups in practical work.

In general, studies of the interchange between small group members is not very encouraging about the quality of mutual peer instruction. All too often there is a dominant or confident member who tends to be the regular handler of the equipment. Other members are quite often passive note takers and record keepers. The less confident are not helped by their peers to get practical experience (Suan, 1976).

The outcome of these uneven patterns of participation is differential learning. Klainin (1984) in a recent study in Thailand, where students work in groups of three in the laboratory, obtained striking evidence of a phenomenon all too familiar to all who have been insecure in laboratories at school or at university. When she tested the groups on their ability to carry out a practical task they had learnt earlier in the year, about 80% of the groups were successful. A little later she retested the class, but this time as individuals, and only about 20% were able to carry out the task correctly! Achievement in a group does not ensure internalization of this learning for all its members.

Accordingly, Science for All seems to face a dilemma. More practical learning is needed yet small group organization which provides realistic cost savings does not seem to lead to effective learning for all.

One alternative that seems to be worthy of exploration is the large group project or class-centred exercise in which, over time, the teacher ensures that proper turns at carrying out critical tasks occur. With fewer groups, the teacher's presence as a resource and task sharer becomes more feasible and may provide cooperative learning that is more effective than our present approaches.

## Inclusive science

There is much talk in the 1980s about education being 'inclusive', and in particular that science education needs to be inclusive. We can therefore take this to mean that science education is at present not inclusive, and if it is not inclusive of some distinguishable groups in society, it is certainly not Science for All.

The commonest use of this term is in association with gender participation and achievement in science education. There are obvious biases - girls more commonly take biology than boys, whereas boys much more commonly study physics than girls, with chemistry somewhere in between. These patterns are common in western countries but in some eastern European countries and some Asian ones the situation is rather different.

Much less attention has been given to other social groups who may be 'excluded' by the way science is taught. For example, the sorts of practical activities (especially at the primary levels) that were advocated in the 'discovery' oriented curriculum projects seem likely to have been biased towards the children of middle and upper class families and to certain cultural groups in multicultural societies. For example, their extensive encouragement of questioning by pupils and the use of open-ended questioning by teachers are not common traits among some Australian families, whereas they are common and even rewarded in others. A contrary example may be the suggestion that the mathematical, abstract quality of much of the conceptual learning in physics and chemistry seems to reduce social class bias in this sort of learning, at least for more able males (Hill *et al.*, 1974).

Among the approaches to lessen gender bias in science education, there is strong encouragement for practical experience, particularly early in primary schooling, that will enable girls to gain experience and confidence in 'tinkering' with scientific equipment (Parker and Rennie, 1986). The extra-school influences concerning confidence in practical science are seen as very strongly biased in favour of boys. Again, cooperative activities in the laboratory which have obviously human-related outcomes are being suggested as more appropriate types of practical work in secondary schooling, if we want girls to be more attracted to science.

Finally, as suggested earlier, it may be that science education will only become truly inclusive or 'for All' when we can use practical work as a vehicle and context for much of its knowledge learning. The use of the practical as a mode for learning science content is a radical thought which is not really recognized in the model in Chapter 1.2 Figure 1.2.1. may be meant to include it, but it would be better represented if the box of Actual Learning Activities (laboratory classroom behaviour) was the only learning activity in that model and if it led on to the two sorts of outcomes that are indicated.

An example of this was a Year 10 class who spent a term of their science lessons building fibreglass canoes to use in a regatta. At an oral level they knew far more about polymerization than the Year 12 elite chemistry group learnt in their specialized Polymer Chemistry option via traditional modes of conceptual learning.

The movement from contemporary Elitist Science towards Science for All requires quite radically new approaches to content and to teaching and learning. Layton (1984), a perceptive historian of science education, has suggested that the prospects for this sort of movement are better now than they have been for a long time. Practical work (its role, and the attitudes of teachers and students to it) is likely to be quite critical if this movement is to have any success.

## REFERENCES


AAAS (1985) *Project 2001: Understanding Science and Technology*. American Association for the Advancement of Science, Washington, DC.

ASE (1979) *Alternatives for Science Education*. Association for Science Education, Hatfield, Herts, England.

Ausubel, D. P. (1968) *Educational Psychology - A Cognitive View*. Holt, Rinehart and Winston, New York.

Bruner, J. S. (1960) *The Process of Education*, Cambridge, Mass., Harvard University Press.

Connell, R. W., Ashenden, D. J., Kossler, S. and Dowsett, G. W. (1982) *Making the Difference*, Allen and Unwin, Sydney.

Fensham, P. J. (1985) Science for all. *Journal of Curriculum Studies*, 17(4), 415-35.

_____ (1986) Science for all. *Educational Leadership*, 44(4), 18-23.

Hill, S. C., Fensham, P. J. and Howden, I. (1974) *Ph.D. Education in Australia: The Making of Professional Scientists*. Canberra, Australian Academy of Science.

Klainin, S. (1984) 'Activity based learning in chemistry', unpublished Ph.D. thesis, Monash University, Clayton, Victoria.

Layton, D. (1984) (Ed.) *The Alternative Road: The Rehabilitation of the Practical*. Centre for Studies in Science and Mathematics Education, Leeds, University of Leeds.

Middleton, M., Brennan, M., O'Neill, M. and Wootten, T. (1986) *Making the Future*, Canberra, Commonwealth Schools Commission.

Owen, J. (1976) Trends in sales of ASEP materials. *Lab. Talk*, 20(3), 31-4.

Parker, L. H. and Rennie, L. J. (1986) A comparison of mixed-sex and single-sex grouping in Year 5 science lessons. Paper presented at AERA Annual Meeting, San Francisco.

Showalter, B. (1974) What is unified science education? *Prism, II*, 1(2), Center for Unified Science Education, Columbus, Ohio State University.

Suan, M. Z. (1976) 'An evaluation of ASEP - a case study approach', M.Ed. thesis, Monash University, Clayton, Victoria.

Tamir, P., Blum, A., Hofstein, A., Sabar, N. (1979) *Curriculum Implementation and its Relationship to Curriculum Development in Science*. Jerusalem, Israel Science Teaching Center, Hebrew University.

UNESCO (1983) *Science for All: Report of a Regional Meeting*, Bangkok, APEID, Unesco Regional Office.

Young, M. F. D. (1971) An approach to the study of curricula as socially organised knowledge, in Young, M. F. D. (Ed.) *Knowledge and*

*Control*: New Directors for the Sociology of Education. London, Collier Macmillan.

# Part 6

# ACTUAL LEARNING ACTIVITIES

# 6.1

# Role of Laboratory Work in a National Junior Secondary Science Project: Australian Science Education Project (ASEP)

John Edwards and Colin Power

## INTRODUCTION

In this chapter we examine the first and probably the most important of the national curriculum development projects undertaken in Australia, the Australian Science Education Project (ASEP). Within Australia, as in the rest of the western world, it is widely accepted that the laboratory is an important agent for developing an understanding of the nature of science, particularly the role of scientific enquiry. In keeping with the recommendations of science teachers, professional associations and the literature, the ASEP materials placed a very heavy emphasis on enquiry on the assumption that the teaching of science should reflect the experimental nature of science. It is also significant in that it was one of the first major science curriculum reforms at the lower secondary level to seek to make laboratory work the centre point of the teaching of science to young adolescents.

The laboratory in the ASEP class was not intended to be a place which would merely illustrate, demonstrate or verify known laws and concepts. It was intended to develop skills and attitudes, and to provide experience in the process of designing investigations and acquiring and interpreting data. All units placed a heavy emphasis on activity-based learning, with simple laboratory-based experiences in virtually every science lesson. ASEP is a particularly interesting case then, of a lower secondary course which sought to follow Klopfer (see Chapter 3.2) and others in making enquiry the heart of science laboratory work. Moreover, it is a project in which the degree to which this intention has been implemented in schools, and the impact of the enquiry approach, has been more systematically evaluated and researched than other comparable courses. As such, it provides a concrete example of what is involved in translating intended learning activities and outcomes in laboratory settings in lower secondary school into actual activities, and indicates some of the conditions needed for the intended outcomes described in the model (Chapter 1.2) to be actualized.

## NATIONAL CURRICULUM REFORM MOVEMENTS

Until the 1960s Australian secondary schools were able to maintain a fairly 'traditional' approach to science teaching. Science in schools sought to equip students with a coverage of the facts of selected science disciplines illustrated by laboratory exercises and demonstrations. Three major factors created pressure for change:

1. Changes in the student population. The actualization of the ideal of a junior secondary education for all precipitated enquiries into the basic purposes of secondary education in all states, each of which recommended the development of new 'general' or, 'integrated' science courses for the junior secondary level.
2. Changes in Australian society and its institutions. In a period of economic growth mingled with concern about the scientific achievements of the USSR, education had come to be viewed as a form of national and private investment. It was generally accepted that the state must assume responsibility for ensuring an adequate supply of well-trained scientists and engineers if continuing progress in the West was to be assured. In the 1960s, Commonwealth Government involvement in secondary schooling, hitherto a State Government responsibility, began. Significantly, this first took the form of Commonwealth grants for the building of school science laboratories (see Chapter 5.2).
3. Changes in science and science education. The rapid

development of science meant that syllabuses had become overcrowded and often out of date. At the senior secondary level, CHEM study, PSSC physics and the Australian adaptation of BSCS biology (*Web of Life*) were beginning to be used in Australian schools. These courses placed less emphasis on the facts of science and more on their conceptual schemes and processes of enquiry. The latter emphasis was reflected in the greater emphasis given to laboratory work, experimentation and the processes of enquiry in each new curriculum package.

These pressures created something of a crisis at the junior secondary level leading to a number of attempts to develop enquiry-based science curriculum materials for use in junior secondary classes. The most important of these was the Victorian-based Junior Secondary Science Project (JSSP). In September 1967 Commonwealth assistance was sought to set up a national curriculum project: ASEP was born. Since 1974, ASEP units have been, and continue to be, used widely throughout Australia.

ASEP represents a centralized attempt to reform the teaching of science at the junior secondary level through the production of activity-based materials. The activity-based approach to teaching science is one of the most widely used variations of laboratory-based teaching, particularly in junior secondary and primary science. Investigation of the theory and practice of incorporating such an approach into a junior secondary science project provides a number of perspectives on the role of the laboratory.

## THE ASEP MATERIALS

The Australian Science Education Project (ASEP), which functioned from late 1969 to early 1974, was Australia's first national curriculum project. At its peak of operation it employed 65 staff, with a maximum of eleven developers (writers) employed at any one time. The Project was jointly funded by the Commonwealth and six State Governments. It resulted in the production of 41 relatively low-cost units for year 7 to 10 science, six service booklets, one handbook *A Guide to ASEP* (ASEP, 1974) and a range of audio-visual and supplementary materials. The materials were designed to fit in with state science curricula in a flexible way. The units were planned to provide teachers with resources which they could integrate with other resources in the design of their own programmes.

Units typically begin with a core section designed to be completed by all students. This leads to a range of options designed to cater for a variety of student ability and interest. Students are invited to choose from the range of options. Most units also contain a self-administered diagnostic test at the end of the core. Students check their own answers and can refer to remedial material before seeking the help of a fellow student or teacher.

Record books are provided for use with most units. However, some of the later units do not have record books, reflecting differences in the views among ASEP staff as to their value. Some unit teams believed that a highly structured approach to activities and the keeping of records was essential, particularly in the early stages of development. Others, particularly those responsible for more controversial units dealing with values and those developing units for students at the formal stage of reasoning, opted for less structured, more open-ended activities, more emphasis on group discussion and less on formal records. For each unit there is a Teacher's Guide which incorporates the student text plus a wide range of background and resource material.

## ASEP PHILOSOPHY

The broad aim of ASEP was to design science experiences which would contribute to the development of children. More specifically, the science experiences were aimed at developing:

1. Some understanding of human beings, their physical and biological environment, and interpersonal relationships,
2. Skills and attitudes important for scientific investigation,
3. Some understanding of the nature, scope and limitations of science, and
4. Some understanding of, and concern for, the consequences of science and technology.

Two statements were considered in conjunction with the aims:

1. The kind of understanding at which the Project aimed

enables children to operate more effectively in their environment.
2. To arouse and foster the interest of children is of prime importance in the development of understanding, skills and attitudes. (ASEP, 1974.)

The main sources of ideas for inclusion in ASEP materials arose from the Project's view of the environment of the child, the nature of science, and the existing state of scientific knowledge. Some ideas were also included to promote particular attitudes, to develop procedures for extending knowledge, or to fit with a particular stage of child development.

The different ASEP units were each developed to match a particular Piagetian stage. Three stages were covered: concrete operational (Stage 1), formal operation (Stage 3), and the transitional phase between these two stages (Stage 2). A range of units were produced at each stage level. From Ginsburg and Opper (1969) a set of Piagetian principles formed a strong basis for ASEP philosophy:

1. New ideas and knowledge should be presented at a level consistent with the child's state of development of thinking and language.
2. A major source of learning is the activity of the child.
3. Classroom practices should be tailored to the needs of individual children and should present moderately novel situations.
4. Children learn by social interaction.
5. Children should have considerable control over their own learning.

The last four of these principles present a clear rationale for the activity-based ASEP materials. At the same time the Project designed its materials to encourage enquiry in a broad sense. Use of a broad range of enquiry types was incorporated since it was felt that no clear research base had been established to favour any particular enquiry technique. Enquiry, in its various forms, was seen as the application of the processes of science. ASEP maintains that this involves the individual in activities which require one to identify problems, observe, measure, classify, order, infer, predict or form hypotheses, search for and discover meaningful patterns, design and perform experiments, interpret and analyse data, and verify the validity of conclusions reached.

An in-house analysis of a sample of ASEP units (Shepherd, 1973) showed that:

1. There is a variety of activity types across a structure spectrum from open to programmed;
2. Most activities are done by students as individuals or in pairs, with whole class activities being rare; and
3. The most common type of activity is one in which students follow detailed instructions, but are free to form their own conclusions from the activity.

The intended approach was that of 'an inductive process in which the students learn by discovery', the degree of guidance offered varying to meet 'their individual needs at each developmental stage' (ASEP, 1974, p. 21).

## DEVELOPMENT OF AN ASEP UNIT

*A Guide to ASEP* (ASEP, 1974, pp. 101-3) describes seventeen steps in the development of an ASEP unit. The following list highlights six particular aspects of the process:

1. Each unit was prepared by a developer who was an experienced, successful teacher. During the development process they had access to: a laboratory assistant; a librarian and a well-stocked library of science and science curriculum materials; a photographer; and an adviser who was a senior member of Project staff.
2. The developer worked from a first specification which had been prepared by senior Project staff following guidelines developed by a meeting of representatives from all States and many areas of education. This was fleshed out into a second specification which clearly outlined: the estimated length; the relation of the material to the Project's aims and philosophy; the links with other ASEP units; a plan of the unit; the objectives; the content and activities; the expected outcomes; suggested references and audio-visuals; and facilities required. Much of the development work was done in the laboratory, developing, trialling and refining activities. Before inclusion in the second specification all activities had to be successfully tried out. This second specification had then to be defended at a meeting of

Project staff. These meetings were extremely rigorous and served to ensure a consistency of standards, philosophy and practice.

3. The production phase then prepared the materials for trialling in local schools. Specialist staff worked as a team with the developer to design, edit, prepare artwork and photographs, and check and correct the draft.
4. Local trials were conducted in a cross-section of schools, usually eight in number. Trial teachers were introduced to the materials and attended regular meetings during the trial. The developer and other Project staff visited trial schools to observe and interact with students and teachers.
5. Following the trial an evaluation summary was prepared. The summary included: comments from outside consultants who were usually experts in the area; an annotated copy of the unit containing comments by teachers and Project observers; collated student diagnostic test results; collated teacher questionnaire returns; detailed ratings and comments from student questionnaires; time taken by students to complete each section of the unit; feedback from the final seminar with trial teachers; and a set of recommendations for improving the unit. The developer, often different from the original developer, again had to defend this document at a meeting, and a re-development plan emerged.
6. The unit was then redeveloped and sent for trial in all States and New Zealand. This usually involved 26 schools, including four nearby schools. Some of the later units were only given this national trial. On completion of this trial a similar process was followed to produce the final development plan. Once this was done the unit went through a final production phase, taking 41 days, to prepare the material for final printing.

Obviously the particular personality, skills and experience of the developer and adviser had the major influence on the unit. However, the process of justification through rigorous meetings and the emphasis on the views of students combined to produce materials with a reasonably consistent philosophy and style. While such an elaborate process is beyond the scope of school-based materials development, the elements of the process provide clear guidelines for those involved in activity-based learning.

When using most ASEP units, it would be difficult to

avoid ensuring that practical work, in or outside a school laboratory, played a central role in the teaching-learning process. Short laboratory-based activities using, for the most part, fairly basic equipment, are an integral part of the course.

## ANALYSIS OF ASEP MATERIALS

There have been four major attempts at analysing ASEP materials. Carss and Clarke (1972) used a content analysis technique to identify major thematic elements and thereby provided a view of the structure of a unit. By plotting the stages when themes emerged and receded throughout a unit, they identified activities in which additional clarification, explanation or linking was needed and other structural weaknesses. They re-wrote sections of the core of the unit on 'Pushes and Pulls' so that it conformed more closely with the principles put forward by Ausubel (1968). Students using the redesigned core achieved better than a control group using the original ASEP trial materials. It is interesting to note the similarity between Carss and Clarke's redesign of the core of one of the ASEP units and the redesign resulting from the ASEP trial process. A cost-benefit analysis of the two procedures suggests the strong potential value of such content analysis in materials development and the value of undertaking such an analysis with other ASEP units. Shepherd (1972) used a count of sentence types to reveal the number of statements, questions or instructions in the materials. This analysis showed that ASEP, and other enquiry-oriented materials such as ISCS and JSSP, contained approximately half as many statements, twice as many questions, and three times as many instructions as a traditional science textbook.

Linke (1972) used a content analysis rating scheme to reveal the environmental emphasis in curriculum materials. A preliminary analysis revealed variation between ASEP units, but a much greater environmental emphasis than in BSCS materials. Fraser (1976) found a very close agreement between the processes of enquiry emphasized in the science education literature and the stated aims of ASEP. Edwards and Dall'Alba (1981) used a Cognitive Demand Scale to reveal the intellectual demand of curriculum materials, and the relationship between the demand level of objectives, learning tasks and evaluation items. A sample analysis of the core of one ASEP unit revealed many mismatches between objectives, learning tasks and evaluation items. It also revealed the level of cognitive demand of the unit (Dall'Alba, 1981). Readability measures are focused procedures for analysis of materials (Edwards, 1983). While developers at ASEP used Gardner's (1972) *Words in Science* and the Flesch readability formula (ASEP, 1974) no measures were made of functional readability. The Cloze procedure (Thelen, 1974) is a powerful and easy-to-use technique which might have been used.

Any teacher or curriculum development team embarking on materials development should look closely at the role of procedures such as content analysis, and readability and cognitive demand measures. Their inclusion in pre-trial evaluation at ASEP would have been a significant adjunct to the methods used.

## SCHOOL-BASED MATERIALS DEVELOPMENT

ASEP serves as a good model for school-based materials development, as well as for an activity-based programme. A large centralized national project such as ASEP usually produces material at a level of generality that cannot possibly take into account the prior experiences of particular groups of students. It is here that school-based materials have their value. In science education there is a wide range of suitable curriculum materials from which to construct the general aspects of a programme. Short, locally produced units should add vitality, interest and relevance. To reproduce materials which already exist is merely adding to curriculum pollution. So what are the lessons of ASEP?

Four of the major characteristics of ASEP are: it is child-centred, it is activity-based, it is self-paced, and it is designed to cater for a range of student ability and interest. For those interested in such materials the ASEP units and *A Guide to ASEP* are good starting points. *A Guide to ASEP* (1974) provides some excellent insights into junior secondary students. Three aspects in the ASEP unit development process ought to inform school-based developers, these are: the use of the community; the involvement of 'experts'; and the role of the laboratory.

Following Piaget and trends in science education, ASEP developers argued that the manipulation of actual objects or situations is essential in order to develop the latent thinking powers of young adolescents. Accordingly, each unit contains a large number of laboratory-type activities designed to encourage enquiry. The activities included demand that students become involved in identifying problems, observing, measuring, classifying, ordering, hypothesizing, designing experiments, interpreting and analysing data and reaching conclusions in the laboratory setting.

Coming up with new ideas for activities was a constant problem at ASEP. One of the major sources was members of the public. Motor mechanics had good ideas for the Petroleum unit, as did oil industry workers, safety officers, industrial chemists, farmers and craft people. Most of these people are not aware of the role they could play in education. The general community is obviously an untapped source of ideas for activities.

At ASEP, feedback from outside 'experts' and from fellow developers was invaluable. Experts in particular are often sensitive to issues that most people do not even notice. Environmentalists, special interest groups and subject matter specialists will pick up things that the developer will miss. To get such advice in the development process provides for more enlightened products. Similarly, using students as critics, editors, or fellow-designers and authors has obvious advantages for relevance and readiness aspects.

Development and refining of activities in the laboratory is essential. To develop hands-on activities in an armchair manner is obviously foolish. In developing activity-based units or worksheets, inexperienced writers will almost always leave out important steps or express instructions in vague terms. To follow verbatim one's own instructions, or to watch without comment while a student does, is usually a chastening experience. One-way communication through the written word is much more difficult for teachers than the two-way communication to which they are accustomed.

At ASEP it was generally accepted that it took six months to train a new developer. There are useful guides such as the *Do It Yourself Curriculum Guide* (South Australian. Education Department, 1977) and *Producing Your Own Curriculum Units* (Edwards, 1985) which can help. Students are quite sophisticated in their tastes so materials must pay close attention to the communication aspects as well as the content. Learning these skills is not difficult but it does take time and commitment.

## ASEP IN ACTION

### School level

Owen (1978) reviewed in detail the impact of ASEP on schools. Owen reported that up to February 1976 approximately 70% of Australian secondary schools had obtained some ASEP units. Heads of science in schools where ASEP had been purchased gave as the major reason its suitability for implementing student-centred learning. In schools where ASEP had not been purchased there were three main barriers: cost, current use of other texts, and inadequate knowledge of ASEP.

There were significant differences between states with respect to ASEP usage. The major factors which favoured introduction of ASEP were:

1. A science syllabus outline more or less congruent with the ASEP philosophy,
2. Usage of a wider range of texts in the state, particularly the use of class sets rather than single texts, and
3. The presence of advocates who provided support for the implementation of ASEP through in-service programmes, curriculum planning guides, and particularly through their work as Science Advisers in schools.

The Head of Science in the school was shown to have a great influence over the activities of the department. Decisions on curriculum materials to be used came mainly from the Head of Science, usually after consultation with staff. Similarly, the Head of Science usually acted as a 'translator' of the ASEP rationale in schools where it appeared to be used effectively. In general, effective use of ASEP occurred where the Head of Science showed strong leadership characteristics. The inexperience of teachers, especially in government schools, accentuated this. The Head of Science is generally the person who writes the science programme for the school, has greater access to externally based experts, and has influence with the administration in provision of facilities.

Owen (1978) found that three major elements led to effective use of ASEP materials: efficient deployment of facilities, group cohesiveness among staff, and a willingness to come to grips with the characteristics of the new materials. Discussions within the Commonwealth Education Department's

Curriculum Development Centre on the future of ASEP reinforce the conclusion that in developing courses which stress enquiry in laboratory settings, careful attention needs to be given to problems of implementation.

## Classroom level

Owen (1978) found that high users of ASEP had a rationale of science teaching more consistent with the rationale propounded by the ASEP developers. Consequently they were more likely to modify the materials for use with different groups and in different circumstances, and to be satisfied with the results.

From the fragmentary data available on what happens in laboratory classes it appears that the importance of enquiry-oriented laboratory investigations described in the professional literature is not often reflected in teaching practices. One explanation for the discrepancy may be the nature of the activities presented in the curriculum materials. It will have been noted that none of the content analyses of the ASEP materials included an analysis of the levels of enquiry or the enquiry processes demanded by ASEP units, although instruments are available which might be used for this purpose (e.g. Herron, 1971; Tamir and Lunetta, 1978). Generally, one suspects that most activities built into ASEP units would be classified as high structure, would be integrated with the text, would demand group activity, and would not often involve students in the planning and design of experiments (e.g. few demand that students formulate questions or problems to be investigated, design observation or measurement procedures, etc.). At the same time, they heavily involve students in performing simple experimental activities and in the analysis and reporting of results (but not formulation of models or raising of questions requiring further investigation, or application of results or procedures to new situations). The pertinent analysis remains to be undertaken.

While, in common with many other 'new curriculum' packages, ASEP may not live up to all our expectations in terms of its emphasis on the processes of enquiry, there is clear evidence that ASEP units do call forth quite different patterns of behaviour in junior science classrooms. Tisher and Power (1973, 1975) developed a scheme for analysing interactions in ASEP-type classrooms which codes the source and target of activities, the nature of the activity, and the intellectual, instructional, non-instructional or affective processes involved. They undertook a series of studies focusing on the ways in which two units ('Light Forms Images' and 'How Many People?') were used. In all, Power and Tisher videotaped and analysed the interaction patterns in 103 ASEP lessons and Power (1973) did the same in 20 conventional junior science lessons. They found that on average the student was the source of activity for only about one-fifth of the time in conventional science classrooms, but for as much as two-thirds of the time in ASEP classrooms. ASEP classes also involved far less fact-stating and a much smaller (one-sixth on average) proportion of time with the whole class as the target of activity. The amount of time spent on laboratory activity varied depending on the unit. In 'Light Forms Images', for example, 34% was spent conducting experiments and discussing procedures, 13% making observations, 7% reading laboratory procedures and 5% writing up results. In 'How Many People?' there was much more small group discussion (as demanded by the activity structure of the unit) and far less laboratory activity (about 7% experimenting and 9% observing).

While activity-based materials do shape the nature and extent of the laboratory activity in science classrooms to a large degree, there are variations from one classroom to the next depending on the educational values of the teacher and the ability level of the class (see Power and Tisher, 1979). Tisher and Power (1973, 1975) found that science teachers whose values were congruent with those of ASEP tended to be more active in working with groups, to spend less classroom time reading and observing, to have better organized systems for distribution of materials and monitoring of student work, and to have classes which spent more time actively engaged in laboratory work, less time waiting and more time attempting to explain results. A high level of emphasis on enquiry was found to be positively associated with student attitude to science, while students in classes characterized by low levels of activity had lower achievement and more negative attitudes.

Suan (1976) undertook a participant observation study of small groups of students using the ASEP unit 'Cells'. He found that only about 50% of each 60-minute lesson was devoted to work-oriented activities, about 30% to social activities like chatting and about 20% to preparing for work and packing up. Suan classified groups according to their orientation and focus of activity: some were Record Book-oriented (they saw the task as filling in the blanks in the

Record Book), or Reading-oriented (students who preferred, and spent more time, reading the materials), or Experiment-oriented (students who enjoyed doing the experiments but who were reluctant to read instructions or complete the task in the Record Book). As reported by Suan (1976), provided they got the right material and had adequate prior instruction on the correct use of the microscope, most groups successfully completed the experiment on onion skin cells. However, only a minority were successful in experiments on check cells and cork cells. In many cases, groups drew cells and parts of cells they had not seen (using text illustrations). In others, there were a variety of drawings from the microscope because students used the wrong material or did not know what the cell would look like and so drew a group of cells as one cell or drew air bubbles as cells.

Fraser (1976) found that, when compared to students in conventional classrooms, students in ASEP classes perceived their classes as more satisfying, more individualized, and as providing more access to resources. Tisher and Power (1975) found that students perceived their ASEP classes as more cohesive, diverse, goal-directed and satisfying, and characterized by less speed, favouritism, disorganization and apathy, and as involving more humour and discussion of interesting ideas. At the same time, the ASEP classes were perceived as more formal and more cliquish. Northfield (1976) reported that ASEP students perceived their learning environment as more goal-directed and individualized, and less difficult and competitive, but less satisfying. Classrooms with a greater amount of student choice were perceived more favourably, particularly in terms of individuality and speed.

After reviewing the research in this area up to 1978, Fraser (1978, p. 26) concluded that 'the use of ASEP materials does promote patterns of interaction and student classroom perceptions which are quite different from and generally more favourable than those in classrooms where conventional materials are used'. Edwards and Marland (1984) found that students in one ASEP classroom spent about 50% of their time involved in wider classroom activities, 27% of their time in interacting with their neighbour, and 23% in their 'private inner-world'. These science lessons, when compared with the students' maths lessons, involved twice as any interactions with the neighbour. The types of interactions with the neighbour(s) included: comparing work, challenging each other's answers, helping in experiments, dividing tasks and cooperative work, discussing results, relaxing each other, reassuring each other, explaining things to each other, and discussing answers to teacher questions. The science lessons studied were at the start of the unit and therefore involved more teacher talk and guidance than would have been the case later in the unit.

## Teaching with ASEP

Part II (pages 39-57) of *A Guide to ASEP* (ASEP, 1974) provides a broad view of teaching with ASEP. The description below presents one view of how such a classroom can operate.

Ideally before starting an activity-based programme students should have had a good introduction to the laboratory and feel at ease with common laboratory equipment and techniques as well as with safety procedures. Many of these experiences can be provided through the ASEP service books. The laboratory should be set up to provide easy student access to equipment without the need to consult the teacher. Becoming trapped in the role of equipment distributor is one of the great dangers in activity-based teaching. Equipment organized on trolleys or benches in small containers such as ice-cream containers is an effective approach to equipment distribution. Indeed, the Power and Tisher (1979) studies showed very clearly that teachers locked into this pattern were ineffective: not only was much time lost, but opportunities to foster learning were missed, resulting in significantly lower achievement and student dissatisfaction.

Students are able to work directly from their units with little teacher direction. Strategies for dealing with students with reading problems must be incorporated where necessary. Similarly students commonly need to be encouraged and/or coerced to read instructions fully before beginning activities. Setting students to work can be accomplished by a quick tour around the room only staying long enough to set each group on-task. Once this has been achieved the teacher has the time to interact with individuals or groups either as a response to expressed needs or in order to achieve certain planned pedagogical goals. This opportunity to individualize instruction through 'escaping from the front of the classroom' is one of the great advantages of activity-based teaching.

Students plan and perform their investigations within the framework provided by the text. The teacher may add enrichment or remedial material as required. Students record their observations and follow-up activities either in record books or on loose sheets which go to make up a unit report. The class is characterized by groups of students working at their own pace. Sometimes for resource reasons the teacher may limit the range of options available at any one time. This freedom of choice within a structured environment is much in keeping with a Montessori approach to education.

Often teachers feel redundant in such a process, particularly teachers who define their role narrowly as that of dispenser of information. As the work of Power and Tisher (1979) shows, these teachers need in-service support of a type which faces the incongruence and respects their professionalism. However the teacher may choose to call the class together occasionally for feedback and discussion sessions to share the generated insights and expertise, or for short periods of instruction by the teacher or fellow students. Tisher and Power (1975) suggested that in the initial stages of use of ASEP materials, teachers need to take extra time at the conclusion of a unit to consolidate what students had learnt from their experiences. Their experimental studies (Power and Tisher, 1979) confirm the need for both advance and post-organizers in laboratory-based science programmes.

## ASEP EFFECTS

The introduction of activity-based curriculum units like ASEP not only alters the nature of the processes in the classroom but also pupil perceptions of the learning environment. Fraser (1978) reviewed his own work and that of others relating to the effects of ASEP on student outcomes. In his studies, the effects of ASEP versus conventional materials in promoting seventeen long-term aims of science education were assessed for 1158 Grade 7 students in Melbourne. The outcome measures included nine scales measuring enquiry skills such as interpreting information and drawing conclusions, and three scales measuring understanding of the nature of science. Fraser found that the performance of ASEP and control students did not differ significantly on sixteen of the seventeen scales: only in the case of enjoyment of science was there a difference, and it was in favour of ASEP. In most cases reported in the literature, favourable changes in enjoyment of science associated with the uses of activity-based materials in the junior high school have been found, but little else. It should also be noted that particular ASEP units excite students and have been preferred by them to more conventional science materials; experience with ASEP suggests that laboratory experience does not guarantee the promotion of favourable attitudes to science.

Within Australia as in the rest of the western world, it is widely accepted that the laboratory is an important agent for developing an understanding of the processes of scientific enquiry. The laboratory in an ASEP class was not intended merely to be a place to illustrate, demonstrate or verify known concepts and laws. It was intended to develop skills and attitudes, and to provide experience in the processes of designing investigations, and acquiring and interpreting data. But the reported evidence suggests that ASEP has not been much more successful than the courses it replaced in the production of measurable effects on 'higher' cognitive processes, attitudes and laboratory skills. It is not sufficient to keep students busy and hope, by providing an activity-based programme, that students will acquire laboratory skills let alone develop deep insights into the ways in which scientists use the laboratory to extend, refine or refute ideas. If one is to enhance the achievement of the broader goals of laboratory work in junior secondary classes, laboratory activities must be systematically selected and integrated into a teaching program directed towards those goals. Any redevelopment of ASEP would need to examine carefully the degree to which activities do, in fact, facilitate the development of the competencies and attitudes being sought.

The appropriateness of the laboratory activities included in the ASEP materials aside, whether any curriculum package has the effects intended depends on how it is used by teachers. In part, as Tisher and Power (1973) found, this depends (a) on the values and beliefs about science and teaching held by the teacher, and (b) on the amount and type of preparation of the teacher. Generally, teachers who were convinced of the value of the ASEP approach were better prepared and better organized, wasted less time during laboratory sessions and spent less time observing pupils and more time discussing the activity with them. For teachers whose beliefs about science teaching were dissonant from those of ASEP (e.g. less laboratory- and enquiry-oriented, more teacher- and discipline-centred) the provision of an in-service course which dealt directly with the practical and ideological problems inherent in using ASEP did help. Teachers not convinced of the ASEP approach undertaking a unit and problem-focused in-service programme did play a more active and productive role in the laboratory setting than those who were simply briefed on ASEP. While qualitatively somewhat different in approach from the ASEP-congruent teachers, like them the more careful preparation and more frequent substantive exchanges with groups during practical activities paid off in improved achievement and attitudes of students. The results also suggested that student achievement of unit objectives in laboratory-based units is enhanced if the teacher plays an active role in consolidating and integrating experiences, data and ideas into a meaningful structure.

There is no such thing as a teacher-proof package. Teachers whose basic view of their role is at variance with that of an externally developed curriculum are likely to respond to the conflict in ways (e.g. inadequate preparation for laboratory activities, passive role while students are working through materials) which make it unlikely that the programme will be implemented as intended. Without an intensive and sensitive professional development programme and strong support within the school, laboratory-based junior secondary science programmes are unlikely to generate the desired student outcomes.

Most of the research on ASEP was undertaken when the materials had just become available. Initially the evidence suggested that few science teachers fully endorsed the position taken by ASEP, and that about one half held views dissonant to those of ASEP developers. Just what the longer term impact of ASEP has been on the views of the role of the laboratory and the importance of the goals towards which such programmes are directed is difficult to judge, but one suspects that activity-based, integrated science has become the norm in many schools. Schibeci (1980) reports that the Western Australian teachers in his study regarded enjoyment of science as the key objective in the junior secondary school and that this was the principal reason for their emphasis on practical work. Teachers did not usually have a clear, coherent view of the contribution laboratory work can make to other science teaching objectives and were not optimistic about the possibility of their students developing an understanding of the nature of science and its processes of enquiry. Power (1984, 1986) has outlined some of the changes needed in pre-service and in-service programmes for this gap between goals and practice to be bridged. In particular, he argued for school based, problem-oriented and action research in-service programmes which build both on contemporary science education research and teachers' professional judgements.

## CONCLUSION

In Chapters 1.1 and 1.2 the importance of frame factors in shaping both the instructional planning system and instruction itself was highlighted. In particular, it raises questions about the reasons for discrepancies between intended and actual learning activities and outcomes. In this chapter, we have examined the role of some of these factors as they relate to the implementation of one enquiry oriented curriculum.

Jackson (1968) in his delightful book *Life in Classrooms* mused that primary teachers are not primarily concerned with learning but with ensuring that students participate in activities. One wonders whether the preoccupation with the transmission of science content and recipe-oriented laboratory exercises in older junior secondary science programmes has not been replaced with a preoccupation with ensuring that students complete activities and response sheets, the meaning and purpose of which are at best only vaguely understood by pupils.

Perhaps the time has come for us to begin to reject much of the mindless ritual that passes for laboratory work in junior secondary schools, and to seek out the types of laboratory activities that do promote the competencies with which we are concerned. This is not just another call for more carefully structured laboratory workbooks in the lower secondary school, but for activities which engage students and which focus their attention on the purposes of science and the role of enquiry in laboratories in achieving these purposes.

In the current context, there is a shortage of adequately trained science teachers in secondary schools. Secondary schools are also struggling to cope with the constant demand to modify their policies and courses to suit the priorities of governments and systems. It may be that as a result science teachers rarely are given the frameworks, skills, opportunity and time to critically examine and reflect on what actually happens when students are working through laboratory activities, and what science means to the average student after three years of laboratory-based 'experience'. Certainly many of the science teachers interviewed by Schibeci (1980) had difficulty in articulating clearly the purpose of science teaching in general, and laboratory work in particular, in the junior secondary school. In a review of the literature relating to scientific attitudes, Gauld and Hukins (1980) point to the vagueness and inconsistency displayed in the literature, a problem shared by curriculum developers. Until some degree of clarity and consensus as to what we hope to accomplish through laboratory work is achieved, we can hardly expect clarity on the part of the teachers.

Elsewhere in this book an indication has been given as to the types of activities which look promising and to how these might be presented so as to more effectively realize the potential of the laboratory in achieving the distinctive goals of science education. But even well designed activities may flounder if their purposes and place in the scientific development of young adolescents is not understood and accepted by those using the materials. Without appropriate laboratory activities and an adequate supply of well trained science teachers, laboratory work in the junior school may degenerate into much ado about nothing, or to stretch a metaphor, a comedy of errors.

## REFERENCES


ASEP (Australian Science Education Project) (1974) *A guide to ASEP*. Melbourne: Victorian Government Printer.

Ausubel, D. (1968) *Educational Psychology: A Cognitive View*. New York, Holt, Rinehart and Winston.

Carss, B. and Clarke, J. (1972) A content analysis of the core of an ASEP unit: 'Pushes and Pulls' examined. *Australian Science Teachers' Journal*, 18(4), 59-61.

Dall'Alba, G. (1981) An analysis of the core of ASEP's Electric Circuits. Unpublished report. School of Education, James Cook University.

Edwards, J. (1983) Producing materials that communicate. *Tesla*, 2(2), 12-19, University of Papua New Guinea.

_____ (1985) *Producing Your Own Curriculum Units*. Unpublished paper. School of Education, James Cook University.

_____ and Dall'Alba, G. (1981) Development of a scale of cognitive demand for analysis of printed secondary science materials. *Research in Science Education*, 11, 158-70.

_____ and Marland, P. (1984) A comparison of student thinking in a mathematics and a science classroom. *Research in Science Education*, 14, 29-38.

Fraser, B. (1976) 'An evaluation of ASEP involving learning outcome, attitudinal and environmental variables.' Unpublished Ph.D. thesis,

Monash University.

_____ (1978) *Review of Research on The Australian Science Education Project*. Canberra: Curriculum Development Centre.

Gardner, P. (1972) *Words in Science*. Melbourne, ASEP.

Gauld, C. and Hukins, A. (1980) Scientific attitudes: a review. *Studies in Science Education*, 7, 129-61.

Ginsburg, H. and Opper, S. (1969) *Piaget's Theory of Intellectual Development: An Introduction*. Englewood Cliffs, New Jersey, Prentice Hall.

Herron, M. (1971) The structure of scientific inquiry. *School Review*, 79(2), 171-212.

Jackson, P. (1968) *Life in Classrooms*. New York, Holt.

Linke, R. (1972) Environmental education research project: Content analysis criteria - Report on the first evaluation trial. Unpublished document, Monash University.

Northfield, J. R. (1976) The effect of varying the mode of presentation of ASEP materials on pupils' perceptions of the classroom. *Research in Science Education*, 6, 63-71.

Owen, J. (1978) *The Impact of the Australian Science Education Project on Schools*. Canberra, Curriculum Development Centre.

Power, C. (1973) The unintentional consequences of science teaching. *Journal of Research in Science Teaching*, 10, 331-9.

_____ (1984) Tinker, tailor, soldier, spy - implications of science education research for teachers. *Science Education*, 68(2), 179-83.

_____ (1986) Providing high quality teachers and support staff for the inter-related teaching of science, technology and mathematics. *Science and Public Policy*, 13(3), 134-46.

_____ and Tisher, R. (1979) A self-paced environment. In H. Walberg (Ed.) *Educational Environments and Effects*. Berkeley, Cal., McCutchen.

Schibeci, R. (1980) Science teachers and science related attitudes. *Research in Science Education*, 10, 159-65.

Shepherd, S. R. (1972) A checklist for analysing the style of instructional materials. *Research*, 28-33. Australian Science Education Research Association.

_____ (1973) Activity Analysis Grid. ASEP (in-house).

South Australian Education Department (1977) *Do-it-yourself curriculum guide for Junior Secondary Science*. Adelaide.

Suan, M. (1976) 'Evaluation of ASEP: a case study approach'. Unpublished M.Ed. thesis, Monash University.

Tamir, P. and Lunetta, V. (1978) An analysis of laboratory inquiries in the BSCS Yellow version. *The American Biology Teacher*, 40(6), 353-7.

Thelen, J. N. (1974) Using the cloze test with science textbooks. *Science and Children*, November/December, pp. 26-7.

Tisher, R. and Power, C. (1973) The effects of teaching strategies where ASEP materials are used. University of Queensland, Report to Australian Advisory Committee on Research and Development in Education.

_____ (1975) The effects of classroom activities, pupils' perceptions and educational values in lessons where self-paced curricula are used. Monash University. Report to Australian Advisory Committee on Research and Development in Education.

# 6.2

# Life in Science Laboratory Classrooms at Secondary Level

**Yael Friedler and Pinchas Tamir**

## INTRODUCTION

On one cold winter morning after the class we were observing had started, a girl rushed in with a red nose and wet eyes saying: 'Actually, I am sick, but I have come only for this class - I would not miss the lab, I like it so much'. In a survey of peak learning experiences in 9th grade in Israeli schools about 40% of peak learning experiences chosen were in biology and of these about two-thirds mentioned experiences in the laboratory (Choppin and Frankel, 1976). When the practical performance of senior high school chemistry students who had observed filmed experiments was compared with that of similar students who had actually performed these experiments in the laboratory there were no statistically significant differences. However, when asked if they had liked their experiences both groups unanimously chose the lab experiences as their preferred mode of learning (Ben Zvi *et al.*, 1976b). The positive attitude of students to practical work is more fully discussed in Chapter 3.4.

What is it that makes the majority of students like their laboratory classes? What is so special in these laboratory experiences? What are the unique characteristics of students' life in their laboratory classes?

As may be seen in Chapter 2.1 the learning laboratory has gradually acquired during the years an increasingly prominent place in the school science curriculum, reaching its peak in the curriculum reform of the late 1950s. One of the major changes advocated by the curriculum reform movement in the USA, in Britain and elsewhere, has been a new conception of the role of the school laboratory; it was not to be a mere illustrative confirmatory adjunct to the learning of science, but, instead, was to become the centre of the instructional process. This conception yielded the Nuffield programmes in the UK, the BSCS, PSSC and CHEM study in the USA and many similar 'laboratory oriented' courses all over the world. Two slogans have become associated with the 'new' school laboratories, namely 'discovery' and 'enquiry' and the people considered as the main conceptual leaders of the curriculum reform, such as Bruner, Gagné, Schwab, Piaget and Ausubel have devoted much of their writing to issues related to learning in the laboratory.

Based on the writings of these and other authors, five major reasons may be offered as a rationale for the school science laboratory:

1. Science involves highly complex and abstract subject matter. Many students fail to comprehend such concepts without the concrete props and opportunities for manipulation afforded in the laboratory (e.g. McNally, 1973; Lawson, 1975; Lawson and Renner, 1975). See also Chapter 3.3.
2. Students' participation in actual investigations developing their enquiry and intellectual skills is an essential component of learning science as enquiry (Schwab, 1960, 1962). It gives students an opportunity to appreciate the spirit of science and it promotes problem solving, analytic and generalizing ability (Ausubel, 1968). It develops important attitudes such as honesty, readiness to admit failure and critical assessment of results and of limitations, better known as scientific attitudes (e.g. Henry, 1975). See also Chapters 3.2, 3.4.
3. Practical experiences, whether manipulative or intellectual, are qualitatively different from other experiences and are essential for the development of skills and strategies with a wide range of generalizable effects (Gagné, 1970; Tamir, 1972, 1975; Olson, 1973). See also Chapter 3.1.
4. The laboratory has been found to offer unique contexts conducive to the identification, diagnosis and remediation of students' misconceptions (e.g. Friedler, 1984; Driver and Bell, 1985;).

5. Students usually enjoy activities and practical work and when offered a chance to experience meaningful and non-trivial experiences they become motivated and interested in science (Selmes *et al.*, 1969; Henry, 1975; Ben Zvi *et al.*, 1976b). See also Chapter 3.4.

The shift to laboratory oriented courses in the 1960s was not based on 'hard' data which show the merits and superiority of such courses, but rather on the biases and convictions of the scientists and educational leaders of the curriculum reform. However, today, we have some evidence which shows that the laboratory can indeed make significant educational contributions (e.g. Pella and Sherman, 1969; Tamir, 1975; Henry, 1975; Lawson, 1975; Ben Zvi *et al.*, 1976b). Yet the controversy concerning the role and the relative emphasis of the laboratory has not been resolved. For example Woolnough and Allsop (1985) argue that 'the imposition of theory on practical work has had a detrimental effect on the development of scientific investigational skills' and advocate that we 'Stop using practical as a subservient strategy for teaching scientific concepts and knowledge' (pp. 38-9).

Another dilemma relates to the optimal proportion of time that should be allocated to learning in the laboratory. While some educators feel that the laboratory merits 80% of class time (FAST, 1976) others suggest that at least half of the class time should be spent on activities and laboratory exercises (Romey, 1968). Some would be satisfied with a third of class time (OECD, 1962), and others have a minimum of one laboratory or field trip per week (Novak, 1970).

Chapters 6.1 and 6.3 deal with the role of the laboratory in an individualized national junior science programme and in tertiary level science, respectively. This chapter touches on some of the same issues raised in these two chapters. However, the main focus of this chapter is on instruction and actual learning activities in the laboratory classrooms of secondary schools. Following Figure 1.2.1 our major concern is to what extent the intended outcomes and activities are actualized in science classrooms.

## TRANSACTIONS IN THE LABORATORY CLASSROOM

### Intended transactions

The nature of classroom transactions is strongly dependent on the curriculum materials used, since these materials, in our case the laboratory manual, to a large extent determine the opportunities to learn offered to the students. Herron (1971) analysed the laboratory manuals of two of the most distinguished enquiry oriented curricula, namely the Physical Science Study Committee (PSSC) and the Biological Science Curricula Study (BSCS). He found that about 80% of the laboratory lessons were at what he called level zero, namely, the problem, the procedure, and even the answer were all given so that there was practically nothing left for the student to find out. About 15% were at level one, namely, problem and procedure given leaving the answer to be found. Only 5% offered the opportunity to the student to design ways and means for the investigation and in no case was the student required to define the problem. Tamir and Lunetta (1978) designed the Laboratory Analysis Inventory (LAI) which allows for a detailed analysis of the opportunities offered by a particular laboratory exercise in terms of the general organization of work (i.e. 'students work on different tasks and pool the results', or 'post-lab discussion required') as well as the enquiry skills called for (i.e. formulating a question, predicting results, hypothesizing, observing, interpreting, explaining, applying, etc.). Using LAI Tamir and Lunetta (1981) found, for example, that the laboratory indeed plays a central role in the PSSC and Project Physics courses. Numerous experiences are provided in which students manipulate materials, gather data, make inferences and communicate the results in a variety of ways. At the same time, however, six important deficiencies were identified:

1. There are no opportunities to identify problems or to formulate hypotheses;
2. There are relatively few opportunities to design observations and measurement procedures;
3. There are even fewer opportunities for students to design experiments and actually follow their own designs;
4. Students are not encouraged to discuss limitations and assumptions underlying their experiments;
5. Students are not encouraged to share their efforts even in

laboratory activities where that is appropriate;
6. There are no provisions for ensuring that teachers carry out post-laboratory discussion, consolidate the findings and analyze their meaning.

The results of the kind of content analysis illustrated above help us to understand why certain lofty goals assigned to the laboratory have not, by and large, been achieved.

## Perceived transactions

### *Students' perceptions*

The most widely used instrument to study students' perceptions of their classroom environment has been the Learning Environment Inventory (LEI) designed by Anderson and Walberg (1974). Typical items are: 'students in the class find the work hard to do', or 'students compete with one another to see who can do the best work'. Fraser (1981) summarized most of the findings based on the use of LEI and similar instruments. Hofstein *et al*. (1980) found that chemistry students in vocational as compared with academic high schools perceived their laboratory classes as more goal directed, more democratic, less competitive and less organized. The students in academic schools, on the other hand, perceived their laboratory classes as more satisfying and more intimate. In general students' perceptions were that laboratory work, compared with regular lessons, required more effort, increased active learning and facilitated the development of social skills such as sensitivity to others and the ability to work in a group.

Two instruments were developed at the University of Texas to gather students' perceptions of their laboratory classes with regard to their enquiry orientation (Barnes, 1967a; Kochendorfer, 1967a). These consist of checklists of statements and the students are asked to indicate whether or not the statements describe accurately what occurs in their classroom. Some statements characterize enquiry oriented laboratories (e.g. 'We often use the laboratory to investigate a problem that comes up in class', or 'The laboratory usually comes before we talk about the specific topic in class'), while others fit conventional confirmatory laboratories (e.g. 'My teacher usually tells us step-by-step what we are to do in the laboratory', or 'Our teacher usually explains exactly what results we should expect from an investigation'). Significant differences between BSCS and non-BSCS classes were found and showed that pupils believed that BSCS laboratories indeed provided more enquiry related experiences (Barnes, 1967b; Kochendorfer, 1967b).

*Teachers' perceptions*

Tamir (1983a) developed the Structured Lesson Report Form (SLRF) which is designed to be completed by the teacher following a given lesson. In this manner, information is obtained on the teacher's perceptions of the lesson's topic, description of students, classroom setting, seating arrangement, students' teaming, use of audio-visual aids, homework, enquiry features, teacher's satisfaction, students' satisfaction and behaviour, learning resources, special features of work in the laboratory, use of teacher guide.

The heart of SLRF is an item which requires the teacher to describe what happened in the lesson.

> Describe briefly, in chronological order, what happened in this class. Include the questions asked (by teacher and students), concepts taught, method of teaching, degree and manner of students' participation, namely: answering teacher's questions, small group discussion, writing, reading, making observations, performing experiments, etc.

SLRF was found to be very useful and quite reliable, thereby contributing an inexpensive substitute for direct observations. It also provides significant feedback to teachers. Using the SLRF to compare junior and senior high school biology classes in Israel, it was found that the teaching of biology at both levels was enquiry oriented and that in their laboratory lessons students conducted investigations involving a variety of organisms. However, several significant differences were identified. For example, junior high classes had, on the average, more varied activities per lesson; while in the senior high there was a sharp division between lab and recitation lessons, in junior high many classes could be characterized as integrated lab-recitation lessons. At the junior high lab students worked in teams of three to four, while at the senior high they worked in pairs. Assistance of laboratory technicians was substantially greater in senior high schools. Senior high teachers were less restricted by the textbook and more adaptive in their use of laboratory investigations. The level of enquiry in junior high classes was not affected by students' scholastic ability while at the senior high level classes of low ability, students tended to be less enquiry-oriented (Tamir, 1983b).

Using teacher interviews and questionnaires designed on the basis of these interviews, Dreyfus, Jungwirth and Tamir (1982) found that enquiry-oriented laboratory activities had become an integral and essential part of high school biology in Israel, well accepted and highly appreciated by the teachers and by the school administrators. In the judgement of the teachers the laboratories provide ample opportunities for student investigations which involve living organisms.

## Implemented transactions

Because of the unique nature of laboratory classes, various specialized observation instruments have been developed mostly low-inference and using short time units (e.g. three seconds) in order to 'capture' the activity. Parakh (1968) found that in a typical laboratory class teacher talk occupied 50% of the time, non-verbal pedagogically relevant teacher activities 40% and pupil talk 10%. He also found that explicit statements about the nature and processes of science occurred only 0.5% of the time. Balzer (1969) found that of all teacher behaviours coded, 47% were teacher-centred content development and only 3% were student-centred content development. The other 50% of teacher behaviours did not pertain to content development or enquiry. Only 6% of the time was devoted to scientific processes most of which related to data interpretation, prediction of results, formation of conclusions and specific questions that posed problems. Behaviour requiring experimental design was rare, while problem identification and hypothesis formulation were virtually non-existent. Note how congruent are these results with what might have been predicted through content analysis of the laboratory manuals.

Smith (1971) made an important contribution by structuring his observation instrument along the dynamics of laboratory lessons in which three distinct stages may be identified, namely pre-lab discussion, actual work and post-lab discussion. This distinction provides useful information. For example, Tamir (1977) found a high positive correlation between enquiry-orientation and the occurrence of post-lab discussion. Friedler (1984) adopted Smith's three stages and added a fourth component labelled 'miscellaneous' since it is not associated necessarily with any one of the three stages. Thus, her Laboratory Observation Schedule (LOS) includes all the identified verbal and non-verbal activities of the teacher and the students; since some of the categories require high inference, a time unit of 30 seconds is used. The pre-lab discussion includes the following categories: identify and formulate problems and/or hypotheses, link to previous lessons, present theoretical background, present facts and problems related to the design of the experiment, talk about techniques, predict results, manipulate data. Performance includes: talk about techniques, present theoretical background, present facts and problems related to experiment, perform experiment, manipulate data, talk about results, interpret and draw conclusions, talk about experiment and limitations, apply and make generalizations, give instructions for further work. The post-lab discussion includes: manipulate data, talk about the results, interpret results/draw conclusions, talk about the experiment (variables, control, limitations), apply and make generalizations, teacher gives instructions for the next experiment. Miscellaneous includes: talk about discipline, talk about management, talk unrelated to lesson, write in note books, read from textbook, illustrate, teacher performs instead of student, student absent.

Using LOS it was found, for example, that as students moved from grade 9 to 12 the level of enquiry supposedly demanded rose considerably. This rise was associated with more time spent on management and performance but very little on discussion. The scarcity of discussion raises serious doubts about the meaningfulness of the observed enquiry oriented laboratories.

A different instrument is the Science Teaching Observation Schedule, STOS (Eggleston, Galton and Jones, 1976). STOS is a very high inference instrument, and therefore employs an extremely long time unit, namely three minutes. An observer using STOS has to make many high inference decisions on the spot. For example, while listening to classroom discourse, the observer has to determine the level of the teacher's questions on a seven category scale ranging from (1) requires recall of facts and principles, to (7) requires inferences from observations or data. Through observations with STOS, it was possible to identify three consistent teaching styles and to substantiate significant differences between experiences offered by physical science laboratories on the one hand and life science laboratories on the other hand. Most biology teachers were found to be 'fact acquirers'. This style is characterized by infrequent use of high level questions as well as high incidence of teachers' statements of facts and directions to sources of fact finding. This is mirrored in the pupils referring to teachers for the purpose of acquiring or confirming facts. On the other hand most physical science teachers fitted the style of 'problem solvers'. This style is characterized by a class in which the initiative is held by the teacher, who challenges his pupils with a comprehensive array of questions, observational, problem solving and speculative in both practical and theoretical contexts. Teachers' statements also reflect these teachers' orientation toward science as a problem solving activity. A third style designated as 'pupil centred enquirers' was also identified. In classes of these relatively few teachers (about 15%), there is clearly a much higher level of pupils' participation, both consultation and referral to the teacher. Pupil-initiated and maintained behaviour is directed to experimental procedure, to inferring, formulating and testing hypotheses as well as to acquiring and confirming facts. Eggleston (1983) reports that further studies in Canada and Australia confirmed the findings described above for England. Moreover, clear associations were found between teaching styles and cognitive and affective outcomes. In physics, pupils of teachers categorized as problem solvers did best, whilst in biology it was pupils of pupil-centred enquirers who did best.

The overall picture obtained from the research cited above using structured observations is that while the teachers are still the dominating figures who initiate and manage classroom transactions, they talk in the laboratory much less than in recitation lessons (40% as compared with 75% of the time), they frequently employ non-verbal instructional behaviours, such as demonstrating, assisting, checking student work, etc. On the average, the students talk about 13% of class time and only for about 6% of the time are they off task. About half of the teacher talk is devoted to management and procedures while the other half is devoted to content associated with the students' work. Relatively little time is devoted to the nature of scientific enquiry, to the processes of science such as problem identification, hypothesis formulation, experimental design and to the assessment of limitations of investigations performed. Enquiry-oriented laboratories are significantly different from conventional ones and may be characterized as follows: the teachers are less direct, more planning takes place, processes of science receive more emphasis, there is more post-lab discussion and teachers give fewer instructions in front of the whole class, but move around more, checking, probing and supporting. Students are usually more active and they initiate ideas more frequently.

## Unstructured observations

The advantage of unstructured observations is that they are open to unpredictable events and they are not constrained by predetermined objectives, categories or instruments. Very few studies report on unstructured observations in laboratory classes. Rubin (1985) found significant behaviour differences between enquiry and confirmatory laboratories. Friedler (1984) supplemented her structured observations with LOS (see above) by unstructured observations which were carried out by two observers who observed simultaneously the pre- and post-lab discussions. During the performance stage one observer concentrated on the teacher and the other observer moved freely from one working group to another, staying with each group for 15 minutes. During the first 10 minutes the observer recorded observations, shifting systematically each minute from one student to the next in the same working group. When all students in the group had been observed, the first student was observed again. After 10 minutes the observer talked with the students informally in order to find out the students' understanding and difficulties in the specific lab.

Using this combined strategy it was found, for example, that the verbal interaction among students in the groups was mainly on a low cognitive level featuring techniques and procedures such as: 'Put a drop of water', 'Bring the test tube', 'What do we have to do now?' Teachers did not interfere with setting up work groups and did not seem to be aware of the importance of the groups' composition. The following difficulties were discovered through the informal conversations and appeared quite frequently: a distorted understanding of fundamental concepts (such as: cell, enzyme); inability to link theoretical knowledge to observed phenomena; inability to distinguish the relevant from the irrelevant in the experiment; misleading associations and deficiencies in knowledge, especially of chemistry related to biological processes.

## FACTORS ASSOCIATED WITH LEARNING IN THE LABORATORY

The four major factors which bear directly on learning in the school laboratory are the subject matter, the student, the teacher and the curriculum (shown in several crucial and different positions in the model, Figure 1.2.1, as commonplaces).

### Subject matter

While research on the differences among laboratories in different disciplines is scarce, such differences obviously exist. In mathematics, for instance, the major purpose of concrete experiences appears to be demonstration of relationships as well as assistance in problem solving and intuitive rule-concept learning. In the physical sciences students make observations, measure and perform experiments. Yet, often, they use instruments which translate phenomena into data without being able to observe the actual phenomena directly. In a chemistry laboratory the students observe changes in colour or in appearance, hear explosions, notice different smells, and feel changes in temperature. Based on their perceptions they have to infer what is happening. Atoms and molecules, electrovalent and covalent bonds can neither be seen nor touched; but nevertheless they constitute the conceptual basis for understanding what is happening.

In the physics laboratory students who work with electrical circuits are expected to explain their observations in terms of the behaviour of electrons which they are not able to see. This lack of direct perception is characteristic of much laboratory work in the physical sciences. While some investigations in biology and geology are similar to those in the physical sciences, biology and geology present more opportunities for direct observations of natural phenomena. Biology has the unique attribute of dealing with life, which implies care for plants and animals as well as employing special precautions and techniques in dealing with living materials. Emotions, such as compassion for animals or reluctance to perform dissections, are unique to laboratory work in biology. As a result of differences such as those described above, different guiding principles of enquiry have evolved for each discipline. Even within a discipline, such as biology, investigations in different subdisciplines, such as anatomy, genetics and physiology, may be guided by different principles, adopt different research strategies and employ different techniques.

## Student

In spite of the increased emphasis on enquiry many students still perceive the laboratory as a place where they do things, but fail to see the meaning of what they do, the relation of what they do to theory, and the place of the laboratory in the larger context of the scientific enterprise (Friedler and Tamir, 1984a; Novak and Gowin, 1984). Woolnough and Allsop (1985) describe the situation as follows: Practical work is designed to enable the student to 'discover and, subsequently, to understand more fully the theory being considered. Yet, somehow, students seem to "discover" the wrong thing, that which seems so obvious to the teacher somehow eludes the student' (p. 10). Driver (1983) has written somewhat sarcastically that 'I do and I understand' should perhaps be replaced by 'I do and I am ever more confused' (p. 9). The Assessment of Performance Unit, having examined hundreds of students, concludes: 'Despite the orientation of science courses to the teaching of content, the results from tests of application of science concepts indicate that it is the minority of 15 year olds who are able to draw on and use some of the most basic scientific concepts' (Assessment of Performance Unit, 1984, p. 190).

Why is it that so often students fail to learn what teachers expect of them from their practical experiences?

One major reason is that children have firmly held views about many science topics and science phenomena prior to being taught science. These ideas can be amazingly tenacious and resistant to change (Driver, 1983; Osborne and Wittrock, 1983). Another important reason is that for many students, enquiry oriented laboratories appear to be too difficult due to their demand for formal reasoning and to the cognitive overload they often impose as a result of the need to apply simultaneously intellectual skills, practical skills and prior subject matter knowledge.

Tasker (1985) carried out extensive observations of practical work of children 11 to 14 years old. Based on his findings he identifies the following as reasons why children retain intuitive views in spite of experiences designed to teach the 'science viewpoint':

1. Lessons are perceived by pupils as isolated events, not as parts of a related series of experiences;
2. The pupil's purpose is different from that of the teacher. Often teachers do not state the purpose. Even when they do they do not make sure that the pupils understand it. The tendency for pupils to construct as a purpose for a scientific classroom task either 'following the set instructions' or 'get the right answer' was found in many classrooms which followed individualized programmes;
3. Pupils fail to understand the relationship between the purpose of the investigation and the design of the experiment which they carry out;
4. Pupils lack prerequisite knowledge;
5. Pupils are unable to grasp the 'mental set' required;
6. The pupils' perceptions relating to the significance of task outcomes are not those assumed by the teacher.

Zilbersztain and Gilbert (1981) who observed 8th grade students (age 13-14) add the following:

7. Even at the basic level of following instructions some are not considered and others are misinterpreted;
8. Interaction with the teacher is an occasion for the presentation of the teacher's knowledge rather than for group discussion of work the class has been doing.

## Teachers

The teacher is undoubtedly the key factor in realizing the potential of the laboratory. In order to be able to accomplish this mission, teachers need to be aware of the goals, potential, merits and difficulties of the school laboratory. Teaching in the laboratory requires a special approach to science (e.g. 'science as enquiry'), special instructional skills (e.g. running pre-lab and post-lab discussion), special management skills (e.g. budgeting time, managing small groups, guarding safety), and special attitudes (e.g. patience, tolerance of uncertainty, readiness to encounter failure, open-mindedness). Careful preparation and planning on the part of the teacher, as well as assessment of performance and understanding of students through observations, diagnostic questioning, adequate homework and practical examinations are all essential. Lastly, the ability to integrate the laboratory with other instructional strategies and to motivate students will both make significant contributions to upgrading learning in the laboratory.

Unfortunately most teachers are ill-prepared to teach effectively in the laboratory. A major reason for this is that 'most science teachers have themselves been brought up on a diet of content dominated cookery book-type practical work and many got in the habit of propagating it themselves' (Woolnough and Allsop, 1985, p. 80). The Learning in Science Project in New Zealand identified two key problems facing science teachers:

1. It is very difficult for busy teachers faced with large classes to explore and establish the meanings of words and views of the world brought to lessons by pupils.
2. Pupils' views of a lesson can be significantly different from the views held by the teacher of that lesson. Pupils on the one hand, and the teacher on the other, often see the purpose of a lesson, the reasons why particular classroom activities are being undertaken and the conclusions that can be drawn from these activities rather differently. While these differences can appear subtle, it was found that if they are not identified and discussed, much of the value of the lesson can be lost. (Osborne, 1985, p. 17)

The issue of teachers' knowledge is explored in detail in Chapter 4.3.

## Curriculum

As much as the teacher is a key figure, curriculum often exerts a decisive effect, since most teachers use commercial textbooks and laboratory manuals.

Content analyses of laboratory manuals reveal that even curricula claiming to be enquiry-oriented such as the BSCS, PSSC and CHEM Study offer, by and large, laboratory exercises which reflect a very low level of enquiry (e.g. Herron, 1971; Tamir and Lunetta, 1981). While some confirmatory laboratory exercises which aim at developing self-confidence as well as basic processes and techniques may be necessary, in our opinion the majority should require students to engage in real problem-solving investigations, reflecting different levels of scientific enquiry according to particular goals and local conditions. Even if high level enquiry investigations cannot realistically be conducted very often, laboratory experiments which require a medium level of enquiry (e.g. Hegarty, 1979) may be most profitable. Another desirable way of facilitating enquiry is by assigning individual research projects which students do on their own under the guidance of the teacher. Examples of excellent high level enquiry investigations may be found in the BSCS Interactions of Ideas and Experiments and in the BSCS Laboratory Blocks (see description by Klinckman, 1970, pp. 83-97, 104-8).

Learning in the laboratory may be supplemented and supported by vicarious experiences such as television, slides, films and computers. Schwab (1960) has rightly argued that many aspects of enquiry may be learned outside the laboratory. Enquiry-oriented materials such as invitations to enquiry (Schwab, 1963), enquiry into enquiry (Connelly *et al.*, 1974) and simulated experiments (Lunetta, 1974; Ben Zvi *et al.*, 1976a) and others have been successfully used. Nevertheless, the bulk of available evidence indicates that while such vicarious experiences may be useful, they *cannot* substitute for the real hands-on experience.

## IMPLICATIONS

How can we improve the life of students and their learning in the laboratory? We can do this mainly through improved instructional materials and through improved instruction.

We certainly improve on the present state-of-the-art by designing a variety of laboratory investigations adequately supplemented by supportive and complementary experiences such as films (Ben Zvi *et al.*, 1976a), computer simulations (Kinnear, 1982), concept mapping and Vee diagrams (Novak and Gowin, 1984). However, our major thesis based on recent research is that special materials are needed to teach enquiry skills. Matthews (1984) developed a guide to teach enquiry skills through an analysis of the story *Pooh Bear*. Rubin (1985) developed an instructional sequence using an advance organizer based on daily experiences, followed by a series of simple enquiry-oriented laboratories, and was able to improve significantly the performance of 9th grade students on all enquiry skills tested. A special module entitled *Basic Concepts of Research in Science* (Friedler and Tamir, 1984b) was developed and used very successfully in Grades 11 and 12. The module opens with an historical-psychological introduction related to the origin and nature of basic concepts such as problem, assumption, hypothesis, control, experiment, etc. This introduction is followed by four chapters, each devoted to one major concept (problem, hypothesis, deduction, experiment). The last chapter provides a synthesis and requires application of the basic concepts by employing them in analysis and design of investigations. The structure of the module, the kinds of activities included and their sequence take into consideration previous research findings in this area. For example, overload is avoided by leading the student step by step from familiar and simple tasks to less familiar and more complex ones, integrating 'dry' thinking exercises, invitations to enquiry and practical investigations, and requiring frequent discussion and reflection on the relationship between theory and the actual research activities. For further details, see Friedler and Tamir (1987). Similarly, Osborne (1985) offered a collection of examples of discrepancies between teachers' and students' viewpoints accompanied by concrete suggestions on how these problems could be overcome.

All the ideas and suggestions described in the previous section will be of no value unless teachers are aware of their existence and know how to use them adequately. This implies special emphasis in pre-service and in-service teacher education with regard to the various issues discussed in this chapter, namely, the nature of enquiry, basic concepts of scientific research, a variety of instructional strategies, i.e. invitations to enquiry, discussion, concept mapping, practical laboratory tests, cooperative learning, and the development of adequate attitudes. (For practical suggestions see Dreyfus, 1983; Novak and Gowin, 1984; Tamir 1983c; Tobin *et al.*, 1984; Woolnough and Allsop, 1985.) See also Chapter 4.3. Special curriculum materials such as the ones described in the previous section should be used extensively in teacher education courses. When these recommended measures are taken seriously and if appropriate physical conditions are ascertained (e.g. well equipped laboratory, easy supply of materials, help of laboratory technicians), the potential of the laboratory for learning and for providing motivating, challenging and satisfying experiences will be better realized and life in laboratory classrooms will be substantially improved.

## REFERENCES

Anderson, G. J. and Walberg, H. J. (1974) Learning environments, in H. J. Walberg (Ed.) *Evaluating Educational Performance*, Berkeley, Calif., McCutchan.

Assessment of Performance Unit (1984) *Science in schools age 15:* Report No. 2, London, Department of Education and Science.

Ausubel, D. P. (1968) *Educational Psychology*, New York, Holt, Rinehart and Winston.

Balzer, L. (1969) Non-verbal and verbal behaviors of biology teachers, *American Biology Teacher, 31*, 226-9.

Barnes, L. W. (1967a) The development of student checklist to determine laboratory practices in high school biology, in A. E. Lee (Ed.) *Research and Curriculum Development in Science Education*, Austin, University of Texas Publications, pp. 90-6.

_____ (1967b) Laboratory instruction in high school biology classes using different curriculum materials, in A. E. Lee (Ed.) *Research and Curriculum Development in Science Education*, Austin, University of Texas Publications, pp. 97-103.

Ben-Zvi, R., Hofstein, A., Samuel, D. and Kempa, R. F. (1976a) The effectiveness of filmed experiments in high school chemical education. *Journal of Chemical Education, 53*, 518-20.

_____ (1976b) The attitudes of high schoold students towards the use of filmed experiments. *Journal of Chemical Education, 53*, 575-77

Bruner, J. S. (1966) Some elements of discovery, Ben Zvi in L. S. Shulman and E. R. Keisler (Eds.) *Learning by Discovery: A Critical Appraisal,* Chicago, Rand McNally.

Choppin, B. and Frankel, R. (1976) The three most interesting things. *Studies in Educational Evaluation, 2*, 56-61.

Connelly, F. M., Feingold, M. and Whalstrom, M. W. (1974) *Science Enquiry and Science Instruction* (Report to the Canada Council).

Dreyfus, A. (1983) Teaching prospective biology teachers to function rationally in the laboratory. *European Journal of Science Education, 5*, 289-98.

Dreyfus, A., Jungwirth, E. and Tamir, P. (1982) An approach to the assessment of teachers' concerns in the context of curriculum evaluation, *Studies in Educational Evaluation, 8,* 87-100.

Driver, R. (1983) *The pupil as a scientist*? Milton Keynes, Open University Press.

_____ and Bell, B. (1985) Students' thinking and the learning of science: a constructivist view, *School Science Review, 67*, 443-56.

Egelston, J. (1973) Inductive vs. traditional methods of teaching high school biology laboratory experiments, *Science Education, 57*, 467-477.

Eggleston, J. (1983) Teaching pupil interactions in science lessons: explorations and theory, in P. Tamir, A. Hofstein and Ben Peretz, M. (Eds.) *Preservice and inservice education of science teachers*, Rehovot, Balaban International Science Services, pp. 519-36.

Eggleston, J.F., Galton, M. J. and Jones, M. (1975) *A Science Teaching Observation Schedule*, Schools Council Research Series, London, Macmillan.

FAST (1976) Description and materials, Curriculum Research and Development Center, University of Hawaii.

Fraser, B. (1981) *Learning environment in curriculum evaluation: a review*, Oxford, Pergamon.

Friedler, Y. (1984) 'Problems and Processes in Learning and Teaching in the Biology Laboratory in Israeli High Schools', unpublished Ph.D. dissertation, the Hebrew University of Jerusalem, Israel (in Hebrew).

_____ and Tamir, P. (1984a) *Basic Concepts in Scientific Research*, Israel Science Teaching Center, The Hebrew University of Jerusalem, Israel (in Hebrew).

_____ and Tamir, P. (1984b) Teaching and learning in high school biology laboratory classes in Israel *Research in Science Education*, *14*, 89-96.

_____ and Tamir, P. (1987) Teaching basic concepts of science to high school students, *Journal of Biological Education*, *20*, 263-70.

Gagné, R.M. (1970) *The Conditions of Learning*, Second Edition, New York, Holt, Rinehart and Winston.

Hegarty, E.H. (1979) 'The Role of Laboratory Work in Teaching Microbiology at University Level', unpublished Ph.D. dissertation, University of New South Wales, Australia.

Henry, N.W. (1975) Objectives for laboratory work, in P.L. Gardner, (Ed.) *The Structure of Science Education*, Hawthorn, Victoria, Longman, pp. 61-75.

Herron, M.D. (1971) The nature of scientific enquiry, *School Review*, *79*, 171-212.

Hofstein, A., Ben-Zvi, R., Gluzman, R. and Samuel, D. (1980) A comparative study of chemistry students' perception of the learning environment in high schools and vocational schools, *Journal of Research in Science Teaching*, *17*, 547-52.

Kinnear, J. (1982) Computer simulation and concept development in students of genetics, *Research in Science Education*, *12*, 89-96.

Klinckman, E. (1970) *Biology Teacher Handbook* (Second Edition), Wiley, New York.

Kochendorfer, L.H. (1967a) The development of a student checklist to determine classroom teaching practices in high school biology, in A.E. Lee (Ed.) *Research and Curriculum Development in Science Education*, Austin, University of Texas, pp. 71-8.

_____ (1976b) Classroom practices of high school biology teachers using different curriculum materials, in A.E. Lee (Ed.) *Research and Curriculum Development in Science Education,* Austin, University of Texas, pp. 79-84.

Lawson, A.E. (1975) Developing formal thought through biology teaching, *American Biology Teacher, 37,* 411-19, 429.

_____ and Renner, J.W. (1975) Piagetian theory and biology teaching, *American Biology Teacher, 37,* 336-43.

Lunetta, V.N. (1974) Computer-based dialogs: A supplement to the physics curriculum, *The Physics Teacher, 12,* 355-56.

Matthews, P. (1984) Processed Pooh, *School Science Review, 66,* 371.

McNally, D.W. (1973) *Piaget, Education and Teaching,* Sydney, Angus and Robertson.

Novak, J.D. (1970) *The Improvement of Biology Teaching,* Ithaca, New York, Cornell University.

_____ and Gowin, D.B. (1984) *Learning How to Learn,* Cambridge University Press, New York.

OECD (1962) *New Thinking in School Biology,* OECD Publication.

Olson, D.R. (1973) What is worth knowing and what can be taught, *School Review, 82,* 27-43.

Osborne, R. (1985) Teachers of science as educational researchers: the Learning in Science Projects, *Australian Science Teachers' Journal, 31,* (2), 14-21.

_____ and Wittrock, M.C. (1983) Learning science: a generative process, *Science Education,* 489-500.

Parakh, J.S. (1968) A study of teacher-pupil interaction in BSCS yellow version biology classes, *American Biology Teacher, 30,* 846-8.

Pella, M.O. and Sherman, J.A. (1969) Comparison of two methods of utilizing laboratory activities in teaching the course of IPS, *School Science and Mathematics, 69,* 303-14.

Romey, W.D. (1968) *Inquiry Techniques for Teaching Science,* Englewood Cliffs, New Jersey: Prentice-Hall.

Rubin, A. (1985) 'Improving Meaningful Learning in the High School Biology Laboratory', unpublished Ph.D. dissertation, The Hebrew University of Jerusalem, Israel (in Hebrew).

Schwab, J.J. (1960) Enquiry - the science teacher and the educator, *The Science Teacher, 27,* 6-11.

_____ (1962) The teaching of science as enquiry, in J.J. Schwab and P.F. Brandwein (eds.) *The Teaching of Science,* Cambridge, Massachusetts, Harvard University Press.

_____ (1963) *Biology Teacher Handbook,* New York, Wiley.

Selmes, C., Ashton, B.G., Meredith, H.M. and Newal, A. (1969) Attitudes to science and scientists, *School Science Review, 51,* 7-22.

Smith, J.P. (1971) The development of a classroom observation instruction relevant to the earth science curriculum project, *Journal of Research in Science Teaching*, *8*, 231-5.

Stake, R.E. (1967) The countenance of educational evaluation, *Teachers' College Record*, *68*, 523-40.

Tamir, P. (1972) The practical mode - a distinct mode of performance in biology, *Journal of Biological Education*, *6*, 175-82.

_____ (1975) Nurturing the practical mode in schools, *The School Review*, *83*, 499-506.

_____ (1977) How are the laboratories used? *Journal of Research in Science Teaching*, *14*, 311-16.

_____ (1983a) Teachers' self reports: An alternative approach to the study of classroom transactions, *Journal of Research in Science Teaching*, *20*, 815-23.

_____ (1983b) A comparsion of biology teaching in junior and senior high schools in Israel, *Journal of Biological Education*,*17*, 65-71.

_____ (1983c) Inquiry and the science teacher, *Science Education*, *7*, 657-72.

_____ and Lunetta, V.N. (1978) An analysis of laboratory activities in the BSCS Yellow Version, *The American Biology Teacher*, *40*, 353-57.

_____ and Lunetta, V.N. (1981) Inquiry related tasks in high school science laboratory handbooks, *Science Education*, *65*, 477-84.

______ Amir, R. and Nussinovitz, R. (1980) High school preparation for college biology in Israel, *Higher Education*, *9*, 399-408.

Tasker, R. (1985) Children's views and classroom experiences, *Australian Science Teachers' Journal*, *27*, 33-7.

Tobin, K., Pike, G. and Lacy, T. (1984) Strategy analysis procedures for improving the quality of activity-oriented science teaching, *European Journal of Science Education*, *6*, 79-89.

Woolnough, B. and Allsop, T. (1985) *Practical Work in Science*, Cambridge: Cambridge, University Press.

Zilbersztain, A. and Gilbert, J. (1981) Does practice in the laboratory fit the spirit? *Australian Science Teachers' Journal*, *27*(3), 39-44.

# 6.3

# Life in Science Laboratory Classrooms at Tertiary Level

**Elizabeth Hegarty-Hazel**

## INTRODUCTION

There's an attractive anecdote at the beginning of the chapter on life in secondary science classrooms (Chapter 6.2) - a 9th grade girl states that she is actually ill but wouldn't miss the lab because she likes it so much. This typified the authors' views that school science labs have novelty value, at the very least are a welcome change from regular classes, and, at the best, can be interesting, challenging, filled with the spirit of enquiry and perhaps be students' preferred learning mode.

That anecdote can be matched with one from a tertiary level science student:

> I found it tremendously upheaving - it made me feel really happy - my spirit felt lifted. You do the project yourself ... and then to find that it all fitted together just as you were shown, was pure happiness really.
>
> (Bliss and Ogborn, 1977, p. 89)

However, in general, university and college science laboratory classes seem to be of a different order from secondary classes. Any novelty value has certainly gone. Students spend long hours in the laboratory (perhaps 20 hours of scheduled classes per week, not counting night or weekend work on projects or write-ups). Preferred learning mode is rarely an issue, at least not for science majors. Classes certainly can be interesting, challenging and filled with the spirit of enquiry. However, terms also likely to be used are necessary, busy, business-like, hard work, tedious, problem prone. One of my own microbiology students, reviewing a videotape of his laboratory activities, stopped the tape to say:

> See the organism I'm working with - it's *Neisseria gonorrhoeae*. Gonorrhoea and sex - you'd expect I'd be thinking about sex! I do all the time away from here! But not in these labs - no time. I notice it each week. I have to spend all my time thinking, planning, organizing - thinking.

In the literature overall, interaction analyses predominate in the descriptions of teacher and student behaviours in science laboratories at tertiary level. These are studies where behaviours are recorded in a systematic way using a structured observation schedule and specific unit analysis. The observer may make the records directly or after reviewing a videotape. Also involving highly structured record keeping, but not using direct observation, are the methods which employ inventories of the psycho-social learning environment. On the other hand, more naturalistic or anthropological methods do involve direct observation but employ a variety of record keeping techniques which attempt to preserve more directly the character of the events observed.

Each of these types of methodology will be briefly described in the chapter. Then, using results obtained in the different traditions, it becomes possible to start piecing together some pictures of life in tertiary science laboratory classrooms. In keeping with the model (Chapter 1.2) the focus will be on the *actual* learning activities - although it is often irresistible to examine the extent to which these mirror the *intended* learning activities.

According to our model (Figure 1.2.1), technical skills, scientific enquiry, scientific knowledge and scientific attitudes (Chapters 3.1 to 3.4) are all intended learning activities in a science curriculum and one way of determining their realization is to examine the actual learning activities which ensue - hence the focus on the behaviour of students and teachers in this chapter.

Research reported in this chapter reflects the dominant methodology (interaction analysis). It is well complemented by

Bliss's report (Chapter 6.4) of one of the few major naturalistic studies of life in laboratory classrooms at tertiary level.

## OVERVIEW OF DIFFERENT RESEARCH TRADITIONS AND METHODOLOGIES

### Interaction analysis in laboratories

Several different observation schemes have been developed independently in the absence of a well-known satisfactory scheme and in the wake of severe criticism (Dunkin and Biddle, 1974) of FIAC - Flanders Interaction Analyses Categories and related schemes. Criticism was focused on ideological bias and the fact that categories demonstrably lacked mutual exclusiveness. In retrospect, this made the results of an earlier interaction study conducted in college chemistry laboratories (Uricheck, 1972) very difficult or impossible to interpret.

One scheme (Tamir, 1977) was arranged on the basis of three common and often-assumed desirable phases of laboratory work : pre-lab discussion (focusing on problem identification and any special instructions), lab-work (carrying out the investigation and evaluation) and post-lab discussion (data analysis and interpretation). In practice, it was found that post-lab discussions, although a feature of secondary science laboratories, were seldom encountered at the tertiary level.

Other schemes focused more on the nature of students' or teachers' behaviour rather than on phases of classwork. Hegarty (1978) used SOLT and SOLS (System for Observation of Laboratory Behaviour: Tutors/Students) and Shymansky and Penick (1979a) used SLIC (Science Laboratory Interaction Categories). SOLT/SOLS and SLIC each contained mutually exclusive categories and shared many common dimensions. Indeed they shared dimensions with schemes which were developed independently at much the same time for observation of school science laboratory classes (Eggleston, Galton and Jones, 1975; Power and Tisher, 1976; Guy, 1982). Major dimensions of SOLT and SLIC were: Cognitive (with separation of talking and listening, of content or procedural orientation and of lower or higher cognitive demands); Laboratory activities (demonstrations, assisting with equipment and resources, watching students working); Laboratory organization (talking, listening and action concerning locating materials, supervision, laboratory management);

Socio-emotional (giving positive and negative reinforcement to students); Unrelated.

In the later sections on behaviour of teachers and students, research results will be reported in terms of these five dimensions in the teacher version shown above or the corresponding student versions.

## Learning environment inventories and laboratories

Learning environment inventories of the psycho-social environment in classrooms use inventories of questions rather than videotapes or any form of description by observers. One of the widely used LEIs has a variety of scales of likely relevance to laboratory classes - friction, competitiveness, cohesiveness, satisfaction, difficulty, material environment, disorganization (Anderson and Walberg, 1976). Other inventories include scales covering involvement, teacher support, and innovation (Moos and Trickett, 1974) and participation, independence and investigation (Rentoul and Fraser, 1979).

Apart from techniques of administration, one of the main distinctions between the use of interaction analysis and LEIs is that the former tend to concentrate on low inference variables (e.g. number of student questions, amount of time practising techniques) whereas the latter tend to concentrate on higher inference variables (e.g. student competitiveness, teacher disorganization) (Rosenshine, 1970; Fraser, 1985, 1986). The relative importance of interactions between low and high inference variables in tertiary science laboratories has been unexplored to date. However, perceptual measures of classroom environment have been found to account for more variance in student learning outcomes than directly observed variables.

In three major reviews (Fraser and Walberg, 1981; Fraser, 1985, 1986) it was concluded that there was considerable support for the predictive validity of students' perception of their learning environments and that the dimensions of the different inventories employed accounted for very significant proportions of achievement variance (beyond that attributable to general ability) and of attitude scores.

However, searches of the literature show that study of science laboratory classes has been almost entirely unaffected by research on LEIs. This is particularly so in tertiary education although one instrument recently developed for use in universities and colleges (Fraser, Treagust and Dennis, 1986) could perhaps be modified for use in the science laboratory setting. An exception is the work of Lawrenz and Munch (1984) who, for one semester, studied the activities in physical science laboratory classes for pre-service elementary teachers at Arizona State University. A shortened form of an LEI was used as one of their measures - including the scales for satisfaction, difficulty, competitiveness, cohesiveness and friction. Choice of scales was predicated on the fact that the major purpose of the study was to investigate student grouping effects. The students questioned perceived their classes as difficult but also as reasonably satisfying and cohesive, approximately neutral in competitiveness and below neutral in friction. No significant differences related to three different methods of grouping students on ability were found.

## Naturalistic descriptions of laboratories

There was a mini methodological boom in the 1960s and 1970s when case studies, anthropological, ethnographic and other naturalistic methods of description produced landmark studies such as *Life in Classrooms* (Jackson, 1968) and *The Complexities of an Urban Classroom* (Smith and Geoffrey, 1968). The harvest for science education was rather meagre though it included one major study, the *Case Studies in Science Education* (Stake and Easley, 1978) with descriptions of the problems encountered in secondary school science laboratories (specifically the failure of the spirit of scientific enquiry to materialize).

At the tertiary level, Parlett and Hamilton (1972) used the term 'illumination' for naturalistic enquiry. In 1969 Parlett described the laboratory classes (digital systems project) for third year students at Massachusetts Institute of Technology. Amongst his results obtained as a participant observer was the description of teaching assistants as 'authoritarian' or 'permissive', but despite the presence of either type of characteristic, the atmosphere of the laboratories featured minimal student discussion and that only about use of equipment. Students acted like total strangers afraid of exposing ignorance in front of others in an extremely competitive environment. The author suggested that 'science and engineering students tend to define their personal worth in terms of technical competence. Admitting confusion in a technical area thus often involves an erosion of self esteem' (Parlett, 1969, p. 1104).

Harris (1977) also described as 'illuminative' his approach to the evaluation of laboratory classes in a first year Human Biology course. His methods included direct observation of the actual functioning of the laboratories as well as formal and informal interviews with students, demonstrators and staff. Results were limited and sometimes contrasting or contradictory. Thus an observer reporting on an exercise on frog cardiac muscle noted (p. 44) that the experiment generated 'a lot of excitement and enthusiasm amongst students'. However, apparently not all students were so enthusiastic and reports of informal interviews included the comment 'not too keen on this experiment - the idea of chopping up a frog does not really appeal to me!' (p. 40). In addition many students apparently did not have a clear idea of what they were doing - as indicated by the following responses to the question 'why do you think you are doing this practical?'

> Someone thinks it's a good idea ... Erm ... let me look at the schedule ... (Harris, 1977, p. 40).

This last finding of student lack of comprehension or concentration was similar to the situation at the University of California, Berkeley, where Reif and St. John (1979) found that only 25% of students leaving 'traditional' physics laboratories were able to account at all coherently for what they had been doing. On the other hand, 80% of students leaving classes composed partly of mini-labs especially designed to enhance 'thinking like a physicist in the laboratory' (St. John, 1980) were able to give a good account of the purpose of their work. Reif and St. John (1979) reported that the latter students performed better than the former in providing verbal descriptions of the essential points of experiments, and on test questions covering both basic skills (estimating statistics) and higher level skills requiring flexible understanding of knowledge gained from experiments (predicting the effect of an error, solving a similar problem).

## BEHAVIOUR OF TEACHERS

In this section, behaviour of teachers is described separately from the behaviour of students (which follows) although of course the two are closely interlinked. Here, and also in the section on students, the headings described above will be used: examples of teachers and students talking and listening (cognitive) will be separated from examples of teachers supervising or students doing laboratory activities. This in turn will be distinguished from organization and management by teachers and students and from reinforcement activities (socio-emotional). It is noted that there were situations where teachers and students were absent altogether from the laboratory or engaged in talk or activities not apparently related to the curriculum.

## Cognitive categories of teachers' behaviour

Laboratory classes have been traditionally thought of as an activities-oriented learning mode. Thus, perhaps the most surprising of all the findings was the sheer amount of time spent by teachers lecturing (and to a lesser extent engaged in discussions). In four studies (Tamir, 1977; Hegarty, 1978; Shymansky and Penick, 1979a; Guy, 1982) there was variation across levels, disciplines, courses or institutions but the range of teachers' classroom time occupied in this manner was about 10-70% and the mean about 50%.

When secondary and tertiary teaching practices were compared, the proportions of laboratory class time occupied by 'teacher talk' were high and very similar (51-54%) for chemistry laboratories at both levels (Guy, 1982). This is at the high end of the range (30-50%) quoted in a major review of school level laboratories (Power, 1977). In another such comparison, university teachers entirely dominated the air waves in both the pre-lab discussion period and during lab work (Tamir, 1977). An index calculated by dividing the sum of enquiry items by verification items revealed low scores (0.3-0.7) in chemistry, biology, histology and physiology at the Hebrew University, Israel, whereas the corresponding index for BSCS biology at high school level was 1.2. This suggests that in the university laboratories there was a large amount of teacher lecturing at low cognitive levels.

In two studies, the distinction was made between lower and higher level talk or questions (the latter labelled 'scientific processes' or 'extended thought'): Hegarty (1978, 1979) and Shymansky and Penick (1979a). Attempts were made to investigate ways in which these processes varied in different contexts, either across institutions or disciplines. Whilst in the Australian study (Hegarty, 1978) proportions of teacher time spent in this presumably desirable manner ranged from 1% to 5%, in the USA study scores were low to non-existent (Shymansky and Penick, 1979a). Across most introductory and advanced laboratory classes at the University of Iowa, results showed 0-0.5% with one only course (introductory botany) with a higher level (1.8%). Comparing subcategories, Hegarty (1979) found other types of differences between microbiology classes at two different universities. At one, the proportion of teacher talk to student talk was 89%, at the other 67%, for the lower cognitive level, and for the higher cognitive level, the corresponding proportions were 12% and 33%. At the first university teachers talked for 56.4% of the time, with 29.6% on procedural knowledge and 26.8% on substantive knowledge, whilst at the other university, the total was lower (29.3%) and the division was 19.5% of this on procedural and 9.8% on substantive knowledge (Hegarty, 1979). Overall, these differences seemed consistent with the aims of the two courses (different emphasis on scientific enquiry and student participation) and with the styles of the laboratory manuals at the two universities.

Patterns of talk were also monitored by Shymansky and Penick (1979a) who concluded:

> Focusing on the subtle implications of the pattern data, it is clear that TAs [teaching assistants] across all courses tend to follow up their questions with 'minilectures' (QLT) rather than with another question (QLQ), and QLT scores are consistently higher than QLQ across the board. Similarly, it would appear that the chemistry TAs have a strong tendency to tell students what to do and an aversion to demonstrating procedures for individual students (SLA).

Raghubir's (1979) criticism of university and school biology courses seems depressingly apt: 'Teachers tell students too much ... In the laboratory, for example, they are likely to tell them just about everything ... to save time and to *save the experiment* ' (p. 13, emphasis added).

Overall, these findings of actual learning activities are in no way congruent with the impression created by Chapters 3.1, 3.2, 3.3 and 3.4. From those discussions of intended learning outcomes it seems reasonable to conclude that (1) scientific enquiry is conceptually the most important issue and can be appropriately taught in the laboratory; (2) technical skills are important but underrated and definitely require a laboratory setting; (3) scientific attitudes and scientific knowledge possibly do not require a laboratory and knowledge can often be more effectively taught elsewhere.

In contrast, from the results presented in this section it will be seen that emphasis on scientific enquiry was practically non-existent (but this will be pursued in a later section) and that talk concerning attitudes was either not studied or non-existent. This leaves the great bulk of scores in the cognitive categories of teacher behaviour divided between substantive and procedural knowledge dimensions in ways which apparently vary greatly from one course to the next.

## Laboratory activities and organization categories of teacher behaviour

These would be the categories expected to occupy the major proportion of teacher time in laboratories. As discussed above, cognitive categories usurped this role. The proportions of time remaining were divided, with 5-20% for activities and 25-30% for organization, in the only two studies from which data were available (Hegarty, 1978; Shymansky and Penick, 1979a).

Again, examination of these categories of behaviour across different contexts revealed interesting patterns. For example, microbiology teachers at one university had about three times as much involvement in laboratory activities as microbiology teachers at another university (Hegarty, 1978). This seemed to be a function of an emphasis on open enquiry-oriented exercises at the former and on controlled exercises at the latter. Shymansky and Penick (1979a) found no consistent differences between introductory and advanced laboratories but within each found differences across disciplines. Some contrasts were in the categories: 'demonstrates procedures' (0.3% and 0.5% for introductory and advanced botany but 12.0% and 14.1% for introductory and advanced physics) and 'shows/illustrates' (0.7% and 2.8% for introductory and advanced chemistry but 24.4% and 7.0% for introductory and advanced zoology). In half of the disciplines observed, figures were startlingly high for 'passive observation' (10-20%). This was defined as 'watching or listening to students in situations where the teacher is not directly or actively involved' and it appears that teachers were not contributing much to the progress of class activities.

### Socio-emotional categories of teacher behaviour

Studies in science classes at all levels reviewed by Hegarty (1979) showed considerable uniformities on the following findings:

> ... more teacher behaviour is 'direct' (that is, stresses lecturing, criticism and direction giving) than 'indirect' (use of praise, questioning, acceptance of students' ideas and feelings). Verbal behaviour in classrooms is predominantly neutral: teachers use praise, criticism and acceptance of students' ideas and feelings for only a small percentage of the time (0-5% each). Students seldom speak for more than 25% of the time. Teachers who are more 'indirect' are associated with students who initiate more questions and discussion (p. 144).

Two studies (Hegarty, 1978; Shymansky and Penick, 1979a) contained relevant dimensions. In neither case did the figure for teachers or students giving or receiving praise or criticism exceed 1%. A non-judgemental interpretation of this situation is 'neutral'. Other observers have used the terms 'sterile', 'remote' and 'an affective desert'. In a review of teaching in higher education, Dunkin and Barnes (1985) expressed it thus: 'It is as though the classrooms of higher education are not only inappropriate contexts in which to display such things as emotions, but that they are also inappropriate places in which to conduct research on emotions' (p. 766). Similarly, little or no classroom observation research on the development of scientific attitudes or attitudes to science (Chapter 3.4) has been found in this category.

Investigating changes in processes across different contexts (microbiology classes at two universities) Hegarty (1979) found that teachers at the first engaged in less verbal behaviour and more non-verbal than those at the second. Teachers at the second talked twice as much as those at the first, whilst on average each teacher at the second talked about seven times as much as each student (compared with twice as much as each student at the first university). At the first university, there were no situations in which a teacher addressed the assembled classroom group as a whole. This activity occupied 12% of teachers' time at the second institution. At the first, teachers gave nearly twice as much attention to individual students (42% vs 25%) but they also spent nearly twice as much time alone in the classroom (23% vs 13%) and nearly twice as much time involved with other teachers as with students (8% vs 5%).

Whereas the second facet of Hegarty's SOLT scheme was the number of students in target grouping, Shymansky and Penick used sex as the second facet of SLIC. In their (1979b) paper on sex bias of teaching assistants they reported that female teaching assistants asked a higher proportion of short-answer and extended-thought questions than the male TAs, appeared slightly better listeners, and tended to provide more assistance with materials and equipment. Rather warily these authors inferred that the female TAs may have taken their teaching responsibilities more seriously than the male TAs (data and effect sizes were not reported). A further finding on possible sex bias was that both female and male TAs tended to interact with single-sex groups of either sex more often than mixed sex groups (14% vs 6%), but much more often 'passively observed' mixed groups than single sex groups (35-45% vs 1-7%).

## Categories of teacher behaviour unrelated to classwork

Again only the studies by Hegarty (1978, 1979) and Shymansky and Penick (1979a) contained relevant categories. In the Australian study, proportions of teacher time spent on class-unrelated tasks or activities were 6-13%. Hegarty (1979) noted this proportion was high by comparison with most school studies (0-5%, Power, 1977) and offered explanations in terms of long laboratory periods (four hours) scheduled without even a coffee break. In the American study, Shymansky and Penick (1979a) reported figures very much higher again (from 3% to a staggering 31%). No explanations were offered but in the latter classes (advanced geology) the following figures were reported - passive observation 11.3%, reading or writing 26.1%, non-lesson related 30.9%, totalling 68.3% of inactive behaviour. One wonders what these teachers *were* doing. It is impossible to invoke the response that it is form not frequency that counts - that one brilliant question could compensate for a day's inactivity. The score for these teachers on 'Asks extended thought questions' was 0.0%.

## BEHAVIOUR OF STUDENTS

Teachers' behaviour has been discussed at some length. So, in the following sections (organized in the same way) where research on students' behaviour tends to be simply the counterpart of teachers' behaviour, it will be reported only briefly here. This is particularly the case with interaction analyses where the schedules for observation of students are in many parts the logical mirror of teacher schedules. For example, if teachers are described as 'supervising student activities', one may assume that students are carrying out laboratory activities.

### Cognitive categories of students' behaviour

One of the few naturalistic behaviour studies was made of first year chemistry students by Mulder and Verdonk (1984) at the University of Utrecht. This was carried out at a stage when course designers were dissatisfied with certain parts of the laboratory course, especially those relating to separation techniques such as recrystallization and distillation - at the end of the course, students had not reached desired levels of proficiency or understanding. The investigation was planned to precede redesign and the introduction of an audio-visual teaching unit.

Target students in the laboratory were supplied with VHF transmitter microphones enabling tape recording of a student's voice as well as that of a lab partner or a laboratory teaching assistant. Some numbers of these recordings (videotapes) were 'observed' and preliminary patterns of behaviour confirmed using more structured observations. The most important finding was that, with practice, students' experimental procedures became coordinated, even automatic. If a student had learned part of the procedure without understanding the reason behind it or without seeing the connection with other parts, then there would be a gap in understanding of the whole automated procedure. These gaps became 'encapsulated' without the students being conscious of their existence. Students were likely to hold unfounded confidence that they had mastered certain steps, when in the eyes of expert observers their performance left much to be desired. One of the basic premises of the course revisions became to prevent students unconsciously learning with 'gaps'.

The observers believed that the problem was three-dimensional:

> *The students* seemed afraid to think about the reasons for the procedures they were performing because they thought they had no time for this. Their fear that tackling their problems would make them desperately short of time apparently caused them to think about their work as little as possible ...
> *The handbook* and other instructions were followed without question; students rarely asked why they were to perform certain tasks or obey assistants' instructions ...
> *Many assistants* tried to preserve the authority that the students had invested in them (at least as far as their knowledge of chemistry was concerned). They would cling desperately to this authority, painstakingly trying to conceal any uncertainty from the student ....
> (Mulder and Verdonk, 1984, p. 451)

A redesigned teaching unit was described which provided for students to watch videotapes of technical procedures and to make choices amongst a number of different possible steps in a procedure. Teaching assistants were required to carry out all steps in the students' exercises and were required to critique students' choices.

In an interaction analysis study of microbiology laboratory classes at two Australian universities, Hegarty (1978, 1979) found little difference across the contexts in the proportion of total student time spent on behaviour labelled as cognitive (26.6% and 30.3%). However, there were very significant differences ($p < 0.001$) for individual categories. At the first university, students spent far more time listening than those at the second - five times as much listening to teachers explaining aspects of content and twice as much listening to teacher talk about procedures. Conversely in microbiology laboratories at the second university there was more student talk overall, with 6.8% vs 3.8% at low cognitive levels and 1.2% vs 0.3% at higher cognitive levels. These students were also more likely to be involved in a discussion than simply listening to teacher talk.

Kyle, Penick and Shymansky (1979) also conducted a process-context study across disciplines and levels at the University of Iowa. Three of their categories covered cognitive aspects - 'transmits information', 'asks questions' and 'listens'. There were few consistent patterns of differences across disciplines or levels for transmitting information. However, it appears that students in physics labs spent more time in this way (7.9% and 10.4%) for introductory and advanced levels than the average (5.8%). Student questioning was low in all cases (1.5%-2.7%) with no significant differences between disciplines. Again, it was found that student listening was a much more prominent feature of some settings than others. Overall it was more common in introductory laboratories than in advanced laboratories, with proportions of time taken in botany, physics, zoology and geology all being above 15% with the latter a high 30.8%. In advanced laboratories no proportion was above 14% and for geology it was the lowest (6.8%).

One wonders if some of these variations, which were unaccounted for by these authors in terms of course aims or instructional design, may not be artefacts of the observation strategy. Only 333 separate 10 minute observations were made across five disciplines at two levels. Across the seven weeks of data collection, students were apparently selected for observation at random each time rather than selected once at random then systematically selected again later.

### Laboratory activities and organization categories of student behaviour

Since carrying out laboratory exercises is the most obvious purpose of having students in class, one would expect this activity to occupy much time and effort. So it has proved, although the proportions of time were not as high as might have been expected with 29% of the time devoted to laboratory activities found by Hegarty (1978) and 36% by Kyle, Penick and Shymansky (1979).

Following up the intentions discussed in Chapter 3.1, one would enquire about the amount of time actually devoted to technical skills. Hegarty's (1979) study showed that in microbiology classes at one university the category of laboratory activities was comprised of technical skills (25.2%) and observations (4.9%). At the other university, proportions were 20.6% and 7.9% respectively. Kyle, Penick and Shymansky (1979) showed differences in time spent on laboratory activities across levels (advanced laboratories with higher percentages than introductory laboratories) and disciplines (range 12.2% for geology to 36.1% for botany at introductory levels and 33.9% for botany to 55.0% for chemistry at advanced levels).

Similarly, laboratory organization is likely to be a major feature - though it proved surprisingly time consuming in both of the studies quoted above (about 30% when fetching supplies, putting away materials and cleaning as well as reading and writing lab notes are included). Examining in detail results using the category 'moves' (gets supplies, moves around the room), Kyle, Penick and Shymansky (1979) found there were major differences across disciplines though not across levels. This was almost the only finding of the study where differences were consistent and may reflect some consistent administrative practices on provision of materials for students or perhaps some inherent feature of the structure of the discipline. So, for example, geology was recorded with 0.6% and 0.8% (at introductory and advanced levels) by comparison with zoology (4.1% and 7.9%), and chemistry (15.4% and 14.0%).

## Socio-emotional categories of student behaviour

Some of the findings of interaction analyses mirrored those for teachers discussed above and will not be repeated here. Few examples of affective behaviour were recorded and little research has been carried out in this area.

When the target of student talk was observed, Hegarty (1979) reported discussion with another student as generally more common than talk to a teacher. This was noticeable for talk at low and higher cognitive levels and for talk on procedures. In both institutions studied, time on talk to teachers was similar (4%) but there was significantly more talk to fellow students in the microbiology classes of one than the other (12% vs 7%). Overall the majority of student behaviour was non-verbal (68% vs 78%) whereas teacher behaviour varied greatly between institutions with 37% vs 72% for verbal and 48% vs 22% for non-verbal. Kyle, Penick and Shymansky (1979) did not report results of any facet of their observation scheme which distinguished student talk with teachers vs other students or with females vs males (though the latter was reported for teachers, Shymansky and Penick, 1979b).

Fortunately there have also been some studies in the traditions of naturalistic research and research on classroom climate. Meredith (1983) developed a 21 item rating scale

('Laboratory Evaluator') for use in assessing the climate of some 200 lab sections in different disciplines at the University of Hawaii. Ten items were identified as having high factor loadings for laboratory impact. Almost all of these items concerned students' feelings about their laboratory teacher (structure and guidance supplied, enthusiasm, communication, well-prepared, open-minded, concerned about student progress). A study of students in a general chemistry laboratory course at a large US mid-western university was instructive although direct behavioural data were strictly limited to records of laboratory class time taken (Kozma, 1982). Results showed that both non-anxious students who lacked ability and able students who were highly anxious were aided by a laboratory programme with increased structure. As it turned out, there were no significant differences in lab time taken. However, the students receiving the additional structure performed better on a laboratory quiz, took less time to solve laboratory problems and felt more satisfied with the knowledge acquired. Both highly conforming and highly motivated students preferred the high structure treatment. Most unusually and commendably amongst such studies, Kozma provided clear examples of high structure and low structure versions of different laboratory exercises.

Making a direct attack on the problem of anxiety amongst beginning chemistry students in the laboratory, Abendroth and Friedman (1983) introduced a programme at Vincennes University called 'Coping with Stress, or, Chemistry, So Who's Worried?' This programme was allocated half-hour slots at the beginning of the first five laboratory classes of the chemistry course (2% of total course time). The programme covered the effect of stress on the body and on academic performance, a discussion of each student's unpleasant past experiences with science or maths, then a variety of exercises on relaxation, stress reduction, present worries about chemistry and preparation for chemistry tests. The students in the special programme became significantly less anxious about chemistry grades, about their lab performance and about their ability to solve mathematical problems. Both the treatment and control students became less anxious about the idea of chemistry itself and about having a teacher who wasn't understanding. The apparently less anxious treatment students outperformed the control group students on the normal chemistry grading procedures. Overall the researchers urged caution in interpretation of these interesting findings since different instructors taught the two chemistry groups and both were aware of the experiment being conducted.

Research on grouping of students in laboratories is very relevant to establishment of the socio-emotional climate. In Chapter 2.1 Layton highlighted the history of the controversy over teacher demonstrations in the laboratory vs individual or small group work an Garrett and Roberts (1982) reviewed the research associated with this controversy throughout the twentieth century. Rather predictably, it seems that demonstrations are less effective than small group work in teaching technical skills but are about as effective (or ineffective) as small group work in encouraging students' learning of science content knowledge in the laboratory.

Hurd and Rowe (1966) conducted their study of small group dynamics and productivity in school level BSCS Laboratory Block programmes. Of interest for tertiary level science was the comparison within the study of the behaviour and performance of college-bound vs non-college-bound classes (college-bound classes were recognized by higher mean scores and smaller standard deviations on the Scholastic Aptitude Test). When a final examination based on content and procedures of the BSCS Laboratory Block was taken as criterion, it was found that the performance and efficiency of college-bound groups increased when group compatibility increased (compatibility measures took into account the controlling behaviour exhibited by a student, the demands made on other students and the extent to which suggested laboratory manual procedures were followed). The reverse findings were reported for non-college-bound students, i.e. performance and efficiency increased as predicted incompatibility increased.

The effects of three different patterns of grouping students in physical science laboratory classes for elementary teachers were investigated at Arizona State University (Lawrenz and Munch, 1984). The patterns were free student choice or allocation to groups which were homogeneous or heterogeneous in performance on Piagetian formal reasoning tests. There were no differences between groups in perception of the laboratory classes measured by the LEI (Learning Environment Inventory, see above), and none of the other findings on performance achieved consistent statistical significance. Nonetheless, the authors felt that, taken overall, the results tended to favour the homogeneously grouped class. This class had the largest gain in formal reasoning ability scores, and was perceived by the students as being the most cohesive, the least competitive and as having the least amount of friction. Within each group, Moreno sociograms were used to account for students' choices of partners. In the homogeneous laboratory section, the highest formal reasoner was chosen at least once in 100% of groups. However, reasons preferred by students for their choices did not refer to formal reasoning nor to any form of critical thinking but rather to helpfulness, personal compatibility, industriousness and prior science background.

### Categories of student behaviour unrelated to classwork

Suffice to say that university and college students spend significant periods of time missing or on task unrelated talk and activities (1-6%) according to Hegarty (1978) and Kyle, Penick and Shymansky (1979). This amount of time is similar to that spent by some teachers but less than others (see the earlier section for teachers). Again, the most likely explanation is the sheer length of the classes which are usually scheduled without an official break. Both students and teachers may be helping themselves to an unofficial break.

## EFFECTS OF ENQUIRY-ORIENTED CURRICULA

Earlier chapters (1.1, 2.1, 2.2 and 3.2) have established the central importance of tertiary science students' understanding the processes of scientific enquiry and actively participating in genuine enquiry in the laboratory. To what extent do enquiry-oriented activities materialize?

Tamir (1977) and Guy (1982) both conducted classroom observation studies comparing behaviour in laboratories at secondary and tertiary level (in Israel and the United Kingdom respectively). The findings were similar. Enquiry-oriented curricula had been implemented in the secondary schools observed and there was evidence of congruent behaviours - investigations carried out, questions and discussion about hypotheses and experimental design. University classes observed were chemistry, biology, physiology and histology in Israel and chemistry in the UK. In both countries, there was little evidence of enquiry-oriented behaviours and, one assumes, little evidence of implementation of enquiry-oriented curricula. There was a dominance of verification rather than investigation or enquiry and a lack of scheduled post-lab time during which any underlying aspects of enquiry could be discussed (Tamir, 1977). There was a far lower incidence of questioning in all categories including the enquiry-oriented group (Guy, 1982).

Hegarty (1978) examined the effects of using different laboratory exercises which were rated from low to high levels of enquiry. The rating scheme used (see Table 6.3.1) was based on a simpler version developed by Herron (1971).

Within a single university microbiology course, it was found that some results were in the desired direction. During higher level exercises there was an increase in time spent by students in questioning and discussions at higher cognitive levels and an increase in peer group interaction. In Piagetian terms, the latter would provide an opportunity for students to compare observations and discuss interpretations, thus aiding the process of self-regulation and the acquisition of abstractions. Enquiry-related behaviours of both students and staff were more noticeable during structured enquiry exercises (level 2A) than during extended projects (level 2B). However, there were some unexpected results for teachers - proportions of time spent on low cognitive talk increased with increase in the level of exercises and time on high cognitive talk decreased during 2B exercises to a point below that found for the level 0 exercises.

**Table 6.3.1** : Levels of scientific enquiry (from Hegarty, 1978)

| Level | Aim | Materials | Method | Answer |
|---|---|---|---|---|
| 0 | Given | Given | Given | Given |
| 1 | Given | Given | Given | Open |
| 2A | Given | Given whole or part | Open or part given | Open |
| 2B | Given | Open | Open | Open |
| 3 | Open | Open | Open | Open |

In addition, there was an increase in time spent by both teachers and students on laboratory management activities (e.g. locating instruments and collecting glassware) during level 2 exercises, but especially during 2B exercises. For teachers, much of this increased time on management activities was spent outside the classroom in research and technical laboratories. This certainly meant that teachers were unavailable for discussions with their students - and this finding could be interpreted in the light of interviews which suggested that the tutors sought a role for themselves as resource persons (especially during the 2B classes) rather than acting as questioners, challengers or promoters of enquiry behaviours. This could perhaps be due to (incorrectly) equating scientific enquiry with the now rather discredited notion of unguided student discovery. It could also be due to teachers simply not possessing a suitable armoury of classroom skills - an issue which was discussed above.

Several authors reported no direct classroom observation data but provided other information from which inferences may be made about the behaviour of students and teachers. Spears and Zollman (1977) compared the effects of structured and unstructured laboratory exercises in physics. When students performed slightly better in the former on a test of understanding science processes, the authors concluded that the structured exercises had revealed to students something of the nature of scientists' activities in scientific enquiry. It was stated that students who were not at the formal operational level in physics could not be expected to design their own laboratory procedures whereby they might also begin to understand the formal processes underlying the activities of scientists. The conclusions of a study by Shymansky and Yore (1980) are in contrast. They reported the effect on outcomes of three different types of enquiry-oriented laboratory exercises: semi-deductive, structured-deductive and hypothetico-deductive. Students were elementary education majors studying general science in a two-course sequence which utilized laboratory activities exclusively. Student measures were (1) grouping according to their cognitive level on tasks for which the criteria ranged from 'all correct with control of all variables' to 'factors confounded in all experiments and subject can only describe actions, not generalizations'; (2) cognitive style. (Student achievement measures combined laboratory performance and quizzes.) It was concluded that structured induction might minimize disadvantages for learners who were concrete-operational and field-dependent. However, hypothetico-deductive activities may have the added benefit of encouraging field-dependent students to use a hypothesis testing approach and of teasing non-formal learners into using a hypothesis-prediction-comparison approach which approximates formal thought and perhaps produces sufficient dissonance to encourage students' cognitive development towards formal operations.

Haukoos and Penick (1983, 1985) conducted an initial study and a replication study on the influence of classroom climate on science process and content achievement of introductory college biology students. The two different curriculum treatments were described briefly and were designated Discovery Classroom Climate (DCC) and Non-Discovery Classroom Climate (NDCC). The replication failed to support the initial finding of superior effects for DCC on students' understanding of science processes nor on a Biology Achievement Test. Scores on gross categories of the SLIC observation scheme (Shymansky and Penick, 1979a) were fairly constant and predictable. Thus, in both studies, in the DCC classes, major types of teacher behaviours included asking extended thought questions, active observation, acknowledgement of student behaviour and listening to students. In NDCC classes teachers did mostly showing and telling together with giving directions.

It seems that a flaw in these two studies, and one which could help explain the equivocal findings, is that the researchers observed teacher behaviours rather than students' behaviours. No doubt teachers have an important mediating effect, but since both groups of students received the same laboratory equipment and materials with the different curriculum treatments embodied in texts and slides, observation of student behaviour was a key omission.

## CONCLUSION

It seems that tertiary science laboratory classes are like the curate's egg - good in parts. Sometimes there is genuine engagement with the materials and methods of enquiry in science, sometimes students gain a sense of accomplishment in technical proficiency, sometimes they grasp a concept and everything clicks. Sometimes an innovation is successfully introduced, sometimes teachers make a difference and sometimes clearly defined objectives help the progress of laboratory activities by having helped the course designer, teachers and students. At other times, the classes are astonishingly stable and looking back to Chapter 2.1 one wonders if there are not some types of low level, illustrative student exercises, carried out mechanically these days, which have been around for a hundred years (literally).

Is this a healthy situation? And if not, why does it persist? From the evidence available to date one can only speculate - but it would seem likely that much is related to teacher coping mechanisms. Important issues for investigation where change is deemed desirable would be teacher training, the role of laboratory teaching in the career paths of tertiary level science staff, reward and support structures.

How should time in the tertiary science laboratory best be allocated for the majority of task-oriented, motivated and academically achieving students? These were identified as 'success' students by Good and Power (1976) who suggested that at least half their time could be spent on individual work without a teacher. Of course in the overall pattern of university life, this could be regarded as time on individual study, assignments, etc. However, in the laboratory setting, it is likely that these students would benefit from two innovations:

1. Learning technical skills and science concepts could take place by means of self-paced, mastery-based learning packages. To date those reported have mostly been in audio-tutorial format in the biological sciences and Keller/PSI (Personalized System of Instruction) in the physical sciences (Dunn, 1986). Dunkin and Barnes (1985) claimed that the latter method was outstanding in the history of innovation in higher education in being consistently superior to all other methods with which it had been compared for student achievement on knowledge-oriented objectives. Whilst other students might not suffer greatly if there was a component of such packages within a laboratory course, success students should both achieve well and gain significant savings of time.

2. Learning many higher level cognitive and scientific enquiry skills could take place by means of experimental investigations and projects (Dunn, 1986). In a laboratory, it is not safe practice to leave students entirely without a teacher. More importantly, all of the studies showing the interaction of one laboratory classroom process with another show that teachers are required as challengers and questioners if there is to

be significant outward behaviour reflecting higher level activity of students. What are the implications for the classroom? McClellan (1971) and Shulman and Tamir (1973) exhorted us to recognize that cognitive rationality is a matter of form and not \frequency, and Power (1977) explained that it would be foolhardy to expect direct relationships between frequency counts and outcomes. Thus, what might feasibly be suggested is that, during experimental investigations and projects, there should be full-time supervision of success students by junior staff for safety reasons. There should also be visits by suitably skilled or trained senior staff for the purposes of probing, prodding, questioning and challenging the students to greater achievements.

## REFERENCES


Abendroth, W. and Friedman, F. (1983) Anxiety reduction for beginning Chemistry students. *Journal of Chemical Education*, 60, 25-6.

Anderson, G. J. and Walberg, H. J. (1976) 'The Assessment of Learning: Environment Inventory and the My Class Inventory' (revised)', unpublished document, Office of Evaluation Research, University of Illinois at Chicago Circle.

Bellack, A. A., Kliebard, H. M., Hyman, R. T. and Smith, F. L. (1966) *The Language of the Classroom*, New York, Columbia University Press.

Bliss, J. and Ogborn, J. (1977) *Students' Reactions to Undergraduate Science*, London, Heinemann Educational Books.

Boud, D. (1986) Introduction, pp. 3-12, in Boud, D., Dunn, J. and Hegarty-Hazel, E. *Teaching in Laboratories*, Guildford, Society for Research in Higher Education/NFER Nelson.

Dunkin, M. J. with the assistance of Barnes, J. (1985) Research on Teaching in Higher Education, pp. 754-77 in Wittrock, M. C. (Ed.) *Handbook of Research on Teaching* (Third Edition), New York, Macmillan.

_____ and Biddle, B. J. (1974) *The Study of Teaching*, New York, Holt Rinehart and Winston.

Dunn, J. (1986) Teaching strategies, pp. 35-6, in Boud, D., Dunn, J. and Hegarty-Hazel, E. *Teaching in Laboratories*, Guildford, Society for Research in Higher Education/NFER Nelson.

Eggleston, J. F., Galton, M. J. and Jones, M. (1975) *A Science Teaching Observation Schedule*, Schools Council Research Series, London, Macmillan.

Fraser, B. J. (1985) Two Decades of Research on Perceptions of Classroom Environment, pp. 1-33, in Fraser, B. J. (Ed.) *The Study of Learning Environments*, Salem, Oregon, Assessment Research.

_____ (1986) *Classroom Environment*, London, Croom Helm.

_____, Treagust, D. F. and Dennis, N. C. (1986) Development of an instrument for assessing classroom psychosocial environments at Universities and Colleges. *Studies in Higher Education*, 11, 43-54.

_____ and Walberg, H. J. (1981) Psychosocial learning environment in science classrooms: a review of research. *Studies in Science Education*, 9, 67-92.

Garrett, R. M. and Roberts, I. F. (1982) Demonstration versus small group practical work in science education: a critical review of studies since 1900. *Studies in Science Education*, 9, 109-46.

Good, T. L. and Power, C. N. (1976) Designing successful classroom environments for different types of students. *Journal of Curriculum Studies*, 8, 45-60.

Guy, J. J. (1982) Quantitative classroom observation illustrated by a study of chemistry practical teaching at university and sixth form level. *Chemical Education Research: Implications for Teaching*. Report of a symposium, Aston University, UK RSC.

Haukoos, G. D. and Penick, J.E. (1983) The influence of classroom climate on science process and content achievement of community college students. *Journal of Research in Science Teaching*, 20, 629-37.

_____ (1985) The effects of classroom climate on college science students: a replication study. *Journal of Research in Science Teaching*, 22, 163-8.

Harris, D. (1977) An illuminative evaluation of a first year laboratory course - a critical appraisal. *Assessment in Higher Education*, 3, 21-48.

Hegarty, E. H. (1978) Levels of scientific enquiry in university science laboratory classes: implications for curriculum deliberations. *Research in Science Education*, 8, 45-57.

_____ (1979) 'The Role of Laboratory Work in Teaching Microbiology at University Level'. Unpublished Ph.D. thesis, University of New South Wales.

Hegarty-Hazel, E. (1986) Research, pp. 129-52, in Boud, D., Dunn, J. and Hegarty-Hazel, E. *Teaching in Laboratories*, Guildford, Society for Research in Higher Education/NFER Nelson.

Herron, M. D. (1971) The nature of scientific enquiry. *School Review*, 79, 171-212.

Hurd, P. D. and Rowe, M. B. (1966) A study of small group dynamics and productivity in the BSCS laboratory block program. *Journal of Research in Science Teaching*, 4, 67-73.

Jackson, P. W. (1968) *Life in Classrooms*, New York, Holt, Rinehart and Winston.

Karplus, R. (1977) Science teaching and the development of reasoning. *Journal of Research in Science Teaching*, 14, 169-75.

Kozma, R. B. (1982) Instructional design in a chemistry laboratory course: the impact of structure and aptitudes on performance and attitudes. *Journal of Research in Science Teaching*, 19, 261-70.

Kyle, W. C., Penick, J. E. and Shymansky, J. A. (1979) Assessing and analyzing the performance of students in college science laboratories. *Journal of Research in Science Teaching*, 16, 545-51.

Lawrenz, F. and Munch, T. W. (1984) The effect of grouping of laboratory students on selected educational outcomes. *Journal of Research in Science Teaching*, 21, 699-708.

McClellan, J. E. (1971) Classroom-teaching research: a philosophical critique, pp. 3-22, in Westbury, I. and Bellack, A. A. (Eds.) *Research into Classroom Processes: Recent Developments and Next Steps*, New York, Teachers' College Press.

Meredith, G. M. (1983) Factor-specific items for appraisal of laboratory and seminar/discussion group experiences among college students. *Perceptual Motor Skills*, 56, 133-4.

Moos, R. H. and Trickett, E. J. (1974) *Classroom Environment Scale Manual*, Palo Alto, Consulting Psychologists' Press.

Mulder, T. and Verdonk, A. H. (1984) A behavioural analysis of the laboratory learning process. *Journal of Chemical Education*, 61, 451-3.

Ogborn, J. (Ed.) (1977) *Practical Work in Undergraduate Science*, London, Heinemann Educational Books.

Parlett, M. (1969) Undergraduate teaching observed. *Nature*, 223, 1102-4.

_____ and Hamilton, D. (1972) *Evaluation as Illumination: a New Approach to the Study of Innovatory Programs* (Occasional Paper No. 9), Edinburgh, Centre for Research in Educational Sciences, University of Edinburgh.

Power, C. (1977) A critical review of science classroom interaction studies. *Studies in Science Education*, 4, 1-30.

_____ and Tisher, R. P. (1976) Relationships between classroom behavior and instructional outcomes in an individualized science program. *Journal of Research in Science Teaching*, 13, 489-97.

Raghubir, K. P. (1979) The laboratory investigative approach to science instruction. *Journal of Research in Science Teaching*, 16, 13-17.

Reif, F. and St. John, M. (1979) Teaching physicists thinking skills in the laboratory. *American Journal of Physics*, 47, 750-7.

Renner, J. W. and Lawson, A. E. (1973) Promoting intellectual development through science teaching. *The Physics Teacher*, 11, 273-6.

Rentoul, A. J. and Fraser, B. J. (1979) Conceptualization of enquiry-based or open classroom learning environments. *Journal of Curriculum Studies*, 11, 233-45.

Rosenshine, B. (1970) Evaluation of classroom instruction. *Review of Educational Research*, 40, 279-300.

Shulman, L. S. and Tamir, P. (1973) Research on teaching in the natural sciences, pp. 1098-148, in Travers, R. M. W. (Ed.), *Second Handbook of Research on Teaching*, Chicago, Rand McNally

Shymansky, J. A. and Penick, J. E. (1979a) Use of systematic observations to improve college science laboratory instruction. *Science Education*, 63, 195-203.

_____ and Penick, J.E. (1979b) Do laboratory teaching assistants exhibit sex bias? *Journal of College Science Teaching*, 8, 222-6.

_____ and Yore, L. D. (1980) A study of teaching strategies, student cognitive development and cognitive style as they relate to student achievement in science. *Journal of Research in Science Teaching*, 17, 369-82.

Smith, L. M. and Geoffrey, W. (1968) *The Complexities of an Urban Classroom*, New York, Holt, Rinehart and Winston.

Spears, J. and Zollman, D. (1977) The influence of structured versus unstructured laboratory on students' understanding of the process of science. *Journal of Research in Science Teaching*, 14, 33-8.

St. John, M. (1980) Thinking like a physicist in the laboratory. *The Physics Teacher*, September, 436-43.

Stake, R. E. and Easley, J. A. (1978) *Case Studies in Science Education*, University of Illinois, Urbana-Champaign: Center for Instructional Research and Curriculum Evaluation and Committee on Culture and Cognition. Volumes I and II (Prepared for National Science Foundation Directorate for Science Education, Office of Program Integration).

Tamir, P. (1977) How are the laboratories used? *Journal of Research in Science Teaching*, 14, 311-16.

Uricheck, M. J. (1972) Measuring teaching effectiveness in the chemistry laboratory. *Journal of Chemical Education*, 49, 259-62.

# 6.4

# Students' Reactions to Undergraduate Science: Laboratory and Project Work

**Joan Bliss**

## INTRODUCTION

The research on students' reactions to undergraduate science was carried out within the Nuffield Foundation-sponsored Higher Education Learning Project HELP(p) (physics), directed by Ogborn. The main emphasis of the project was to help university lecturers in science in a number of areas of their work. The project chose to look at laboratory work, small group work and individual learning.

It was felt by all the teachers participating in the project that it was important to look at how students experienced learning at the undergraduate level, so as to gain some insights into the impact of different teaching methods. This chapter looks in particular at the results of this study that have to do with laboratory work and projects.

Before actually describing the research study, some brief details about the HELP(p) study of laboratory work are mentioned because they are relevant to the theme of the present book. The area of laboratory work was examined because it was felt that it was an interesting period in the development of practical work (see Ogborn, 1977a). There had been a great deal of questioning of the value of such work, as well as a great deal of experimenting with new methods and patterns of organization, but there was, at that time, no clear or agreed way forward. The HELP(p) book about laboratories was based on the obser- vations of a team of physicists who paid an extensive series of visits to teaching laboratories throughout England. Eight physics departments were involved, covering four first year and four second year laboratories. In parallel another eight institutions were visited to look mainly at third year project work.

Thus, the HELP(p) project was carried out during the era of development of the 'new' curricula at secondary levels (see Chapters 6.1 and 6.2) and, occasionally, tertiary levels. Since that era, there have been a variety of other research studies on laboratory work at the tertiary level, some using questioning techniques or direct observation for data collection, as in HELP(p), others using more unstructured techniques. These are discussed in detail in Hegarty-Hazel's Chapter 6.3.

In the conclusions to the HELP(p) study, Ogborn (1977a) stressed the importance of choosing between aims, and of the necessity of choosing both amongst ways of looking at them and amongst levels of thinking about them. He concluded that at these levels one is trading off meaning against perspective; the wider the point of view the harder it is to interpret or translate it in terms of action, while the more definite the point of view, the harder it is realistically to weigh up its general value.

The model (Figure 1.2.1) differentiates intended and actual transactions and many of the case studies in the HELP(p) laboratory book show that to achieve these aims is much trickier than it may look. Aims and actions may be in conflict, with students detecting the inconsistencies. Lastly the study stressed the significance of the teachers involved in laboratory work, for, as Ogborn suggested: 'It is important to ask what they can do well. It is no less important to ask what might lead them to do better. Change of this kind can have as great an effect as any amount of improvement to hardware, to forms of instruction, or to administration, and without it the latter changes may have little effect'.

## THE STUDY

In the research project on students' motivation to learn, it was decided to ask students to tell us a story about how they reacted to a real situation in which they were trying to learn some physics. Students were asked to tell us about some time when learning felt particularly good or particularly bad. The idea of approaching students in this manner came from the work of Herzberg, Mausner and Snyderman (1959) who looked at the motivation to work of engineers and accountants. Herzberg adapted the 'critical incident' technique from Flanagan's (1954) studies of factors of importance in the content or performance of a job.

The interview, as it was developed after a pilot study, lasted between 35 and 45 minutes with each student, and had three important stages for each story. First, an elicitation from the student of a story about a definite event with enough detail so as to know why the story was significant to the student. Second, a description by the student of his or her reaction to the situation and reasons for this reaction. Finally, the student was asked whether or not the situation had had an effect on his or her work. Students were asked to tell one 'bad' and one 'good' story, in whichever order they preferred. During the pilot study it was noticed that often after two stories had been told a third extremely revealing story came to mind, and in the main study students were given the opportunity to recount more than two stories if they wished. We aimed to interview 120 students in ten representative universities, taking four from each undergraduate year, and obtained 115.

## THE ANALYSIS

Various classifications were tried but none seemed to capture the intricate and detailed nature of students' experiences. At this point we turned to constructing a novel method of analysis, which entailed organizing a large set of descriptive categories into a structured system called a network. It is not the goal of this chapter to discuss this method in any detail. Descriptions of it can be found in Bliss and Ogborn (1977, 1979), Bliss, Monk and Ogborn (1983) and Ogborn (1977b, 1980). The idea is to develop a strategy for elaborating categories, using a notation derived from systematic linguistics (Halliday 1973, 1975). The notation sets out the categories in such a way as to show their interdependences. The notion is simple and rigorous, but also flexible, generating a network-like structure. Examination of a network will show relationships such as those categories which belong within others, those which are independent and those which are conditional on the choice of others.

## LTS

115 interviews produced 307 stories in all. In order for a y to be included in the data it had to be able to answer three uestions described previously: that is, to be about a definite earning situation, to contain clearly identifiable feelings about that situation and to contain an identifiable set of reasons associated with the feelings. Details of coding were:

| *Stories coded* | |
|---|---|
| 'Good' events with clear feelings and reasons | 129 |
| 'Bad' events with clear feelings and reasons | 142 |
| Stories with 'conflicting' feelings with clear and identifiable reasons | 14 |
| total of coded stories | 285 |

| *Stories not coded* | |
|---|---|
| 'Good' and 'bad' stories with unclear feelings and reasons | 16 |
| Mixture of 'good' and 'bad' events | 3 |
| Stories about non-science subjects | 3 |
| total number of stories | 307 |

In the checking process involving two researchers 50% of the stories were left unchanged, 40% had minor changes (adding, omitting or altering up to two lines of code) and 10% suffered major changes. 12% of the identified stories were not used in the analysis, including all those with 'conflicting' feelings.

## IMPORTANCE OF AREAS OF LEARNING

The students' stories covered an extremely wide range of learning situations. Of the total 271 stories:

| | | |
|---|---|---|
| 128 | (47%) | were about lectures |
| 43 | (16%) | were about laboratories |
| 40 | (15%) | were about individual work |
| 26 | (9.5%) | were about tutorials |
| 23 | (8.5%) | were about project work |
| 11 | (4%) | were about other areas, like examinations |

Do the frequencies for laboratories and projects reflect what might be expected? In what sense does the preponderance of stories about lectures actually represent what is happening to the student when learning physics at the university? It is possible that many teachers in English universities would see work in laboratories and on projects or in tutorials as important if not crucial aspects of a science course.

## LABORATORY AND PROJECT STORIES

### Projects good, laboratories bad?

The scene will be set for us by an extract from a story of a first year student about a 'bad' laboratory experience:

> ... the experiment - it's the thought of being there for five hours at a time and when you come out you don't feel you've done anything. It's also the thing that thousands of undergraduates have done it before - it's so stereotyped you could almost use a tape recorder instead of a person ... All right, you've got to build up from basics, but surely you must feel inside that you're doing something useful, because if you're not, what's the point of it? In my mind it leads to a lot of apathy.

As shown earlier of the 43 stories told about laboratory work 27 were about 'bad' events, and 16 about 'good' events, whereas for projects 21 stories were about 'good' situations and only 2 were about 'bad' ones. On the surface, students' stories would seem to paint a fairly black picture for laboratory work as opposed to project work. However, when these figures are looked at for first year and later year students it can be seen that the picture is not quite so simple:

| | 'good' stories | 'bad' stories |
|---|---|---|
| Laboratories (first year) | 7 | 21 |
| Laboratories (later years) | 9 | 6 |

The picture of laboratory work is very different, dependent on whether the student is a first year student or a second or final year student, with the stories of the later group of students even favouring laboratory work.

This raises four questions: What makes laboratory work so bad? What are good laboratories like? Why is project work mainly good? How do laboratory work and project compare?

## What makes laboratory work so bad?

Overall there were two features that distinguished bad laboratory stories from many of the other bad stories in the study and these were both to do with negative interpersonal feelings. Firstly, students felt a greal lack of security in the laboratory and this feeling was often associated with negative feelings about themselves, like self-doubt or dread. Secondly, students felt annoyance either towards the teacher or the situation. The reasons for these feelings were:

1. Some difficulty with the experiment, too hard, too long, apparatus did not work
2. Unable to understand the ideas or to cope with apparatus
3. Unable to get help or guidance from the teacher, or not getting recognition for work done

There were clear differences between the first year laboratory stories and stories told about later years. The differences can be best illustrated by looking at students' reactions and grouping them into three categories: 'turned off', 'upset' and 'cross'. 'Turned off' feelings were about such things as: lack of interest, time-wasting, being bored, fed-up, pointlessness and indifference. The number of such feelings in the first year and in later stories were in the same ratio as the total numbers of feelings in the two sets of stories. So the two groups were equally 'turned off'.

At another level, however, first year and later year students reacted very differently to laboratory work. The first year students had serious doubts about themselves as science students, saying they felt inferior, inadequate, or, as they put it, 'I'm not good at it'. They also felt very upset with feelings of depression, bewilderment, and being lost which were often accompanied by a set of feelings of foreboding such as panic or anxiety. However, the feelings of second or third year students could be characterized as about 'being cross', for example, annoyance with apparatus, annoyance with teacher, resentment, hatred and frustration. The following quotations contrast the learning of students in different years:

Comment from a first year student:

> I was on the first experiment for three or four weeks - you get bogged down and worried. You're afraid to break anything ... At first you're afraid to go and ask, and the demonstrator at that time wasn't coming to ask me. You don't know anybody really, and you're afraid to show yourself up in front of people. And it's just very bewildering. You need someone to help you to adjust at first, someone to come round and say you're all right. It was all rather depressing.

Comments from second and third year students:

> ... you spend a really phenomenal amount of time just waiting for somebody to explain what you're supposed to do. It's really very frustrating and if anything goes wrong you have to wait again. It kills off any enthusiasm. And little technical things go wrong, with only one person who can fix them. A waste of time.

> We now find there are some ready-made filters on the third floor we could have borrowed instead of making our own. It diverted attention from the real problem and we lost a bit of interest.

Thus, in the first year, students are nervous, anxious and somewhat intimidated by the hardware of the laboratory and by their own inexperience and gaucheness faced with a range of new experiments and, as it appears to them, very little help. With second and third year laboratories, apparatus still goes wrong and help is still not available but the students have become a little immune to the situation. They do not see failures in the laboratory work as due to their own inadequacy but perhaps as an intrinsic part of the work itself; they have become used to laboratories and so problems do not get under their skin in quite the same way as for first year students.

**What are good laboratories like?**

There are essentially two kinds of stories about times when learning was good in a laboratory. Firstly, quite a large number of the stories are about experiments which work. Students talk about being able to cope, about learning something and about succeeding in doing the task. Secondly, there is another kind of 'good' laboratory story which has to do with independence and freedom. An example of coping in the laboratory was as follows:

> I sort of settled down with a piece of paper and came up with an idea which I thought would work, and apart from one modification it did, which was quite pleasing. I sort of showed the circuit to one of the demonstrators and he said it should work. We sort of got everybody's circuits together and the whole bench was covered with them and it worked, it was pleasing. For the rest of term we did logic circuits [and when someone got stuck] you knew it was a matter of, 'Well you've got to this state'. [the tutor would say] 'Go and ask him' [me], 'he did that one last week or the week before.'

A story which reflects feelings of freedom is:

> I like the practical in the second year because you don't have someone breathing down your neck - you are given a very open-ended approach to experiments, you are just given a few essentials and you go away and read up background literature and you are left on your own to make of it what you want. [It's good] ... because you have so much freedom, for once, you are allowed to pick for yourself rather than have everything presented to you.

In both these categories the teacher's role is a difficult one. In the first he or she is expected to give help when it is needed, but to let well alone when it is not. In the second category the teacher is seen either to allow or to encourage the student to work independently, whether or not the policy is a deliberate one. The next example illustrates this balancing act of the former category:

> If you get any problems you have demonstrators to consult, but most of it is up to you. If they think the experiment is difficult they'll give you more help and more or less tell you how to do it word for word, but if they think you might be able to do with a bit of thinking by yourself, they let you think it out by ourself, which is in some ways good and in some ways bad ... If you can

think for yourself it's all right, but if you can't you're having problems all the time.

In the following story, the sense of goodness and achievement from having done something independently is striking, with the tutor playing both a significant and at the same time marginal role:

> ... I can't remember his exact words, but he looked at it and sort of said, 'You know something, that's right. It works - it's bound to - come here, Derek, have a look at this - he's done it.' It was probably the first time anyone had done that bit of the practical ... it was supposed to be theoretical - just design the circuit, rather than construct it.

So not all stories about laboratories are bad. Goodness comes from the confrontation of student and apparatus, with the student winning the battle. A feeling of goodness also emerges when students feel they are doing a proper experiment, responsibly and independently, and not just repeating a series of five-finger exercises 'that thousands of other undergraduates have done before'.

## What makes projects mainly good?

In many ways project stories mirror good laboratory stories but they are very much more intense. There are four essential features of good project stories: success or achievement, a feeling of responsibility or independence, the sense of coping and understanding and lastly a somewhat unusual intense satisfaction, rarely found in any of the other stories of the study. Feelings of success were found in about three-quarters of the stories and, in the following story, feelings of achievement and independence are linked, as a student recounts what it was like working with a partner on a third year project:

> The good thing was that we were able to work on our own when we wanted to and the supervisor only came in when we were in difficulty. So most of the time we were working on our own and we felt a sense of achievement really. I think that it is very important that when we hit problems he was fairly readily available to come and help us. This way we didn't waste too much time just thinking

about the problem not getting anywhere with it, he put us on the right line. He never actually told us what was wrong but he sort of gave us pointers to think about, that was very interesting.

More than half the stories talk about freedom, responsibility or independence. These types of feelings would appear to be important for students because they see themselves as capable people and possibly as emerging scientists, as the following example illustrates:

> ... at last you were being trusted with it [expensive apparatus] - at last you were being treated as if you knew what you were doing, and didn't have to be watched over the whole time ... If you make a mistake, you take the responsibility for it. You get the credit when you do it right, and you get moaned at when you do it wrong, but at least you know, 'Well that was all mine'.

All the good stories about projects contained something about coping and understanding, about facing and surmounting a challenge, as the following student recounts:

> ... that got us really thinking about what we were going to be doing in practice ... Things were rather slow to begin with - we had various hold-ups, but we were able to do some preliminary experiments, and after that we were able to make a fair bit of progress ... I suppose it was being able to carry through an idea right from hearing about it in the first place, then finding out about heat pipes in general, then narrowing it down to what would be feasible.

Perhaps the following story illustrates both the student's involvement in project work and the tremendous satisfaction, 'thrill', he got out of it:

> When finding out about it, I found a few anomalous phenomena really. And I think that sort of gave me the first taste of doing something, thinking about it, rather than just regurgitating what was shoved down my throat. I spent a lot of time on it, I actually rekindled my own interest in physics in four weeks. So I worked very hard, I was for the first time in three years interested in physics. I was really interested in something and it really annoyed

me that I couldn't think of reasons why certain things happened. In the end it was difficult because I'd not done any magnetic theory before. It meant reading quite a lot. Finally I got a picture of what was actually happening in my mind and it gave me a kind of thrill really.

## How do laboratory work and project work compare?

In the report on this study (Bliss and Ogborn, 1977), we suggested that differences between projects, 'good' laboratory and 'bad' laboratory stories, might well be understood in terms of the tasks set to students. Laboratory work, unlike lectures, nearly always presents the student with a task of work to do. Faced with this, the student can very quickly measure his or her own success or failure. First year students who find it hard to cope with experiments feel insecure and anxious. The successes of the later year students lead to feelings of achievement and satisfaction but they never mention security, possibly because as soon as one task is finished there is yet another to do.

Also the third year students seem to have come to terms with laboratory work, recognizing that it can be frustrating, that often apparatus does not work, and that frequently people are not around to help, but, that in some sense, this is the nature of laboratory work. So, rather than feeling depressed and inadequate, they now look for what might be going wrong, the organization, the apparatus, etc. and get annoyed at this, with realism being now more characteristic of their reactions.

Common to both 'good' project and 'good' laboratory stories are feelings of independence, freedom and responsibility. Yet it is not easy to achieve situations of this kind because too little help can soon plunge students into feelings of getting nowhere with their tasks, so lack of freedom and responsibility does not get a mention when students are faced with difficult laboratory experiments. Thus the problem for the teacher who wishes to promote these sorts of feelings is to find experiments which are meaningful, neither too overwhelming nor too trivial (for further analysis of this problem see Ogborn, 1977b).

The intensity of the satisfactions produced by projects, by contrast with the more muted pleasure and enjoyment associated with 'good' experiments, is perhaps a challenge to teachers planning laboratory work. That is, is there any way laboratories can provide the powerful and sustained involvement a project can provide? Let one student who really experienced a tremendous satisfaction from his project finish this section:

> ... the beauty of Saturday morning, when it melted! That I'd got a fluctuation big enough to cause melting. I'd worked through the maths behind the fluctuation terms and I'd got an answer, and it was right. There's a beauty behind it all - there's a reason.

## CONCLUSION

One of the most striking and frequent reactions in the study in general is that of 'cutting off', of disengaging with the subject. One possible interpretation is that this is how students protect themselves from potential failure. While this form of defence may work well for lectures and tutorials, it is not helpful for laboratory work because students must attempt to do 'the experiments' and so they can realistically measure their own success or failure, this latter situation leaving them with feelings of depression and anxiety. Third year project work, however, is an exception to this trend. Projects are one of the rare learning situations where students do feel totally involved.

A second perhaps puzzling feature of many stories of the general study is the rather limited mention of success, whether in the form of inner personal achievement or as external public recognition. Even when students do talk about inner success it is often couched in words such as 'getting on top of it', 'coming to terms with it' - hardly the sort of way in which one would generally think of achievement. While this feature is true of laboratory work in the first year of university, we showed that in third year laboratory work there were occasions when students really felt they were understanding and coping successfully with experiments. Again third year project work presented students with a real challenge and often they experienced a tremendous sense of achievement, particularly in situations where they felt responsible and independent. In other words, because of its nature, work in the laboratory (and this more than many other kinds of study in the university) whether in the form of projects or well-thought-out experiments can provide students with meaningful tasks where success is a possible real outcome.

A third feature of the results of the study which is rather worrying is the tremendous amount of self-doubt and insecurity expressed by students, particularly in the first year and in laboratory work. If the challenges of learning at university are considered a virtue it is important that these challenges are not made too great.

The results of this study together with some of the conclusions of the HELP(p) work on laboratories indicated a number of points about what determines what happens in laboratories.

1. The mythology of the laboratory - that it is a place where one learns experimenting by direct contact with those skilled in experimental research - needs to be called into question. There is less contact than is usually imagined and when it happens, it is not often of this kind. But it does and can happen, from time to time, and deserves recognition.

2. Laboratories are in many ways enormously stable, and innovation often makes much less difference than people might suppose or hope for. In other words, change cannot be undertaken lightly and a great deal of care and planning is needed to make any change effective. However, such innovations could introduce a sense of purpose to the more traditional set experiment laboratory. It should also be remembered that the laboratory time scale must necessarily be a long one. What happens in any one experiment is rarely enough plausibly to support the 'educational or other hopes' that might be embedded in it.

3. The students' task in the laboratory is not an easy one. They are often in situations where they have to switch between doing physics and keeping the teacher happy. It is important to understand this dual but implicit role of the student, and the conflict it can set up. Also by the time students reach university they have very different backgrounds and so differ enough for it to be reasonable to attempt to match practical work with the individual's own needs.

4. Who people are, what they are like and how they behave make a big difference, perhaps more important than any planned organization. However, staff differ and it is not plausible to imagine that any one pattern of laboratory teaching can be operated by anybody. It is crucial to find ways of allowing staff to be 'at their best' and to do what they are individually good at within the design of a scheme for laboratory teaching.

## REFERENCES

*HELP publications*:

Bliss, J. and Ogborn, J. (1977) *Students' Reactions to Undergraduate Science*, Heinemann Educational Books Ltd., London.

Ogborn, J. (Ed.) (1977a) *Practical Work in Undergraduate Science*, Heinemann Educational Books Ltd., London.

_____ (Ed.) (1977b) *Small Group Teaching in Undergraduate Science*, Heinemann Educational Books Ltd., London.

*Other publications*:

Bliss, J. and Ogborn, J. (1979) The analysis of qualitative data. *European Journal of Science Education*, 1, 427-40.

_____ Monk, M. and Ogborn, J. (1983) *Qualitative Data Analysis for Educational Research*, Croom Helm, London.

Flanagan, J. (1954) The critical incident technique. *Psychological Bulletin*, 51, 327-58.

Halliday, M. A. K. (1973) *Explorations in the Function of Language*, Edward Arnold, London.

_____ (1975) *Learning How to Mean*. Edward Arnold, London.

Herzberg, F., Mausner, B. and Snyderman, B. (1959) *The Motivation to Work*, John Wiley, New York.

Ogborn, J. (1980) Some uses of networks of options for describing qualitative data, in W. F. Archenhold *et al.* (Eds.) *Cognitive Development Research in Science and Mathematics Education*, University of Leeds.

# Index